SOLUTIONS MANUAL

THERMAL ENVIRONMENTAL ENGINEERING
THIRD EDITION

Thomas H. Kuehn

James W. Ramsey

James L. Threlkeld

PRENTICE HALL, Upper Saddle River, NJ 07458

Acquisitions Editor: Laura Curless
Supplement Editor: Lauri S. Friedman
Special Projects Manager: Barbara A. Murray
Production Editor: Maria T. Molinari
Supplement Cover Manager: Paul Gourhan
Supplement Cover Designer: PM Workshop Inc.
Manufacturing Buyer: Dawn Murrin

© 2000 by Prentice Hall
Upper Saddle River, NJ 07458

All rights reserved. No part of this book may be
reproduced, in any form or by any means,
without permission in writing from the publisher.

Printed in the United States of America

10 9 8 7 6 5 4 3 2 1

ISBN 0-13-917238-6

Prentice-Hall International (UK) Limited, London
Prentice-Hall of Australia Pty. Limited, Sydney
Prentice-Hall Canada, Inc., Toronto
Prentice-Hall Hispanoamericana, S.A., Mexico
Prentice-Hall of India Private Limited, New Delhi
Pearson Education Asia Pte. Ltd., Singapore
Prentice-Hall of Japan, Inc., Tokyo
Editora Prentice-Hall do Brazil, Ltda., Rio de Janeiro

2.1

a) $28.75 \text{ in Hg} \left(\dfrac{1.450 \times 10^{-4} \text{ psia}}{2.961 \times 10^{-4} \text{ in Hg}} \right) = 14.08 \text{ psia}$

b) $28.75 \text{ in Hg} \left(\dfrac{4.019 \times 10^{-3} \text{ in H}_2\text{O}}{2.961 \times 10^{-4} \text{ in Hg}} \right) = 390.2 \text{ in H}_2\text{O}$

c) $\dfrac{390.23 \text{ in H}_2\text{O}}{0.8} \left(\dfrac{1 \text{ m}}{39.37 \text{ in}} \right) = 12.39 \text{ m fluid}$

d) $28.75 \text{ in Hg} \left(\dfrac{1 \text{ Pa}}{2.961 \times 10^{-4} \text{ in Hg}} \right) = 97.10 \text{ kPa}$

2.2

a) Let x represent dry air and y represent water vapor. By Eqns. (2.18 and 2.19)

$$\frac{P_y}{P_x} = \frac{m_y}{m_x} \frac{R_y}{R_x}$$

$$m_y/m_x = 0.012$$

$$\frac{R_y}{R_x} = \frac{85.76 \text{ ft lb}_f/\text{lb}_m \,°R}{53.352 \text{ ft lb}_f/\text{lb}_m \,°R} = 1.607$$

$$P_x = P - P_y$$

Thus

$$P_y = \frac{P}{1 + \frac{m_x R_x}{m_y R_y}} = \frac{14.32 \text{ psia}}{1 + \frac{1}{(0.012)(1.607)}} = 0.27 \text{ psia}$$

b) $\rho = \frac{m}{V} = \frac{m_x + m_y}{V} = \frac{m_x}{V}\left(1 + \frac{m_y}{m_x}\right)$

By Eq. (2.18), $\rho = \frac{P_x}{R_x T}\left(1 + \frac{m_y}{m_x}\right)$

$$\rho = \frac{(14.32 - 0.27) \text{lb}_f/\text{in}^2 (144 \text{in}^2/\text{ft}^2)}{53.352 \frac{\text{ft lb}_f}{\text{lb}_m \,°R}(70°F + 459.67°R)}(1 + 0.012) = 0.0725 \frac{\text{lb}_m}{\text{ft}^3}$$

c) $R = \frac{PV}{mT} = \frac{(14.32 \text{ lb}_f/\text{in}^2)(144 \text{ in}^2/\text{ft}^2)}{(0.0725 \text{ lb}_m/\text{ft}^3)(70 + 459.67°R)} = 53.70 \frac{\text{ft lb}_f}{\text{lb}_m \,°R}$

2.2 Cont'd

d) Based upon $m = m_x$, by Eq. (2.23)

$$c_p = c_{px} + \frac{m_y}{m_x} c_{py}$$

From Table A.5E, $c_{px} = c_{pa} = 0.2403 \text{ Btu/lb}_{m_a}{}^\circ F$

From Table A.6E, $c_{py} = c_{p_v} = 0.450 \text{ Btu/lb}_{m_v}{}^\circ F$

$$c_p = 0.2403 + (0.012)\, 0.450 = 0.246 \frac{\text{Btu}}{\text{lb}_{m_a}{}^\circ F}$$

2.3

Procedure follows that outlined in Prob. 2.2

a) $m_y / m_x = 0.01 \ kg_v / kg_a$

$$P_y = \frac{95 \ kPa}{1 + \frac{1}{(0.01)(1.607)}} = 1.50 \ kPa$$

b) $\rho = \dfrac{(95 - 0.01) \ kPa}{\left(0.2870 \ \frac{kJ}{kg \ K}\right)(20 + 273.15 K)} = 1.13 \ kg/m^3$

c) $R = \dfrac{(95 \ kPa)}{(1.13 \ kg/m^3)(20 + 273.15 K)} = 0.287 \ kJ/kg \ K$

d) From Table A.5SI, $c_{Px} = c_{Pa} = 1.006 \ kJ/kg_a \ K$

From Table A.6SI, $c_{Py} = c_{Pv} = 1.882 \ kJ/kg_v \ K$

$c_P = 1.006 + (0.01)(1.882) = 1.025 \ kJ/kg_v \ K$

2.4

a) let ρ_1 be the correct density and ρ_2 the density computed using the perfect gas law

By Eq. (2.24), $\dfrac{\rho_2}{\rho_1} = Z$

By Fig. 2.1 at $-20°F$, 10 atm, $Z = 0.9892$

$\text{Error} = (100)\dfrac{(\rho_1 - \rho_2)}{\rho_1} = 100(1 - 0.9892) = 1.08\%$

b) $P = 1\text{ MPa}\left(\dfrac{10^3 \text{ kPa}}{1 \text{ MPa}}\right)\left(\dfrac{1 \text{ atm}}{101.325 \text{ kPa}}\right) = 9.87 \text{ atm}$

By Fig. 2.1 at $-30°C$, 9.87 atm, $Z = 0.9895$

$\text{Error} = 100(1 - 0.9895) = 1.05\%$

2.5

By Table A.1E, $P = 0.95034 \text{ psia}$, $v = 349.98 \text{ ft}^3/\text{lb}_m$

$R_v = 85.76 \text{ ft lb}_f/\text{lb}_m \, °R$

Thus

$$Z = \frac{Pv}{RT} = \frac{(0.95034 \text{ lb}_f/\text{in}^2)(144 \text{ in}^2/\text{ft}^2)(349.98 \text{ ft}^3/\text{lb}_m)}{(85.76 \text{ ft lb}_f/\text{lb}_m \, °R)(100 + 459.67 \, °R)}$$

$Z = 0.998$

2.6

The energy balance on the radiator becomes

$$|\dot{Q}_{1-2}| = \dot{m}_{st}(h_1 - h_2)$$

where $|\dot{Q}_{1-2}| = 5000$ Btu/hr

at ①, by Table A.1E, $h_f = 184.64$, $h_g = 1151.89 \frac{Btu}{lbm}$

$$h_1 = h_f + x_1(h_g - h_f) = 184.64 + 0.97(1151.89 - 184.64)$$

$$h_1 = 1122.87 \text{ Btu/lbm}$$

at ②, by Table A.1E, $h_2 = h_f(200°F) = 168.34 \frac{Btu}{lbm}$

Thus

$$\dot{m}_{st} = \frac{|\dot{Q}_{1-2}|}{h_1 - h_2} = \frac{5000 \text{ Btu/hr}}{(1122.87 - 168.34) \text{ Btu/lbm}}$$

$$\dot{m}_{st} = 5.24 \text{ lbm/hr}$$

2.7

The energy balance on the evaporator is

$$\dot{Q}_{1-2} = \dot{m}_{ref}(h_2 - h_1)$$

at ①, by Table A.2 SI, $h_f = 88.73$, $h_g = 1417.81 \frac{kJ}{kg}$

$$h_1 = 88.73 + 0.3(1417.81 - 88.73) = 487.45 \frac{kJ}{kg}$$

at ②, by Table A.2 SI, $h_2 = h_g = 1417.81 \frac{kJ}{kg}$

Thus

$$\dot{Q}_{1-2} = 1\frac{kg}{s}(1417.81 - 487.45)\frac{kJ}{kg} = 930.4 \text{ kW}$$

2.8

By Eq. (2.42)

$$\dot{I} = \dot{m}_{st}(s_2 - s_1) - \frac{\dot{Q}}{T_{air}}$$

at ①, by Table A.1E, $s_f = 0.31869$, $s_g = 1.7494 \frac{Btu}{lbm \cdot °R}$

$$s_1 = s_f + x_1(s_g - s_f) = 0.31869 + 0.97(1.7494 - 0.31869)$$

$$s_1 = 1.7065 \; Btu/lbm \cdot °R$$

at ②, by Table A.1E, $s_2 = s_f(200°F) = 0.29430 \frac{Btu}{lbm \cdot °R}$

Thus

$$\dot{I} = 5.24 \frac{lbm}{hr}(0.29430 - 1.7065)\frac{Btu}{lbm \cdot °R} - \frac{(-5000) \; Btu/hr}{(70 + 459.67 \; °R)}$$

$$\dot{I} = 2.04 \frac{Btu}{hr \cdot °R}$$

2.9

By Eq. (2.42)

$$\dot{I} = \dot{m}_{ref}(s_2 - s_1) - \frac{\dot{Q}_{1-2}}{T_{air}}$$

at ①, by Table A.2 SI, $s_f = 0.3644$, $s_g = 5.6145 \frac{kJ}{kg\,K}$

$s_1 = s_f + x_1(s_g - s_f) = 0.3644 + 0.3(5.6145 - 0.3644)$

$s_1 = 1.9394 \; kJ/kg\,K$

at ②, by Table A.2 SI, $s_2 = s_g(-20°C) = 5.6145 \frac{kJ}{kg\,K}$

Thus

$$\dot{I} = 1\,\frac{kg}{s}\left(5.6145 - 1.9394\right)\frac{kJ}{kg\,K} - \frac{930.4\,kW}{(-10 + 273.15\,K)}$$

$\dot{I} = 0.1395 \; kW/K$

2.10

By Eq. (2.86)

$$R_t = R_i + \left(\frac{l_1}{k_1} + \frac{l_2}{k_2} + \frac{l_3}{k_3}\right) + R_o$$

$$R_i = 1/h_i = 1/1.5 \text{ Btu/hr ft}^2\text{°F} = 0.667 \frac{\text{hr ft}^2\text{°F}}{\text{Btu}}$$

$$\frac{l_1}{k_1} = \frac{0.5 \text{ in}}{0.1 \text{ Btu/hr ft °F}} \left(\frac{\text{ft}}{12 \text{ in}}\right) = 0.417 \frac{\text{hr ft}^2\text{°F}}{\text{Btu}}$$

$$\frac{l_2}{k_2} = \frac{3.5 \text{ in}}{0.03 \text{ Btu/hr ft °F}} \left(\frac{\text{ft}}{12 \text{ in}}\right) = 9.722 \frac{\text{hr ft}^2\text{°F}}{\text{Btu}}$$

$$\frac{l_3}{k_3} = \frac{4.0 \text{ in}}{0.75 \text{ Btu/hr ft °F}} \left(\frac{\text{ft}}{12 \text{ in}}\right) = 0.444 \frac{\text{hr ft}^2\text{°F}}{\text{Btu}}$$

$$R_o = 1/h_o = 1/6 \text{ Btu/hr ft}^2\text{°F} = 0.167 \frac{\text{hr ft}^2\text{°F}}{\text{Btu}}$$

Thus

$$R_t = 0.667 + 0.417 + 9.722 + 0.444 + 0.167 = 11.417 \frac{\text{hr ft}^2\text{°F}}{\text{Btu}}$$

$$U = 1/R_t = 1/11.417 \text{ hr ft}^2\text{°F/Btu}$$

$$U = 0.0876 \text{ Btu/hr ft}^2\text{°F}$$

By Eq. (2.83), $Q/A = U(t_i - t_o)$

$$\frac{Q}{A} = 0.0876 \frac{\text{Btu}}{\text{hr ft}^2\text{°F}} (70-0)\text{°F} = 6.13 \frac{\text{Btu}}{\text{hr ft}^2}$$

2.11

We will use Eq. (2.52) to calculate h

$$\vec{V} = \frac{\dot{V}}{A} = \frac{2000 \text{ cm}^3/\text{s}}{\pi (3.0 \text{cm})^2/4} = 2.83 \text{ m/s}$$

From Table A.6 SI
- $k = 0.6154$ W/m°C
- $\mu = 7.97.7 \times 10^{-6}$ kg/m s
- $Pr = 6.24$

From Table A.1 SI
- $v_f = 0.001005$ m³/kg

$$Re_D = \frac{(2.83 \text{ m/s})(0.03 \text{ m})}{(797.7 \times 10^{-6} \text{ kg/ms})(0.001005 \text{ m}^3/\text{kg})} = 1.059 \times 10^5$$

Flow is Turbulent so Eq. (2.52) is valid with the coefficients listed

$$h = \frac{k}{D} C Re_D^{0.8} Pr^{0.4}$$

$$h = \frac{0.6154 \text{ W/m°C}}{0.03 \text{ m}} (0.023)(1.059 \times 10^5)^{0.8} (6.24)^{0.4}$$

$$h = 10,270 \text{ W/m}^2\text{°C}$$

2.12

We will use Eq. (2.54) neglecting the viscosity ratio term to determine the heat transfer coefficient

$$Re_D = \vec{V}D/\nu = \rho\vec{V}D/\mu$$

From Table A.5E at the free stream temp. of 70°F,

$\rho = 0.07493 \; lb_m/ft^3$
$\mu = 0.04400 \; lb_m/hr\,ft$
$k = 0.01491 \; Btu/hr\,ft\,°F$
$Pr = 0.709$

$$Re_D = \frac{(0.07493 \; lb_m/ft^3)(10 \; ft/s)(5 \; in)}{(0.04400 \; lb_m/hr\,ft)(12 \; in/ft)}\left(\frac{3600 \; s}{hr}\right) = 2.554 \times 10^4$$

$$h = \frac{k}{D}\left(0.4 \; Re_D^{0.5} + 0.06 \; Re_D^{2/3}\right) Pr^{0.4}$$

$$h = \frac{0.01491 \; Btu/hr\,ft\,°F}{5 \; in \; (1 \; ft/12 \; in)}\left(0.4(2.554\times10^4)^{0.5} + 0.06(2.554\times10^4)^{2/3}\right) 0.709^{0.4}$$

$$h = 3.62 \; Btu/hr\,ft^2\,°F$$

$$\dot{Q} = hA(t_p - t_{air})$$

For 1 ft of pipe

$$\frac{\dot{Q}}{ft} = \frac{3.62 \; Btu}{hr\,ft^2\,°F}\left(\pi \; 5 \; in\right)\left(\frac{ft}{12 \; in}\right)(150-70)°F \; (1 \; ft)$$

$$\frac{\dot{Q}}{ft} = 379 \; Btu/hr$$

2.13

The natural convection heat transfer coefficient will be determined from Eq. (2.57)

$$Ra_D = Gr_D Pr = \frac{g\beta D^3 \Delta t}{\nu^2} Pr = \frac{g\beta D^3 \Delta t}{(\mu/\rho)^2} Pr$$

From Table A.5E at the free stream temp. 70°F
- $\rho = 0.07493$ lbm/ft³
- $\mu = 0.04400$ lbm/hr ft
- $k = 0.01491$ Btu/hr ft °F
- $Pr = 0.709$

For gases, $\beta = \frac{1}{T} = 1/(70+459.67°R) = 1.888 \times 10^{-3}$ °R⁻¹

$$Ra_D = \frac{\left(32.174 \frac{ft}{s^2}\right)\left(1.888 \times 10^{-3} \frac{1}{°R}\right)(5 in)^3 (150-70)°F (0.709)}{\left(\frac{0.04400 \text{ lbm/hr ft}}{0.07493 \text{ lbm/ft}^3}\right)^2 \left(\frac{1728 \text{ in}^3}{ft^3}\right)\left(\frac{hr}{3600 s}\right)^2}$$

$$Ra_D = 9.37 \times 10^6$$

By Eq. (2.57)

$$h_m = \frac{2(0.01491 \text{ Btu/hr ft °F})/(5 in)(ft/12 in)}{\ln\left[1 + \frac{2}{\left[\left(0.518(9.37\times 10^6)^{1/4}\left[1+\left(\frac{0.559}{0.709}\right)^{3/5}\right]^{-5/12}\right)^{15} + \left(0.1(9.37\times 10^6)^{1/3}\right)^{15}\right]^{1/15}}\right]}$$

$$h_m = 0.848 \text{ Btu/hr ft}^2 °F$$

$$\frac{\dot{Q}}{ft} = \frac{0.848 \text{ Btu}}{\text{hr ft}^2 °F}(\pi \cdot 5 in)(1 ft)\left(\frac{ft}{12 in}\right)(150-70)°F = 88.8 \text{ Btu/hr}$$

2.14

We will use Eq. (2.60) to determine the natural convection heat transfer coefficient in the window cavity with $L = 1$ cm, $H = 1$ m

$$Ra_L = \frac{g L^3 \Delta t \, Pr}{T (\mu/\rho)^2}$$

From Table A.SSI at 10°C
- $\rho = 1.248$ kg/m³
- $\mu = 1.765 \times 10^{-5}$ kg/m s
- $k = 0.02493$ W/m K
- $Pr = 0.712$

$$Ra_L = \frac{(9.8 \, m/s^2)(10^{-2} m)^3 (15°C)(0.712)}{(283.15°K)\left(\frac{1.765 \times 10^{-5} \, kg/ms}{1.248 \, kg/m^3}\right)^2} = 1848$$

$$Nu_{L_{conv}} = \frac{1}{2}\left[\left(0.67\left[1848(0.01)\right]^{1/4}\left[1+\left(\frac{0.559}{0.712}\right)^{3/5}\right]^{-5/12}\right)^{15} + \left(0.1[1848]^{1/3}\right)^{15}\right]^{1/15}$$

$$Nu_{L_{conv}} = 0.619$$

$$h_m = \frac{k}{L}\left[1 + \left(Nu_{L_{conv}}\right)^{15}\right]^{1/15}$$

$$= \frac{0.02493 \, W/m K}{0.01 \, m}\left[1 + (0.619)^{15}\right]^{1/15} = 2.493 \, W/m^2 K$$

$$\dot{Q} = h_m A (t_h - t_c)$$

$$= 2.493 \frac{W}{m^2 K} (1 m^2)(15°C) = 37.4 \, W$$

2.15

From Fig. 2.8b

150 kPa → 10 kW/m² → h

$h \approx 1500 \text{ W/m}^2\text{K}$

$q = h \Delta t$

$\Delta t = q/h = \dfrac{10 \text{ kW/m}^2}{1500 \text{ W/m}^2\text{K}} = 6.67 \text{ K}$

From Table A.2 SI, the saturation temperature at 150 kPa is -25.23 C

Therefore the surface temperature becomes

$t_{surface} = t_{sat} + 6.67 \text{ C}$

$= -25.23 + 6.67 = -18.56 \text{ °C}$

2.16

a) We will use Eq. (2.61) to determine the heat transfer coefficient

From Table A.8E at 10°F
$$\mu_\ell = 0.581 \text{ lbm/ft hr}$$
$$k_\ell = 0.0586 \text{ Btu/hr ft °F}$$

From Table A.3E at 10°F
$$h_{fg} = h_g - h_f = 105.4 - 13.1 = 92.3 \text{ Btu/lbm}$$

$$Re_\ell = \frac{1.0 \text{ lbm/min}}{\frac{\pi (0.5 in)^2}{4}} \left(\frac{12 in}{ft}\right) \frac{(0.5 in)}{(0.581 \text{ lbm/ft hr})} \left(\frac{60 min}{hr}\right) = 3156$$

Let $L = 10 ft$

$$K = \frac{778 \text{ ft lbm}}{Btu} \frac{0.8 (92.3 \text{ Btu/lbm})}{10 \text{ ft}} = 5745$$

$$h_m = \frac{(0.0586 \text{ Btu/hr ft·°F}) \, 0.0082 \left(3156^2 (5745)\right)^{0.4}}{(0.5 in)(ft/12 in)}$$

$$h_m = 231 \text{ Btu/hr ft}^2 \text{°F}$$

(different choice of L will affect values for K and h_m)

b) The length necessary will depend on the temperature difference maintained between the tube and the fluid

For our choice of 10 ft in part a,

$$\dot{Q} = h_m A \Delta t = \dot{m} \Delta x h_{fg}$$

or $\Delta t = \dot{m} \Delta x h_{fg} / (h_m A)$

$$\Delta t = \frac{(1.0 \, lb_m/min) \, 0.8 \, (92.3 \, Btu/lb_m)}{\left(\frac{231 \, Btu}{hr \, ft^2 \, °F}\right)(\pi \, 0.5 in \, 10 \, ft)} \left(\frac{60 \, min}{hr}\right)\left(\frac{12 \, in}{ft}\right)$$

$\Delta t = 15 \, °F$

if the temperature difference is one-half this value, the necessary length would be doubled to 20 ft

2.17

We will use Eq. (2.63)

From Table A.8E at 100°F
- $c_{p_\ell} = 0.3162$ Btu/lbm °F
- $\mu_\ell = 0.338$ lbm/ft hr
- $k_\ell = 0.0466$ Btu/hr ft °F
- $Pr_\ell = 2.29$

From Table A.3E at 100°F
- $\rho_\ell = 71.24$ lbm/ft³
- $\rho_v = 3.89$ lbm/ft³
- $h_{fg} = 72.80$ Btu/lbm

$$Re_c = \frac{(1.0 \text{ lbm/min})(0.5 \text{ in})}{\frac{\pi (0.5 \text{ in})^2}{4} \cdot \frac{0.338 \text{ lbm}}{\text{hr ft}}} \left(\frac{71.24}{3.89}\right)^{0.5} \left(\frac{12 \text{ in}}{\text{ft}}\right) \left(\frac{60 \text{ min}}{\text{hr}}\right) = 23,214$$

$$M = \frac{72.80 \text{ Btu/lbm}}{\frac{0.3162 \text{ Btu}}{\text{lbm °F}}(8°F)} = 28.78$$

$$Nu_\ell = 0.1 (2.29)^{1/3} (28.78)^{1/6} (23,214)^{2/3} = 188$$

$$h = \frac{188 (0.0466 \text{ Btu/hr ft °F})}{0.5 \text{ in}}\left(\frac{12 \text{ in}}{\text{ft}}\right) = 210 \frac{\text{Btu}}{\text{hr ft}^2 \text{°F}}$$

2.18
Using Eq. (2.74) with $\varepsilon_1 = 1$, $\varepsilon_2 = 1$, and $F_{1-2} = 1.0$

$$\frac{\dot{Q}_{1-2}}{A_1} = \sigma (T_1^4 - T_2^4)$$

$$= 0.1713 \times 10^{-8} \frac{Btu}{hr\ ft^2\ {}^\circ R^4} \left[(10+459.67)^4 - (-40+459.67)^4\right] {}^\circ R^4$$

$$\frac{\dot{Q}_{1-2}}{A_1} = 30.22 \frac{Btu}{hr\ ft^2}$$

2.19

We will use the following equation assuming the surroundings behave as a black body at 70°F

$$\dot{Q}_{1-2} = \varepsilon_1 A_1 \sigma (T_1^4 - T_2^4)$$

for a length of 1 ft

$$\frac{\dot{Q}_{1-2}}{ft} = 0.85 \; \frac{\pi \sin(1\,ft)}{(12\,in/ft)} \; 0.1713 \times 10^{-8} \frac{Btu}{hr\,ft^2\,°R^4} \left(609.67^4 - 529.67^4\right)°R$$

$$\frac{\dot{Q}_{1-2}}{ft} = 113 \; \frac{Btu}{hr}$$

2.20

We will use Eq. (2.74) with $F_{1-2} = 1.0$ and $A_1/A_2 = 1.0$

$$\dot{Q}_{1-2} = \frac{A_1 \sigma (T_1^4 - T_2^4)}{\frac{1}{\varepsilon_1} + \frac{1}{\varepsilon_2} - 1}$$

$$= \frac{1 m^2 (5.670 \times 10^{-8} W/m^2 K^4)(288.15 K^4 - 273.15 K^4)}{\frac{1}{0.9} + \frac{1}{0.9} - 1}$$

$$\dot{Q}_{1-2} = 61.6 \, W$$

2.21

Let the radiative heat transfer be written as

$$\dot{Q}_r = h_r A(t_1 - t_2)$$

$$h_r = \frac{\dot{Q}_r}{A(t_1 - t_2)}$$

$$= \frac{113 \text{ Btu/hr } (12 \text{ in/ft})}{\pi \, 5 \text{ in } (1 \text{ ft})(150 - 70)°F} = 1.079 \frac{\text{Btu}}{\text{hr ft}^2 °F}$$

The combined convective/radiative heat transfer coefficient becomes

$$h = h_c + h_r$$

$$= 0.848 + 1.079 = 1.927 \frac{\text{Btu}}{\text{hr ft}^2 °F}$$

2.22

Let the radiative heat transfer coefficient be written as

$$h_r = \frac{\dot{Q}_r}{A(t_1 - t_2)}$$

$$= \frac{61.6 \text{ W}}{1 \text{m}^2 (15-0)°C} = 4.11 \frac{\text{W}}{\text{m}^2 °C}$$

The combined convective/radiative heat transfer coefficient becomes

$$h = h_c + h_r$$

$$= 2.493 + 4.11 = 6.60 \frac{\text{W}}{\text{m}^2 °C}$$

Chapter 3

3.1
Properties at states 1, b, 3 and d are taken from Table A.3SI, properties at state c are taken from Fig. E.2SI. State points refer to Fig. 3.2.

State	Phase	t °C	P MPa	h kJ/kg	v m3/kg	s kJ/kg °K
1	sat. liquid	30	1.192	81.26	0.00085	0.3004
b	liq./vap. mix	10	0.681	80.00	0.00485	0.3004
3	sat. vapor	10	0.681	253.43	0.03472	0.9129
c	sup. vapor	30	1.0	263	0.025	0.9129
d	sat. vapor	30	1.192	259.12	0.01975	0.8872

determine the quality at state b from the known entropy
$x_b = (s_b-s_f)/(s_g-s_f) = (0.3004-0.2173)/(0.9129-0.2173) = 0.1195$

$v_b = v_f + x_b(v_{fg}) = 0.0008 + 0.1195(0.03472-0.0008) = 0.00485$ m^3/kg
$h_b = h_f + x_b(h_{fg}) = 56.46 + 0.1195(253.43-56.46) = 80.00$ kJ/kg
C.O.P. = $_bq_3/[_3w_c + _cw_d - _1w_b]$
C.O.P. = $(h_3 - h_b)/[(h_c - h_3) + T_c(s_c - s_d) - (h_c-h_d) - (h_b-h_1)]$
$=(253.43-80.00)/[(263-253.43) +303.15(0.9129-0.8872)-(263-259.12) -(80.00-81.26)]=14.19$
C.O.P. = $T_3/(T_1-T_3) = 283.15/20 = 14.16$

3.2
States 1,a,3,b and c have been determined in Problem 3.1. States 2 and 4 are determined here from Table A.3SI and Fig. C.2SI respectively.

State	Phase	t °C	P MPa	h kJ/kg	v m3/kg	s kJ/kg °K
1	sat. liquid	30	1.192	81.26	0.000852	0.3004
a	liq./vap. mix	10	0.681	80.00	0.00485	0.3004
2	liq./vap. mix	10	0.681	81.26	0.00507	0.3049
3	sat. vapor	10	0.681	253.43	0.03472	0.9129
b	sup. vapor	30	1.0	263	0.025	0.9129
4	sup. vapor	41	1.192	269	0.021	0.9129
c	sat. vapor	30	1.192	259.12	0.01975	0.8872

determine the quality at state 2 from the known enthalpy
$x_2 = (h_2-h_f)/(h_g-h_f) = (81.26-56.46)/(253.43-56.46) = 0.1259$

$v_2 = v_f + x_2(v_{fg}) = 0.0008 + 0.1259 (0.03472-0.0008) = 0.00507$ m^3/kg
$s_2 = s_f + x_2(s_{fg}) = 0.2173 + 0.1259(0.9129-0.2173) = 0.3049$ kJ/kg°K
A1 = (269-259.12) - 303.15(0.9129-0.8872) = 2.089 kJ/kg
A2 = 81.26-80.00 = 1.26 kJ/kg

3.3
Properties for R-22 are obtained from Table A.3E and Fig. C.2E.

State	Phase	t °F	P psia	v ft3/lbm	h Btu/lbm	s Btu/lbm°R
1	sat. liquid	90	183.05	0.01375	36.2	0.0739
2	liq./vap. mix	0	38.64	0.38514	36.2	0.0797
3	sat. vapor	0	38.64	1.373	104.5	0.2281
4	sup. vapor	139	183.05	0.35	121.2	0.2281

$x_2 = (h_2-h_f)/h_{fg} = (36.2-10.4)/(104.5-10.4) = 0.2742$

$v_2 = v_f + x_2(v_{fg}) = 0.01193+0.2742(1.373-0.01193) = 0.38514$ ft^3/lbm
$s_2 = s_f + x_2(s_{fg}) = .0236 + 0.2742(0.2281-0.0236) = 0.0797$ Btu/lbm °R
(a) C.O.P. = $(104.5-36.2)/(121.2-104.5) = 4.09$
(b) $\eta_R = 4.09(90)/459.6 = 0.801$
(c) $\dot{m} = (15)(200)/(104.5-36.2) = 43.92$ lbm/min
(d) Hp = $(43.92)(121.2-104.5)/42.4 = 17.3$
(e) C.D. = $43.92(1.373) = 60.30$ ft^3/min

Figure E-2sI   Pressure-Enthalpy Diagram for R22

Problem 3.2

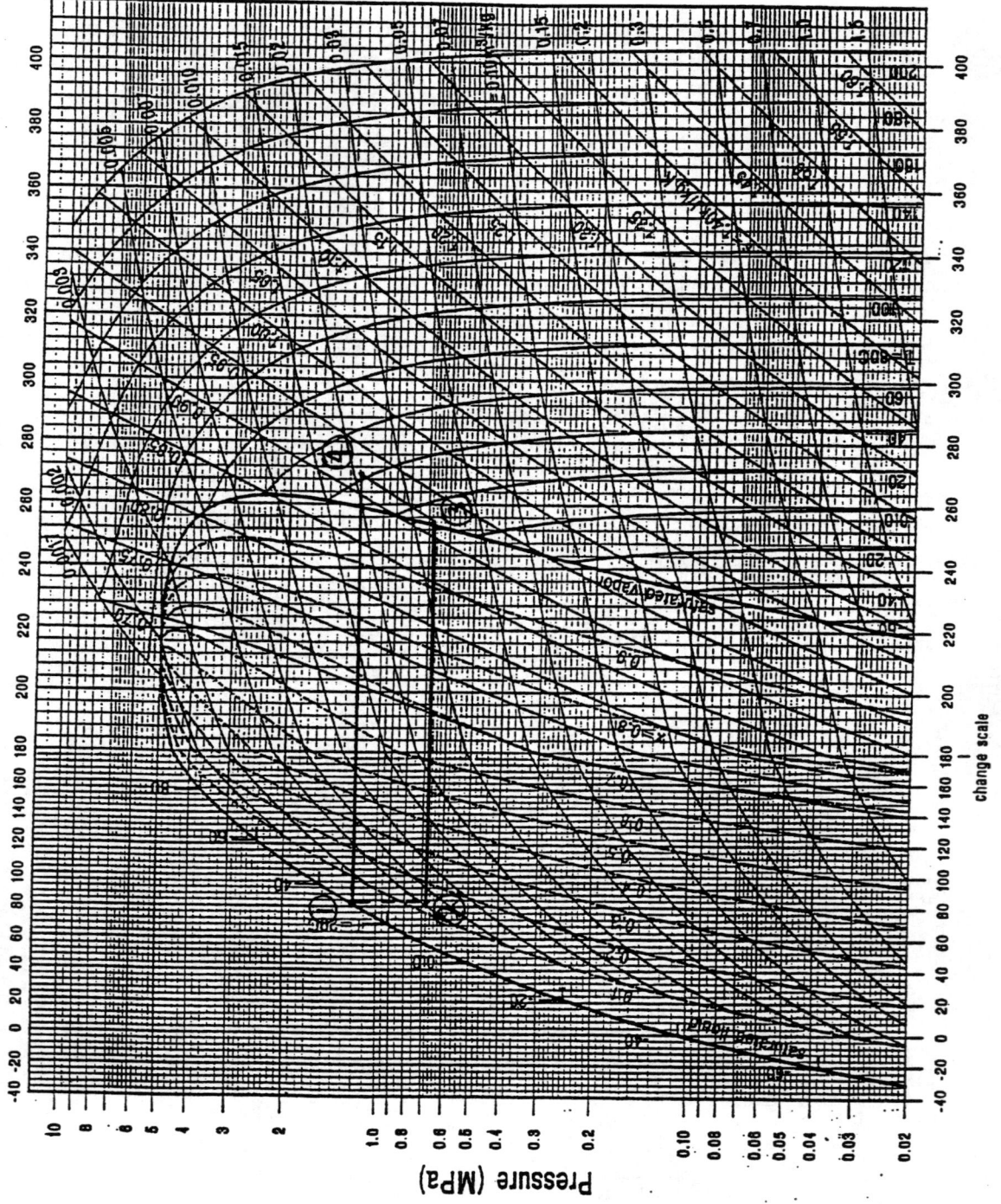

Figure E-2E Pressure-Enthalpy Diagram for R22

Problem 3.3:

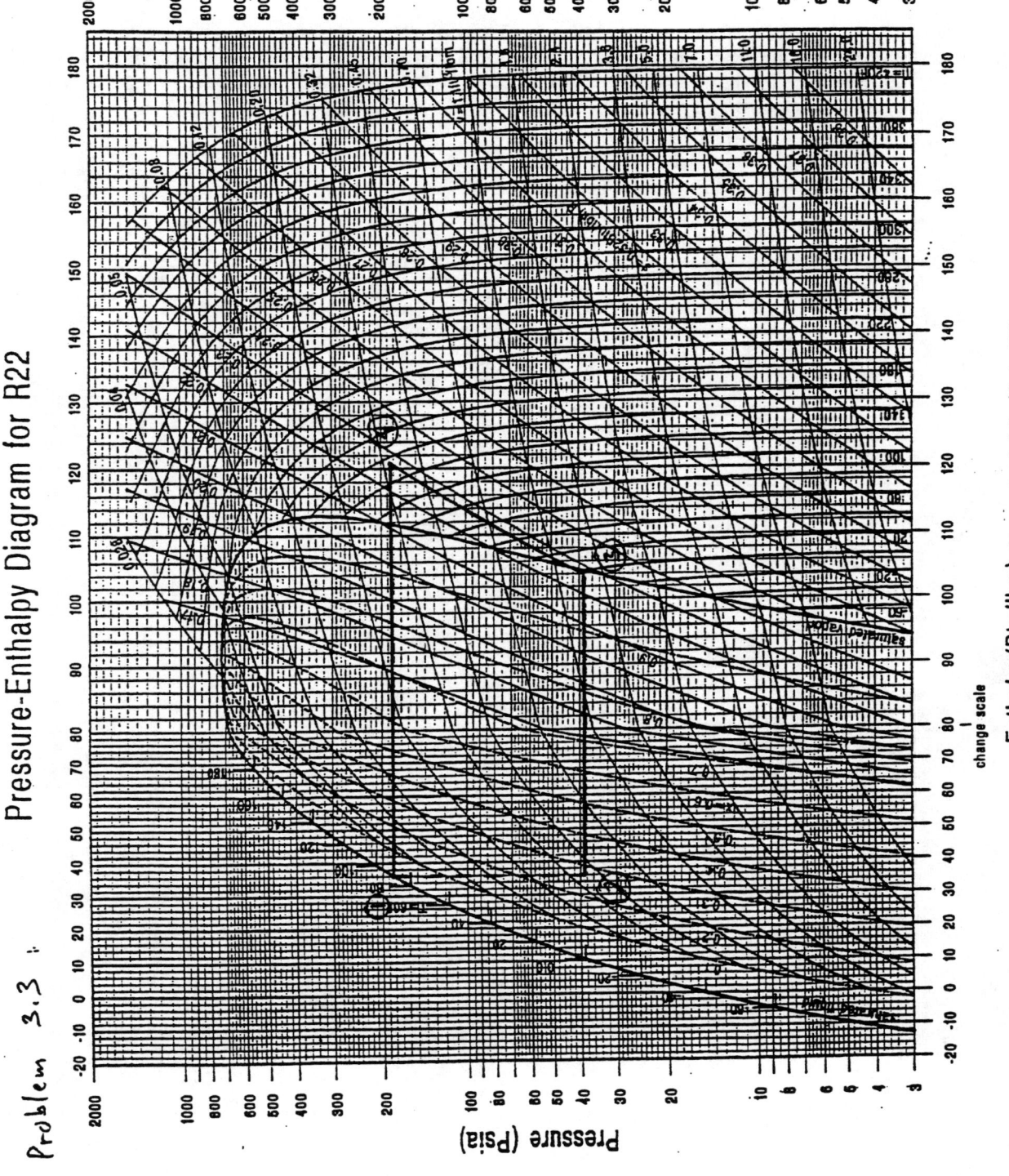

3.4
Solution procedure follows that outlined in Example 3.3.
State leaving the condenser is compressed liquid and state leaving the evaporator is superheated vapor from observing Table A.2E. State points are as shown in Figure 3.8.
(a) $h_1 = h_2 = h_f(104°F) = 159.7$ Btu/lbm, $h_3 = 615$ Btu/lbm, $h_4 = 802$ Btu/lbm
   C.O.P. = $(615-159.7)/(802-615) = 2.43$
(b) $\dot{m} = 25(200)/(615-159.7) = 10.98$ lbm/min
   C.D. = $10.98(16.0) = 176$ ft^3/min

3.5
Property data obtained from Table A.3E and Fig. C.2E.

State	Phase	t °F	P psia	h Btu/lbm	s Btu/lbm°R
1	sat. liquid	100	210.55	39.3	
2	subcooled liq.	80	210.55	33.1	
3	liq./vap. mix		31.15		
4	sat. vapor	-10	31.15	103.5	
5	superheated vap.		31.15	109.9	0.245
6	superheated vap.		210.55	133.5	0.245

Energy balance on heat exchanger to determine h_5.
$h_5 = h_1 + h_4 - h_2 = 39.3 + 103.5 - 33.1 = 109.7$ Btu/lbm
$s_5 = s_6 = 0.245$ Btu/lbm °R, $h_6 = 133.7$ Btu/lbm
Hp/ton = $[(133.5-109.9)/(103.5-33.1)](1.341 \times 10^{-3}$ hp$/2.844 \times 10^{-4}$ ton$) = 1.58$

3.6
Solution procedure follows problem 3.5.

Cycle #	t °F	performance hp/ton
1	60	1.56
2	40	1.56
3	20	1.55
4	0	1.54

Figure E-2E

Prob. 3.5

Pressure-Enthalpy Diagram for R22

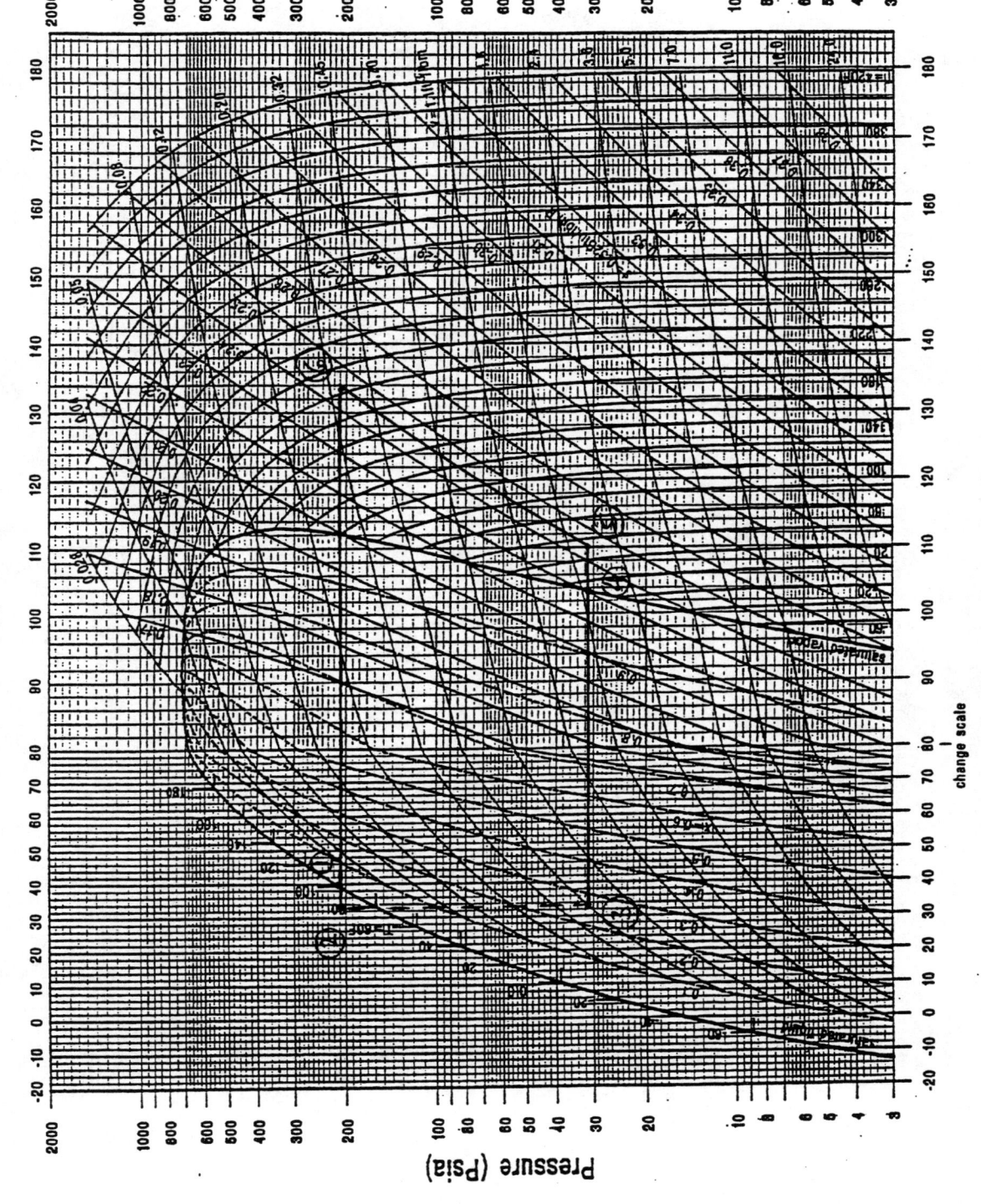

3.7 Solve Ex. 3.2 using ammonia (Table A.2SI & Fig. C.1S)
$T_o = 30°C$, $T_R = 0°C$

(a) $w_{carnot} = 30K(5.3301 - 1.2008) \, kJ/kg\,K = 123.9 \, kJ/kg$

(b) $q_{carnot} = 273.15K(4.1293) = 1127.9 \, kJ/kg$

(c) By Eq. (3.7), $A_1 = (1580 - 1466.17) - (303.15)(5.3301 - 4.9732)$
$= 222.0 \, kJ/kg$

(d) $x_a = \dfrac{1.2008 - 0.7102}{5.3301 - 0.7102} = 0.1062$

$h_a = 180.07 + 0.1062(1442.0 - 180.07) = 314.09 \, kJ/kg$

$A_2 = 322.58 - 314.09 = 8.49 \, kJ/kg$

(e) $A_3 = A_2 = 8.49 \, kJ/kg$

(f) $\eta_R = \dfrac{(1442.0 - 322.58)}{(1580 - 1442.0)} \dfrac{30}{273.15} = 0.891$

3.8

(a) $\text{C.O.P.}_{\text{Carnot}_A} = \dfrac{499.67°R}{50°R} = 9.993$

$\dot{W}_{A_{\text{Carnot}}} = \dfrac{\dot{Q}_A}{\text{C.O.P.}_A} = \dfrac{10\ \text{Tons}}{9.993} = 1.001\ \text{Tons}$

$\text{C.O.P.}_{\text{Carnot}_B} = \dfrac{459.67°R}{90°R} = 5.107$

$\dot{W}_{B_{\text{Carnot}}} = \dfrac{\dot{Q}_B}{\text{C.O.P.}_B} = \dfrac{5\ \text{Tons}}{5.107} = 0.979\ \text{Tons}$

$\text{C.O.P.}_{\text{Carnot}} = \dfrac{\dot{Q}_A + \dot{Q}_B}{\dot{W}_{A_{\text{Carnot}}} + \dot{W}_{B_{\text{Carnot}}}} = \dfrac{(10+5)\ \text{Tons}}{(1.001+0.979)\ \text{Tons}} = 7.576$

(b)

state	t (°F)	P (psia)	h (Btu/lbm)
1	90	180.73	143.3
2	0	30.42	143.3
3	0	30.42	610.8
8		180.73	724

$\text{C.O.P.} = \dfrac{\dot{Q}_A + \dot{Q}_B}{\dot{m}_1 (h_8 - h_3)}$

$\dot{m}_1 = \dot{m}_2 + \dot{m}_5 = \dfrac{\dot{Q}_A}{h_3 - h_2} + \dfrac{\dot{Q}_B}{h_3 - h_2} = \dfrac{\dot{Q}_A + \dot{Q}_B}{h_3 - h_2}$

$\text{C.O.P.} = \dfrac{(\dot{Q}_A + \dot{Q}_B)/(h_8 - h_3)}{(\dot{Q}_A + \dot{Q}_B)/(h_3 - h_2)} = \dfrac{h_3 - h_2}{h_8 - h_3}$

$= \dfrac{610.8 - 143.3}{724 - 610.8} = 4.130$

$\eta_R = \dfrac{\text{C.O.P.}}{\text{C.O.P.}_{\text{Carnot}}} = 4.130/7.576 = 0.545$

Prob. 3.8 cont'd.

(c)

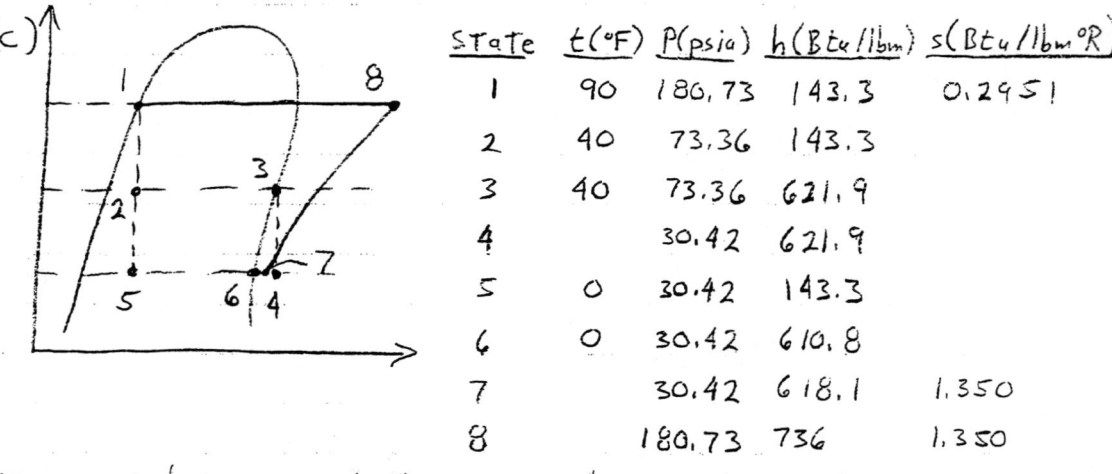

State	t(°F)	P(psia)	h(Btu/lbm)	s(Btu/lbm °R)
1	90	180.73	143.3	0.2951
2	40	73.36	143.3	
3	40	73.36	621.9	
4		30.42	621.9	
5	0	30.42	143.3	
6	0	30.42	610.8	
7		30.42	618.1	1.350
8		180.73	736	1.350

Mass and energy balance to determine state 7

$$\dot{m}_4 + \dot{m}_6 = \dot{m}_7$$

$$\dot{m}_4 h_4 + \dot{m}_6 h_6 = \dot{m}_7 h_7$$

$$h_7 = (\dot{m}_4 h_4 + \dot{m}_6 h_6)/\dot{m}_7$$

$$\dot{m}_4 = \frac{\dot{Q}_A}{h_3 - h_2} = \frac{10 \text{ Tons } (200 \text{ Btu/min ton})}{(621.9 - 143.3) \text{ Btu/lbm}} = 4.179 \text{ lbm/min}$$

$$\dot{m}_6 = \frac{\dot{Q}_B}{h_6 - h_5} = \frac{5 \text{ Tons } (200)}{610.8 - 143.3} = 2.139 \text{ lbm/min}$$

$$h_7 = (4.179 \times 621.9 + 2.139 \times 610.8)/(4.179 + 2.139) = 618.1 \text{ Btu/lb}$$

$$C.O.P. = \frac{\dot{Q}_A + \dot{Q}_B}{\dot{m}_7 (h_8 - h_7)} = \frac{15 \text{ Tons } (200 \text{ Btu/min ton})}{6.318 \frac{\text{lbm}}{\text{min}} (736 - 618.1) \frac{\text{Btu}}{\text{lbm}}} = 4.027$$

$$\eta_R = \frac{C.O.P.}{C.O.P._{Carnot}} = \frac{4.027}{7.576} = 0.532$$

Prob. 3.8 cont'd.

(d)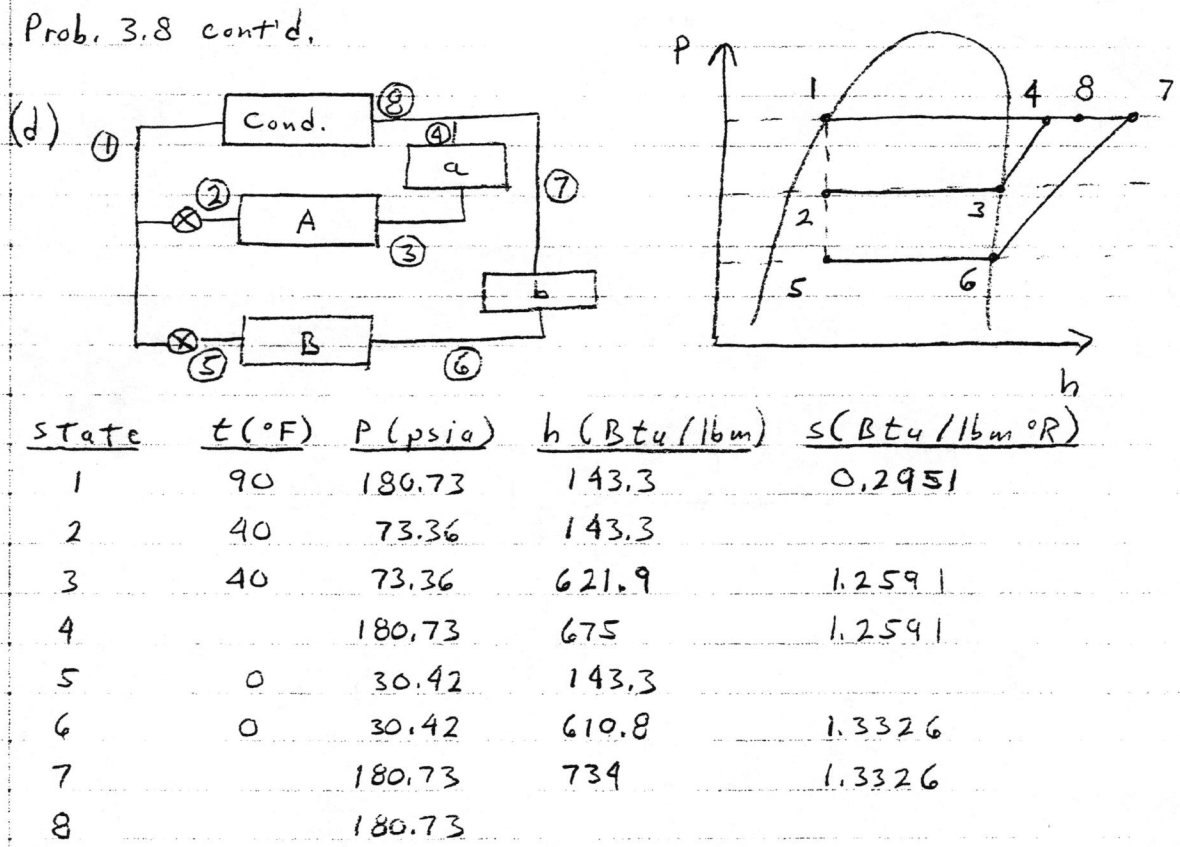

State	t (°F)	P (psia)	h (Btu/lbm)	s (Btu/lbm °R)
1	90	180.73	143.3	0.2951
2	40	73.36	143.3	
3	40	73.36	621.9	1.2591
4		180.73	675	1.2591
5	0	30.42	143.3	
6	0	30.42	610.8	1.3326
7		180.73	734	1.3326
8		180.73		

Evaporator mass flow rates same as part (c)

$$C.O.P. = \frac{\dot{Q}_A + \dot{Q}_B}{\dot{m}_4 (h_4 - h_3) + \dot{m}_6 (h_7 - h_6)}$$

$$= \frac{15 \text{ Tons} (200 \text{ Btu/min ton})}{4.179 \frac{lbm}{min}(675 - 621.9)\frac{Btu}{lbm} + 2.139 \frac{lbm}{min}(734 - 610.8)\frac{Btu}{lbm}}$$

C.O.P. = 6.180

$$\eta_R = \frac{C.O.P.}{C.O.P._{Carnot}} = \frac{6.180}{7.576} = 0.816$$

(e) Not much difference between parts (b) and (c) with a single compressor, considerably better performance with 2 compressors shown in part (d)

3.9

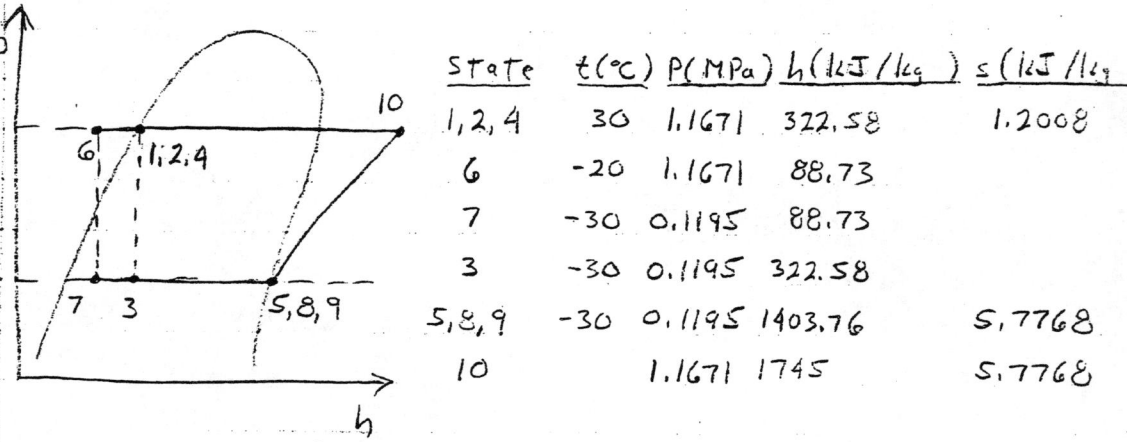

State	t(°C)	P(MPa)	h(kJ/kg)	s(kJ/kg K)
1,2,4	30	1.1671	322.58	1.2008
6	-20	1.1671	88.73	
7	-30	0.1195	88.73	
3	-30	0.1195	322.58	
5,8,9	-30	0.1195	1403.76	5.7768
10		1.1671	1745	5.7768

(a) with liquid cooler

energy balance on cooler to determine mass ratios

$$\dot{m}_3 (h_5 - h_3) = \dot{m}_6 (h_4 - h_6)$$

$$\frac{\dot{m}_3}{\dot{m}_6} = \frac{h_4 - h_6}{h_5 - h_3} = \frac{322.58 - 88.73}{1403.76 - 322.58} = 0.2163$$

$$\dot{m}_{10} = \dot{m}_6 + \dot{m}_3 = \dot{m}_6 + 0.2163\, \dot{m}_6 = 1.2163\, \dot{m}_6$$

$$C.O.P. = \frac{\dot{Q}_{7-8}}{\dot{W}_{9-10}} = \frac{\dot{m}_6 (h_8 - h_7)}{1.2163\, \dot{m}_6 (h_{10} - h_9)}$$

$$= \frac{(1403.76 - 88.73)}{1.2163 (1745 - 1403.76)} = 3.168$$

without liquid cooler

$$C.O.P. = \frac{h_8 - h_3}{h_{10} - h_8} = \frac{1403.76 - 322.58}{1745 - 1403.76} = 3.168$$

Prob. 3.9 cont'd.

(b) There is no change in cycle performance when a liquid cooler is introduced so from a theoretical view, liquid coolers are not useful.

(c) If there is a considerable pressure drop between the condenser exit (state 1) and the inlet to the expansion valve (state 6) caused by a long line or elevation change, it would be useful to include a liquid cooler to ensure no vapor enters the valve. If vapor exists in the line upstream of the valve, the valve will not function properly and will not reach its rated capacity.

3.10 R-22

State	t (°F)	P (psia)	h (Btu/lbm)	s (Btu/lbm·R)
1	90	183.05	36.2	0.0739
2	20	57.71	36.2	
3	20	57.71	15.8	
4	0	38.64	15.8	
5	0	38.64	104.5	0.2281
6		57.71	108.3	0.2281
7	20	57.71	106.4	
8		57.71	107.9	0.2275
9		183.05	121.2	0.2275

Mass and energy balance on flash tank to determine mass flow ratios

$$\dot{m}_2 = \dot{m}_3 + \dot{m}_7, \quad \dot{m}_2 h_2 = \dot{m}_3 h_3 + \dot{m}_7 h_7$$

$$\dot{m}_3 (h_2 - h_3) = \dot{m}_7 (h_7 - h_2)$$

$$\frac{\dot{m}_7}{\dot{m}_3} = \frac{h_2 - h_3}{h_7 - h_2} = \frac{36.2 - 15.8}{106.4 - 36.2} = 0.291, \quad \frac{\dot{m}_8}{\dot{m}_3} = \frac{\dot{m}_2}{\dot{m}_3} = 1.291$$

Find state 8 from mass and energy balances of mixing

$$\dot{m}_8 = \dot{m}_3 + \dot{m}_7, \quad \dot{m}_8 h_8 = \dot{m}_3 h_6 + \dot{m}_7 h_7$$

$$h_8 = \frac{\dot{m}_3}{\dot{m}_8} h_6 + \frac{\dot{m}_7}{\dot{m}_8} h_7 = \frac{h_6}{\dot{m}_8/\dot{m}_3} + \frac{\dot{m}_7/\dot{m}_3}{\dot{m}_8/\dot{m}_3} h_7$$

$$h_8 = \frac{108.3}{1.291} + \frac{0.291}{1.291} \cdot 106.4 = 107.9 \text{ Btu/lbm}$$

Locate state 9 at correct pressure, $S_8 = S_9$

Problem 3.10 cont'd

(a) $\text{C.O.P.} = \dfrac{\dot{Q}_{4-5}}{\dot{W}_{5-6} + \dot{W}_{8-9}} = \dfrac{\dot{m}_3 (h_5 - h_4)}{\dot{m}_3 (h_6 - h_5) + \dfrac{\dot{m}_8}{\dot{m}_3}(h_9 - h_8)}$

$$\text{C.O.P.} = \frac{104.5 - 15.8}{(108.3 - 104.5) + 1.291\,(121.2 - 107.9)} = 4.23$$

(b) when $\dot{m}_7 \to 0$, $h_4 \to h_1$ and $s_9 \to s_5$

$h_9^* = 121.5$ Btu/lbm

$$\text{C.O.P.} = \frac{h_5 - h_1}{h_9^* - h_5} = \frac{104.5 - 36.2}{121.5 - 104.5} = 4.02$$

(c) The use of the flash Tank increases the C.O.P. of this cycle from 4.02 to 4.23, a 5% increase.

3.11 Ammonia (see Fig. 3.14)

State	t (°F)	P (psia)	h (Btu/lbm)	s (Btu/lbm °R)	v (ft³/lb)
1	90	180.73	143.3		
2	15	43.15	143.3		
3	25	180.73	69.7		
4	-40	10.40	69.7		
5	-40	10.40	597.0	1.4222	24.867
6,7		43.15	677	1.4222	
8	15	43.15	615.3	1.3035	6.556
9		180.73		1.3035	

Assuming equal volumetric efficiencies

$$\dot{m}_8 / \dot{m}_5 = \frac{C.D._8 / v_8}{C.D._5 / v_5}$$

solving for $C.D._5$; $C.D._5 = C.D._8 (v_5/v_8)(\dot{m}_5/\dot{m}_8)$

Find mass flow rate ratio from mass and energy balance on shell-and-coil intercooler & expansion valve

$\dot{m}_1 = \dot{m}_8$, $\dot{m}_3 = \dot{m}_7$

$\dot{m}_1 h_1 + \dot{m}_7 h_7 = \dot{m}_3 h_3 + \dot{m}_8 h_8$

$\dot{m}_3 (h_7 - h_3) = \dot{m}_8 (h_8 - h_1)$

$$\frac{\dot{m}_3}{\dot{m}_8} = \frac{h_8 - h_1}{h_7 - h_3} = \frac{615.3 - 143.3}{677 - 69.7} = 0.777$$

$$C.D._5 = 100 \text{ ft}^3/\text{min} \left(\frac{24.867 \text{ ft}^3/\text{lbm}}{6.556 \text{ ft}^3/\text{lbm}} \right) 0.777 = 295 \text{ ft}^3/\text{min}$$

3.12 R-22

State	t(°C)	P(MPa)	h(kJ/kg)	s(kJ/kg K)
1	35	1.5	87.72	
2		0.5	87.72	
3	5	1.5	50.48	
4		0.2	50.48	
5	-20	0.2	243.1	0.985
6		0.5	266.5	0.985
7	5	0.5	254	
8		0.5	264	
9		1.5	294	

Energy balance on intercooler to obtain mass ratios

$$\dot{m}_3 h_1 + \dot{m}_7 h_2 = \dot{m}_3 h_3 + \dot{m}_7 h_7$$

$$\dot{m}_3 (h_1 - h_3) = \dot{m}_7 (h_7 - h_2)$$

$$\frac{\dot{m}_3}{\dot{m}_7} = \frac{h_7 - h_2}{h_1 - h_3} = \frac{254 - 87.72}{87.72 - 50.48} = 4.47$$

Mixing of streams 6 and 7 to determine h_8

$$\dot{m}_3 + \dot{m}_7 = \dot{m}_1, \quad 4.47\dot{m}_7 + \dot{m}_7 = 5.47\dot{m}_7 = \dot{m}_1, \quad \frac{\dot{m}_1}{\dot{m}_7} = 5.47$$

$$\dot{m}_3 h_6 + \dot{m}_7 h_7 = \dot{m}_1 h_8$$

$$h_8 = \frac{\dot{m}_3 h_6 + \dot{m}_7 h_7}{\dot{m}_1} = \frac{h_6}{\dot{m}_1/\dot{m}_3} + \frac{h_7}{\dot{m}_1/\dot{m}_7}$$

$$h_8 = \frac{h_6}{(1 + \dot{m}_7/\dot{m}_3)} + \frac{h_7}{\dot{m}_1/\dot{m}_7} = \frac{266.5}{(1 + 1/4.47)} + \frac{254}{5.47} = 264 \frac{Btu}{lbm}$$

Locate state 9 at 1.5 MPa, $S_9 = S_8$

Prob. 3.12 cont'd.

$$C.O.P. = \frac{\dot{Q}_{4-5}}{\dot{W}_{5-6} + \dot{W}_{8-9}} = \frac{\dot{m}_3(h_5 - h_4)}{\dot{m}_3(h_6 - h_5) + \dot{m}_1(h_9 - h_8)/\dot{m}_3}$$

$$C.O.P. = \frac{243.1 - 50.48}{(266.5 - 243.1) + 1.22(294 - 264)}$$

$$C.O.P. = 3.21$$

3.13 Ammonia

State	t (°F)	P (psia)	h (Btu/lbm)	s (Btu/lbm °R)	v (ft³/lbm)
1	100	211.95	155.0		
2,9	12	40.32	155.0		
3	20	211.95	64.2		
4	-50	7.66	64.2		
5	-50	7.66	593.1	1.4479	33.087
6		40.32	688	1.4479	
7	100	40.32	664.5		
8,10,11	12	40.32	614.5	1.3092	6.989
12		211.95		1.3092	

$$\frac{C.D._{LP}}{C.D._{HP}} = \frac{\dot{m}_5 \, v_5 / \eta_{v_{LP}}}{\dot{m}_{11} \, v_{11} / \eta_{v_{HP}}} \quad \text{(assume equal volumetric efficiencies)}$$

Need the mass ratio $\dot{m}_5 / \dot{m}_{11}$

Given: $\dot{m}_9 (h_{10} - h_9) = 4 \, \dot{m}_5 (h_5 - h_4)$

$$\frac{\dot{m}_9}{\dot{m}_5} = \frac{4(h_5 - h_4)}{h_{10} - h_9} = \frac{4(593.1 - 64.2)}{614.5 - 155.0} = 4.60$$

Energy balance on shell and coil intercooler to determine additional mass flow ratios

$\dot{m}_8 = \dot{m}_2 + \dot{m}_5$

$\dot{m}_5 h_1 + \dot{m}_2 h_2 + \dot{m}_5 h_7 = \dot{m}_5 h_3 + \dot{m}_8 h_8$

$\dot{m}_5 (h_1 + h_7) + \dot{m}_2 h_2 = \dot{m}_5 h_3 + (\dot{m}_2 + \dot{m}_5) h_8$

$\dot{m}_2 (h_2 - h_8) = \dot{m}_5 (h_3 - h_1 - h_7 + h_8)$

3.13 Cont'd.

$$\frac{\dot{m}_2}{\dot{m}_5} = \frac{h_3 - h_1 - h_7 - h_8}{h_2 - h_8}$$

$$= \frac{64.2 - 155.0 - 664.5 + 614.5}{155.0 - 614.5} = 0.306$$

Mass balance on mixing of streams 8 and 10

$$\dot{m}_8 + \dot{m}_9 = \dot{m}_{11} \quad \text{or} \quad (\dot{m}_2 + \dot{m}_5) + \dot{m}_9 = \dot{m}_{11}$$

dividing through by $\dot{m}_5$

$$\frac{\dot{m}_2}{\dot{m}_5} + 1 + \frac{\dot{m}_9}{\dot{m}_5} = \frac{\dot{m}_{11}}{\dot{m}_5} \quad ; \quad 0.306 + 1 + 4.60 = 5.91$$

Then:

$$\frac{C.D._{LP}}{C.D._{HP}} = \frac{v_5/v_{11}}{\dot{m}_{11}/\dot{m}_5} = \frac{33.087/6.989}{5.91} = 0.801$$

3.14 R-22

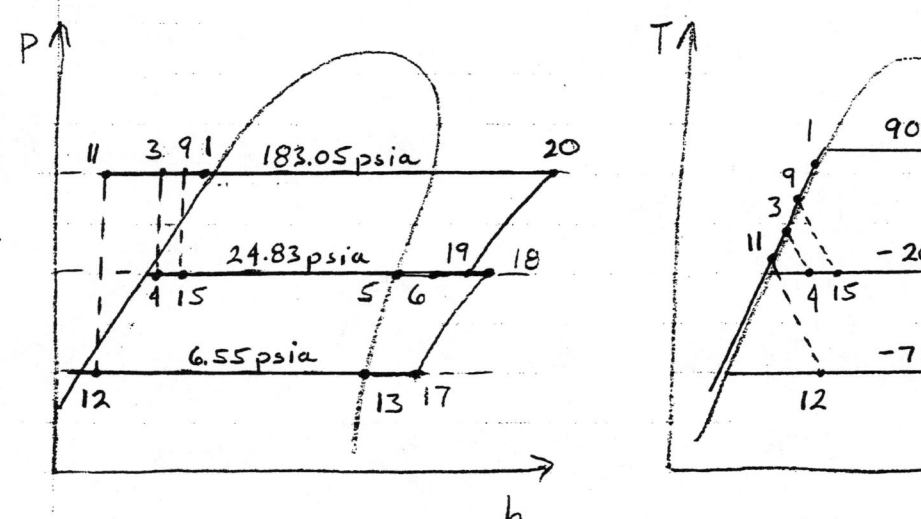

States 1, 2 & 8 are the same
States 9, 10 & 14 "   "   "
States 6, 7 & 16 "   "   "

a) C.O.P. = $\dfrac{\dot{Q}_A + \dot{Q}_B}{\dot{m}_{17}(h_{18} - h_{17}) + \dot{m}_{19}(h_{20} - h_{19})}$

need mass flow rates and enthalpy values

Combining Superheater A and Evaporator A

$\dot{m}_2 = \dot{m}_6$, $\dot{m}_2 = \dfrac{\dot{Q}_A}{h_6 - h_2}$

$\dot{m}_2 = \dfrac{5 \text{ Tons } (200 \text{ Btu/min ton})}{(104.5 - 34.3) \text{ Btu/lbm}} = 14.25 \text{ lbm/min}$

For Evaporator B

2/3

3.14 Cont'd

$$\dot{m}_{12} = \frac{\dot{Q}_B}{h_{13} - h_{12}} = \frac{10 \text{ Tons}(200 \text{ Btu/min ton})}{(96.9 - 15.8) \text{ Btu/lbm}} = 24.66 \frac{\text{lbm}}{\text{min}}$$

Combining Superheater B, Evaporator B and the Liquid Cooler

$$\dot{m}_8 = \dot{m}_{16} + \dot{m}_{17}$$

$$\dot{m}_8 h_8 + \dot{Q}_B = \dot{m}_{16} h_{16} + \dot{m}_{17} h_{17}$$

$$(\dot{m}_{16} + \dot{m}_{17}) h_8 + \dot{Q}_B = \dot{m}_{16} h_{16} + \dot{m}_{17} h_{17}$$

$$\dot{m}_{16} = \frac{\dot{Q}_B - \dot{m}_{17}(h_{17} - h_8)}{h_{16} - h_8}$$

$$= \frac{10 \text{ Tons}(200 \text{ Btu/min ton}) - 24.66 \text{ lbm/min}(114.8 - 34.3) \text{ Btu/lbm}}{(104.5 - 34.3) \text{ Btu/lbm}}$$

$$\dot{m}_{16} = 0.212 \text{ lbm/min}$$

$$\dot{m}_8 = 0.212 + 24.66 = 24.872 \text{ lbm/min}$$

$$\dot{m}_7 = \dot{m}_6 + \dot{m}_{16} = 14.25 + 0.212 = 14.462 \text{ lbm/min}$$

$$\dot{m}_{19} = \dot{m}_7 + \dot{m}_{18} = 14.462 + 24.66 = 39.12 \text{ lbm/min}$$

Energy Balance on mixing tee, 7, 18, 19

$$\dot{m}_{19} h_{19} = \dot{m}_7 h_7 + \dot{m}_{18} h_{18}$$

$$h_{19} = \frac{\dot{m}_7}{\dot{m}_{19}} h_7 + \frac{\dot{m}_{18}}{\dot{m}_{19}} h_{18}$$

3.14 Cont'd

$$h_{19} = \frac{14.462 \cdot 104.5}{39.12} + \frac{24.66 \cdot 132}{39.12} = 122 \text{ Btu/lbm}$$

$$h_{20} = 151 \text{ Btu/lbm}, \quad v_{19} = 2.9 \text{ ft}^3/\text{lbm}$$

$$\text{C.O.P.} = \frac{(5+10)(200)}{24.66(132-114.8) + 39.12(151-122)} = 1.92$$

b) By Eq. 3.6 (assuming 100% volumetric efficiency)

C.D. = $\dot{m} \, v$

for Compressor A

$$\text{C.D.}_A = \dot{m}_{19} \, v_{19} = 39.12 \frac{\text{lbm}}{\text{min}} \cdot 2.9 \frac{\text{ft}^3}{\text{lbm}} = 113 \text{ ft}^3/\text{min}$$

for Compressor B

$$\text{C.D.}_B = \dot{m}_{17} \, v_{17} = 24.66 \frac{\text{lbm}}{\text{min}} \cdot 10 \frac{\text{ft}^3}{\text{lbm}} = 247 \text{ ft}^3/\text{lbm}$$

c) for Compressor A

$$HP_A = \frac{\dot{m}_{19}(h_{20} - h_{19})}{42.4} = \frac{39.12 \frac{\text{lbm}}{\text{min}} (151 - 122) \frac{\text{Btu}}{\text{lbm}}}{42.4 \text{ Btu/min hp}} = 26.8 \text{ hp}$$

for Compressor B

$$HP_B = \frac{\dot{m}_{17}(h_{18} - h_{17})}{42.4} = \frac{24.66(132 - 114.8)}{42.4} = 10.0 \text{ hp}$$

3.14 Pressure-Enthalpy Diagram for R22

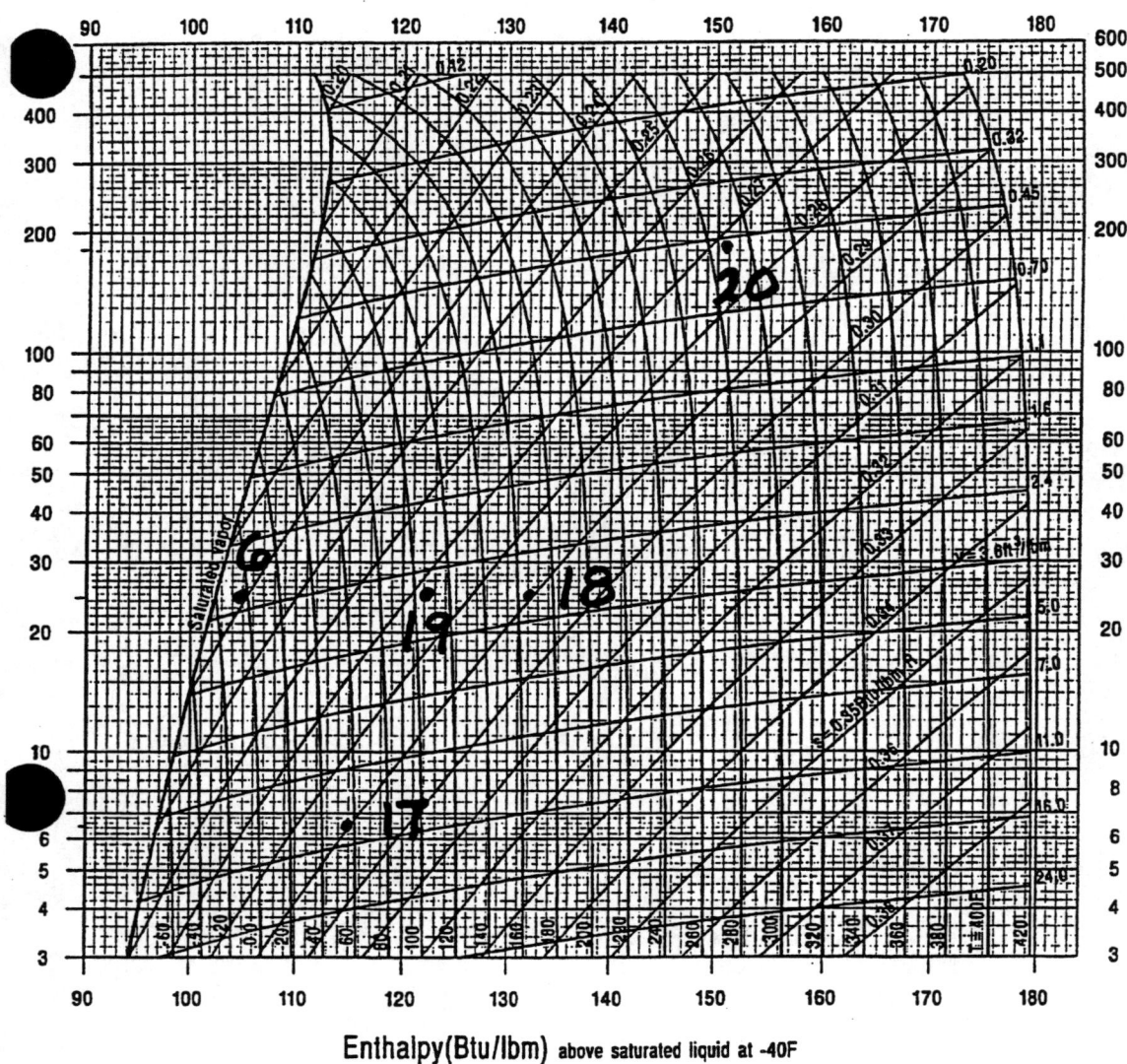

Enthalpy(Btu/lbm) above saturated liquid at -40F

Chart C-2E *(continued)*

4.1 R-22, C = 0.04, Capacity = 20 tons

$\dot{Q}_{2-3} = \dot{m}(h_3 - h_2)$

$\dot{m} = \eta_v (C.D.) / v_3$

$\eta_v = [1 + C - C(P_4/P_3)^{1/n}]$

$n = 1.12$

From Table A.3E
$h_2 = h_1 = 30.1$ Btu/lbm, $h_3 = 104.5$ Btu/lbm

$P_4 = 136.08$ psia, $P_3 = 38.64$ psia, $v_3 = 1.373$ ft^3/lbm

$\eta_v = [1 + 0.04 - 0.04(136.08/38.64)^{1/1.12}] = 0.917$

$\dot{m} = \dot{Q}_{2-3}/(h_3 - h_2) = \left(20 \text{ Tons} / (104.5 - 30.1)\dfrac{\text{Btu}}{\text{lbm}}\right)\left(\dfrac{200 \text{ Btu}}{\text{min ton}}\right)$

$\dot{m} = 53.8$ lbm/min

C.D. $= \dfrac{\dot{m} v_3}{\eta_v} = \dfrac{(53.8 \text{ lbm/min}) 1.373 \text{ ft}^3/\text{lbm}}{0.917} = 80.5 \dfrac{\text{ft}^3}{\text{min}}$

a) increase the condensing temperature to 100°F, determine the capacity

$P_4' = P_{sat}(100°F) = 210.55$ psia (Table A.3E)

$h_2' = h_f(100°F) = 39.3$ Btu/lbm (Table A.3E)

$\eta_v' = [1 + 0.04 - 0.04(210.55/38.64)^{1/1.12}] = 0.858$

47

4.1 cont'd.

$$\dot{m}' = 0.858\,(80.5\text{ ft}^3/\text{min})/1.373\text{ ft}^3/\text{lbm} = \underline{50.3\frac{\text{lbm}}{\text{min}}}$$

$$\dot{Q}_{2-3} = 50.3\frac{\text{lbm}}{\text{min}}(104.5-39.3)\frac{\text{Btu}}{\text{lbm}}\left(\frac{\text{ton min}}{200\text{ Btu}}\right) = \underline{16.4\text{ Tons}}$$

b) When the capacity $\to 0$, $\dot{m} \to 0$ as Δh is not zero, therefore $\eta_v \to 0$

$$0 = \left[1 + c - c\left(P_4/P_3\right)^{1/n}\right]$$

solve for P_3 with P_4, c and n fixed.

$$\frac{P_4}{P_3} = \left(\frac{1+c}{c}\right)^n$$

$$P_3 = P_4\left(\frac{1+c}{c}\right)^{-n}$$

substituting values from part a)

$$P_3 = 210.55\text{ psia}\left(\frac{1+0.04}{0.04}\right)^{-1.12} = \underline{5.48\text{ psia}}$$

This corresponds to a minimum evaporating temperature of about $-75°F$.

4.2 ammonia compressor with cooling water

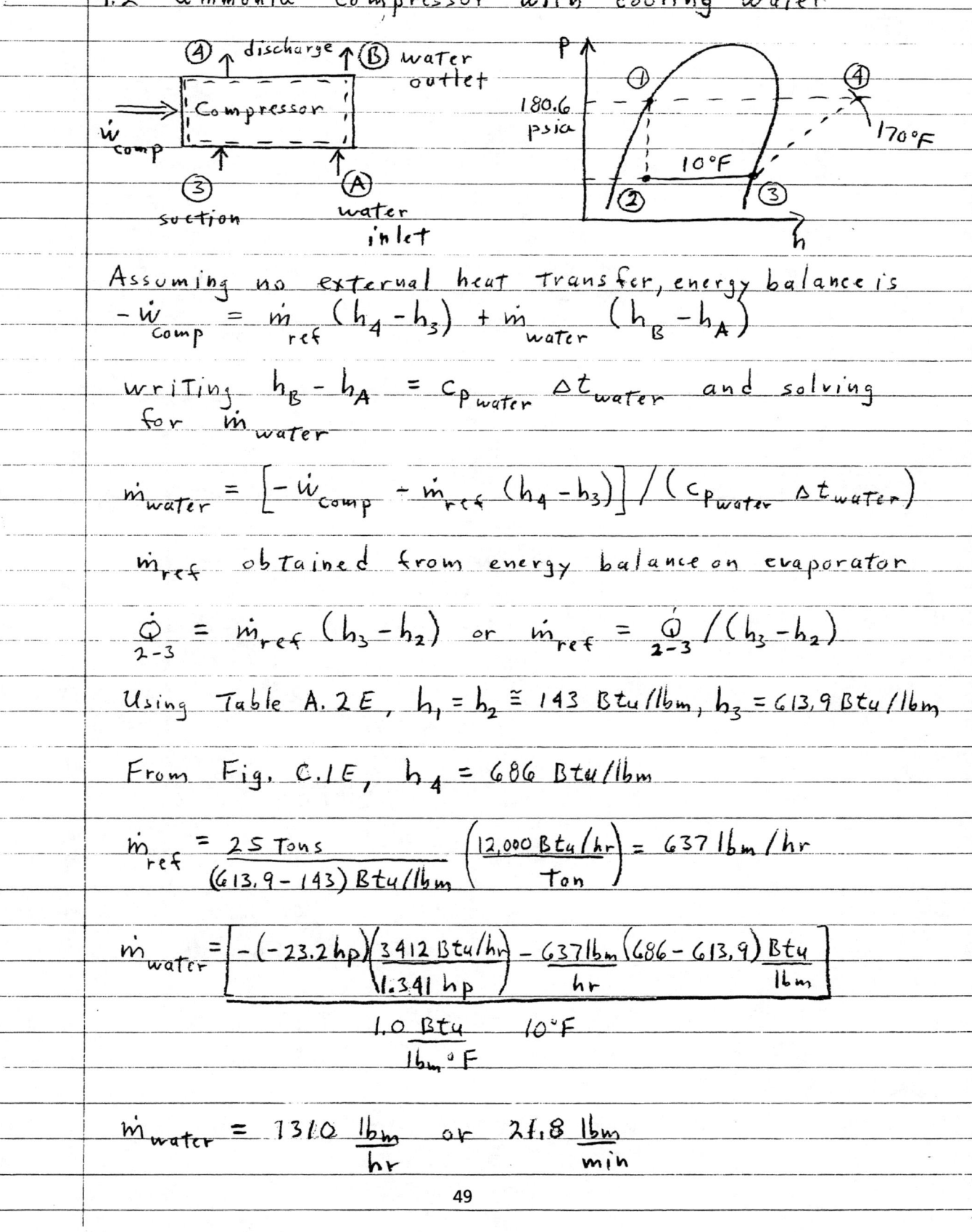

Assuming no external heat transfer, energy balance is

$$-\dot{W}_{comp} = \dot{m}_{ref}(h_4 - h_3) + \dot{m}_{water}(h_B - h_A)$$

writing $h_B - h_A = c_{p\,water} \Delta t_{water}$ and solving for $\dot{m}_{water}$

$$\dot{m}_{water} = \left[-\dot{W}_{comp} - \dot{m}_{ref}(h_4 - h_3)\right] / (c_{p\,water} \Delta t_{water})$$

$\dot{m}_{ref}$ obtained from energy balance on evaporator

$$\dot{Q}_{2-3} = \dot{m}_{ref}(h_3 - h_2) \quad \text{or} \quad \dot{m}_{ref} = \dot{Q}_{2-3}/(h_3 - h_2)$$

Using Table A.2E, $h_1 = h_2 \cong 143$ Btu/lbm, $h_3 = 613.9$ Btu/lbm

From Fig. C.1E, $h_4 = 686$ Btu/lbm

$$\dot{m}_{ref} = \frac{25 \text{ Tons}}{(613.9 - 143) \text{ Btu/lbm}} \left(\frac{12{,}000 \text{ Btu/hr}}{\text{ton}}\right) = 637 \text{ lbm/hr}$$

$$\dot{m}_{water} = \frac{\left[-(-23.2 \text{ hp})\left(\frac{3412 \text{ Btu/hr}}{1.341 \text{ hp}}\right) - 637 \frac{\text{lbm}}{\text{hr}}(686 - 613.9) \frac{\text{Btu}}{\text{lbm}}\right]}{\frac{1.0 \text{ Btu}}{\text{lbm} \cdot °F} \quad 10°F}$$

$$\dot{m}_{water} = 1310 \frac{\text{lbm}}{\text{hr}} \quad \text{or} \quad 21.8 \frac{\text{lbm}}{\text{min}}$$

4.3 R-22

$\Delta P_{suct} = 0.02$ MPa
$\Delta P_{disch} = 0.005$ MPa
$C = 0.05$
$t_b - t_a = 5°C$
$C.D. = 0.005$ m³/s
$\eta_m = 0.85$
$\eta_{motor} = 0.9$

find $\dot{W}_{elec}$ & $\dot{Q}_{2-3}$

From Table A.3SI and Fig. C.2SI

State	t °C	P MPa	v m³/kg	h kJ/kg
1	35	1.3544		87.72
2	0	0.4974		87.72
3	0	0.4974	0.04715	249.95
a	-1	0.4774		249.95
b	4	0.4774	0.051	254
c,d		1.3594		
4		1.3544		

$$\dot{m} = \frac{\eta_v \, C.D.}{v_3} = \left[1 + C - C\left(\frac{P_c}{P_b}\right)^{1/n}\right] \frac{C.D.}{v_b}$$

$$\dot{m} = \left[1 + 0.05 - 0.05\left(\frac{1.3594}{0.4774}\right)^{1/1.12}\right] \frac{0.005 \, m^3/s}{0.052 \, m^3/kg} = 0.0904 \, kg/s$$

$$\dot{Q}_{2-3} = \dot{m}(h_3 - h_2) = 0.0904 \frac{kg}{s}(249.95 - 87.72) \frac{kJ}{kg} = 14.67 \, kW$$

$$\dot{W}_{elec} = \frac{\dot{m} \, n \, P_b \, v_b}{(n-1) \eta_m \eta_{motor}} \left[\left(\frac{P_c}{P_b}\right)^{\frac{n-1}{n}} - 1\right]$$

4.3 Cont'd.

$$\dot{W}_{elec} = \frac{0.0904 \text{ kg/s} \cdot 1.12 \cdot 0.4774 \text{ MPa} \cdot 0.051 \text{ m}^3/\text{kg}}{(1.12-1) \cdot 0.85 \cdot 0.9} \left[\left(\frac{1.3594}{0.4774}\right)^{\frac{1.12-1}{1.12}} - 1 \right]$$

$$\dot{W}_{elec} = 3.19 \text{ kW}$$

4.4

Solution procedure follows that for Prob. 4.3

a) $C = 0.04$

$\dot{m} = 0.0902$ kg/s

$\dot{Q}_{2-3} = 14.63$ kW

$\dot{W}_{elec} = 3.24$ kW

b) $C = 0.03$

$\dot{m} = 0.0917$ kg/s

$\dot{Q}_{2-3} = 14.88$ kW

$\dot{W}_{elec} = 3.30$ kW

4.5

Solution procedure follows that for Prob. 4.3

a) $\Delta P_{suct} = 0.015$ MPa
$\Delta P_{disch} = 0.03$ MPa

$P_a = P_b = P_3 - \Delta P_{suct} = 0.4974 - 0.015 = 0.4824$ MPa
$P_c = P_d = P_4 + \Delta P_{disch} = 1.3544 + 0.03 = 1.3844$ MPa

$v_b = 0.050$ m³/kg

$\dot{m} = 0.0922$ kg/s

$\dot{Q}_{2-3} = 14.96$ kW

$\dot{W}_{elec} = 3.24$ kW

b) $\Delta P_{suct} = 0.01$ MPa, $\Delta P_{disch} = 0.015$ MPa

$P_a = P_b = 0.4974 - 0.01 = 0.4874$ MPa
$P_c = P_d = 1.3544 + 0.015 = 1.3694$ MPa

$v_b = 0.050$ m³/kg

$\dot{m} = 0.0924$ kg/s

$\dot{Q}_{2-3} = 14.99$ kW

$\dot{W}_{elec} = 3.20$ kW

1/3

4.6 Solution procedure follows that of Prob. 4.3
a) $t_2 = t_3 = 10°C$

State	t °C	P MPa	v m³/kg	h kJ/kg
1	35	1.3544		87.72
2	10	0.6805		87.72
3	10	0.6805	0.03472	253.43
a	9	0.6605		253.43
b	14	0.6605	0.037	257
c,d		1.3594		
4		1.3544		

$\dot{m} = 0.1290$ kg/s
$\dot{Q}_{2-3} = 21.38$ kW
$\dot{W}_{Elec} = 3.09$ kW

b)

State	t °C	P MPa	v m³/kg	h kJ/kg
1	35	1.3544		87.72
2	-10	0.3542		87.72
3	-10	0.3542	0.06536	246.15
a	-11	0.3342		246.15
b	-6	0.3342	0.075	249
c,d		1.3594		
4		1.3544		

$\dot{m} = 0.0583$ kg/s

$\dot{Q}_{2-3} = 9.23$ kW

$\dot{W}_{elec} = 2.89$ kW

4.6 Cont'd

c)

State	t °C	P MPa	v m³/kg	h kJ/kg
1	35	1.3544		87.72
2	-30	0.1634		87.72
3	-30	0.1634	0.1359	237.73
a	-31	0.1434		237.73
b	-26	0.1434	0.16	240
c,d		1.3594		
4		1.3544		

$\dot{m} = 0.0212$ kg/s

$\dot{Q}_{2-3} = 3.18$ kW

$\dot{W}_{elec} = 1.62$ kW

Problem 4.6

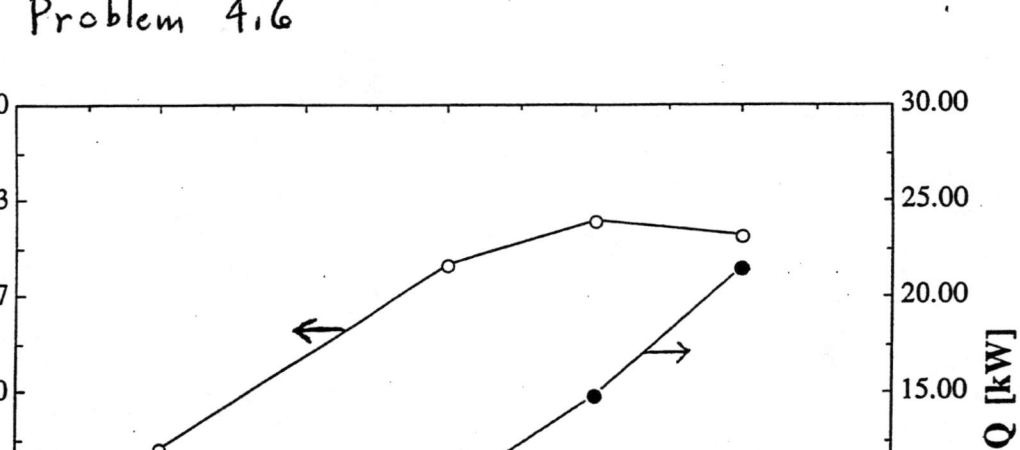

4.7

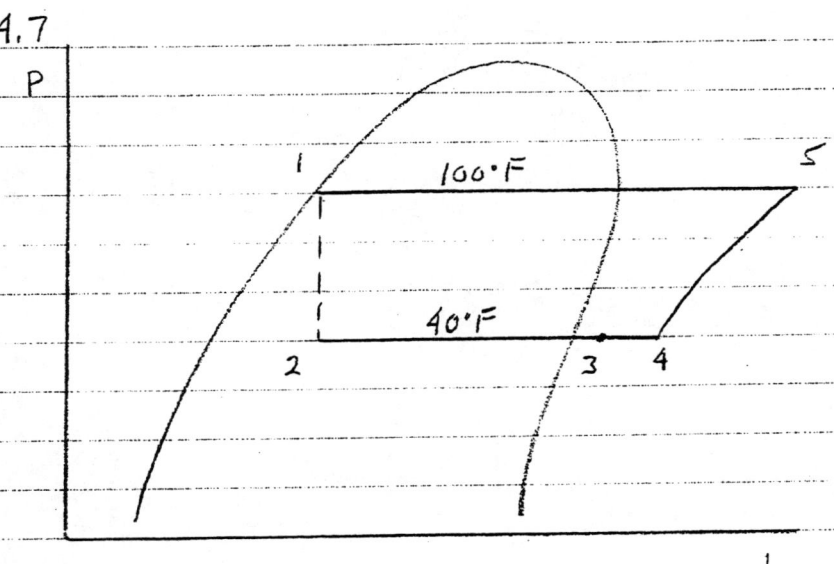

$$\dot{Q}_{2-3} = \dot{m}(h_3 - h_2)$$

$h_2 = h_1 = 39.3 \text{ Btu/lbm}, \quad h_3 = 109 \text{ Btu/lbm}$

By Eq. 4.9, $\dot{m} = \eta_v \, C.D. / v_4$

$$C.D. = \frac{4}{4}\pi(3in)^2 \, 4in \, \frac{800}{min} \left(\frac{ft^3}{1728 in^3}\right) = 52.36 \text{ ft}^3/\text{min}$$

$v_4 = 0.69 \text{ ft}^3/\text{lbm}$

$$\dot{m} = \frac{0.7 \, (52.36) \, ft^3/min}{0.69 \, ft^3/lbm} = 53.1 \text{ lbm/min}$$

$$\dot{Q}_{2-3} = \frac{53.1 \, lbm}{min} (109 - 39.3) \, Btu/lbm = 3700 \, \frac{Btu}{min}$$

$$= 3700 \, \frac{Btu}{min} \left(\frac{ton}{200 \, Btu/min}\right) = 18.5 \text{ Tons}$$

4.8

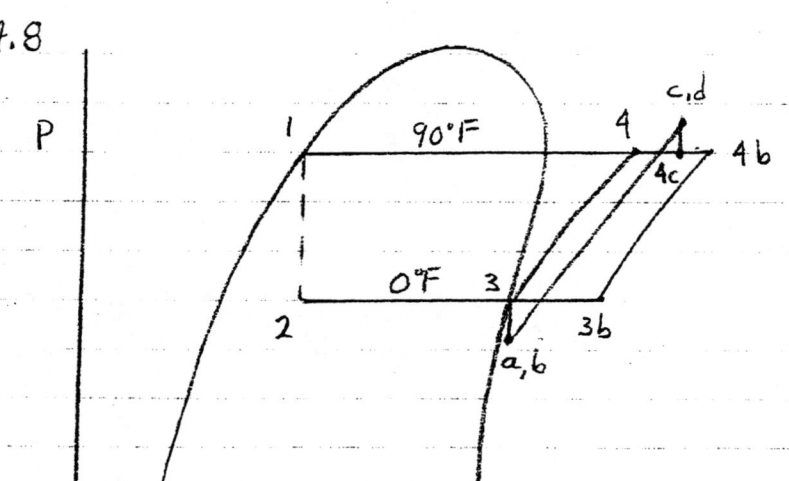

a) see Example 3.1

$$C.O.P. = \frac{h_3 - h_2}{\omega_{3-4}}$$

by Eq. 4.7

$$\omega_{3-4} = \frac{n}{n-1} P_3 v_3 \left[\left(\frac{P_4}{P_3}\right)^{\frac{n-1}{n}} - 1 \right]$$

$$= \frac{1.24}{0.24} \; \frac{30.42 \, lb_f}{in^2} \; \frac{9.110 \, ft^3}{lbm} \left[\left(\frac{180.73}{30.42}\right)^{\frac{0.24}{1.24}} - 1 \right] \left(\frac{Btu}{778 \, ft\,lb_f}\right)\left(\frac{144\,in^2}{ft^2}\right)$$

$$= 109 \; Btu/lbm$$

$$C.O.P. = \frac{(610.8 - 143.3) \, Btu/lbm}{109 \, Btu/lbm} = 4.29$$

From Ex. 3.1, C.O.P. = 4.15 which is 3% less

4.8 Cont'd

b) only change is to v_3 in the equation for w_{3-4}

$$\text{Decrease in C.O.P.} = 100\% \left(1 - \frac{v_3}{v_3'}\right)$$

$$= 100\% \left(1 - \frac{9.110}{10.5}\right) = 13.24\%$$

c) by Eq. 4.7, $w_{3-4c} = \frac{n}{n-1} P_b v_b \left[\left(\frac{P_c}{P_b}\right)^{\frac{n-1}{n}} - 1\right]$

$$= \frac{1.24}{0.24} \cdot \frac{26.42 \text{ lb}_f}{\text{in}^2} \cdot \frac{10.7 \text{ ft}^3}{\text{lb}_m} \left[\left(\frac{186.73}{26.42}\right)^{\frac{0.24}{1.24}} - 1\right] \frac{144}{778} = 124 \frac{\text{Btu}}{\text{lb}_m}$$

$$\text{Decrease in C.O.P.} = 100\% \left(1 - \frac{w_{3-4}}{w_{3-4'}}\right)$$

$$= 100\% \left(1 - \frac{109}{124}\right) = 12.4\%$$

4.9

Part b) $\dot{Q}_{2-3} = \dot{m}(h_3 - h_2)$, only $\dot{m}$ changes

by Eq. 4.4, $\dot{m} = \eta_v \, C.D./v_3$, only v_3 changes

Decrease in capacity $= 100\% \left(1 - \dfrac{v_3}{v_{3b}}\right) = 13.24\%$

(same answer as Prob. 4.8, part b)

Part c) only η_v changes in equation for $\dot{m}$

$$\eta_{v_a} = 1.05 - 0.05\left(\dfrac{180.73}{30.42}\right)^{1/1.24} = 0.84$$

$$\eta_{v_b} = \left[1.05 - 0.05\left(\dfrac{186.73}{26.42}\right)^{1/1.24}\right]\dfrac{9.110}{10.7} = 0.688$$

Decrease in capacity $= 100\%\left(1 - \dfrac{\eta_{v_b}}{\eta_{v_a}}\right)$

$= 100\%\left(1 - \dfrac{0.688}{0.84}\right) = 18.1\%$

4.10 ammonia

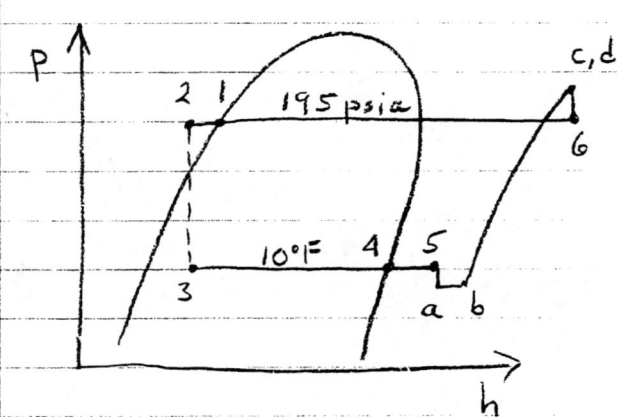

$t_5 - t_4 = 20°F$
$C = 0.04$
$\Delta P_{suct} = 4 psi$
$t_b - t_a = 15°F$
$n = 1.27$
$\Delta P_{disch} = 6 psi$
$\eta_m = 0.8$
$bore = 4 in$
$stroke = 5 in$

Determine whether this compressor can provide a refrigerating capacity of 18.5 tons

a) From Eqn. 4.3
$$\eta_v = \left[1 + C - C\left(\frac{P_c}{P_b}\right)^{1/n}\right] \frac{v_5}{v_b}$$

From Table A.2E and Fig. C.1E

State	t °F	P psia	v ft³/lbm	h Btu/lbm
1	95	195		
2	76	195		127.1
3	10	38.51		127.1
4	10	38.51		613.9
5	30	38.51	8	
a	28	34.51	8.5	
b	43	34.51	9	
c		201		

$$\eta_v = \left[1 + 0.04 - 0.04\left(\frac{201}{34.51}\right)^{1/1.27}\right] \frac{8}{9} = 0.78$$

From Eqn. 4.4
$$\dot{m} = \frac{\eta_v \, C.D.}{v_5}$$

4.10 cont'd

$$C.D._{max} = \left(\frac{\pi \, bore^2}{4}\right)(stroke)(rpm)_{max} \times 4 \, cylinders$$

$$= \frac{\pi (4in)^2}{4} \times 5in \times 600 \, rpm \left(\frac{ft^3}{1728 \, in^3}\right) \times 4 = 87.3 \, \frac{ft^3}{min}$$

$$\dot{m}_{max} = \frac{0.78 \times 87.3 \, ft^3/min}{8 \, ft^3/lbm} = 8.5 \, lbm/min$$

$$\dot{Q}_{3-4 \, max} = \dot{m}_{max}(h_4 - h_3)$$

$$= \frac{8.5 \, lbm}{min}(613.9 - 127.1)\frac{Btu}{lbm}\left(\frac{ton \, min}{200 \, Btu}\right) = 20.7 \, Tons$$

Capacity is sufficient as 20.7 tons > 18.5 Tons required

$$actual \, rpm = rpm_{max}\left(\frac{18.5 \, Tons}{20.7 \, tons}\right) = 600 \, rpm \left(\frac{18.5}{20.7}\right)$$

actual rpm = 536

b) From Eqn 4.11

$$w_{comp} = \frac{n}{n-1} P_b v_b \left[\left(\frac{P_c}{P_b}\right)^{\frac{n-1}{n}} - 1\right]$$

$$= \frac{1.27}{0.27} \, 34.51 \frac{lb}{in^2} \, \frac{9 \, ft^3}{lbm}\left(\frac{144 \, in^2}{ft^2}\right)\left(\frac{Btu}{778 \, ft\,lb_f}\right)\left[\left(\frac{201}{34.51}\right)^{\frac{0.27}{1.27}} - 1\right]$$

$$w_{comp} = 123 \, Btu/lbm$$

$$\dot{W}_{comp} = \dot{m}_{max}(18.5/20.7) \, w_{comp}/\eta_m$$

$$= \frac{8.5 \, lbm}{0.8 \, min}\left(\frac{18.5}{20.7}\right)\frac{123 \, Btu}{lbm}\left(\frac{60 \, min}{hr}\right)\left(\frac{1 \, hp \, hr}{2544 \, Btu}\right) = 27.5 \, hp$$

4.11 Displacement of rolling piston rotary compressor, $A = 10\text{cm}$, $B = 8\text{cm}$, $W = 4\text{cm}$, $\omega_{shaft} = 60\text{ rps}$

By Eqn. 4.19

$$C.D. = \frac{\pi W}{4}(A^2 - B^2)\omega_{shaft} \bigg/ \left[\left(\frac{A}{B} - 1\right)\frac{A}{B}\right]$$

$$= \frac{\pi \cdot 4\text{cm}}{4}(10^2 - 8^2)\text{cm}^2 \cdot 60\text{ rps} \bigg/ \left[\left(\frac{10}{8} - 1\right)\frac{10}{8}\right]$$

$$C.D. = 108{,}570 \text{ cm}^3/\text{s} = 0.1086 \text{ m}^3/\text{s}$$

4.12

By Eqns. 4.21 and 4.22

$A = 3$ in., $B = 2.5$ in., $w = 1$ in., $t = 0.25$ in.

$$c = \sqrt{2.5(2\cdot 3 - 2.5)} = 2.96 \text{ in.}$$

$$V_s = \frac{1}{4}\left[\pi\left(3^2 - \frac{2.5^2}{2}\right) - 3^2 \cos^{-1}\left(\frac{3-2.5}{3}\right) + 2.96(3-2.5) - 2\cdot 0.25(2.96-2.5)\right]$$

$$V_s = 1.77 \text{ in}^3 \left(\frac{\text{ft}^3}{1728 \text{ in}^3}\right) = 1.024 \times 10^{-3} \text{ ft}^3$$

$$C.D. = 2(1.024 \times 10^{-3} \text{ ft}^3)\, 3450 \text{ rpm} = 7.07 \text{ ft}^3/\text{min}$$

4.13

Use R-22 refrigerant as in Fig. 4.10

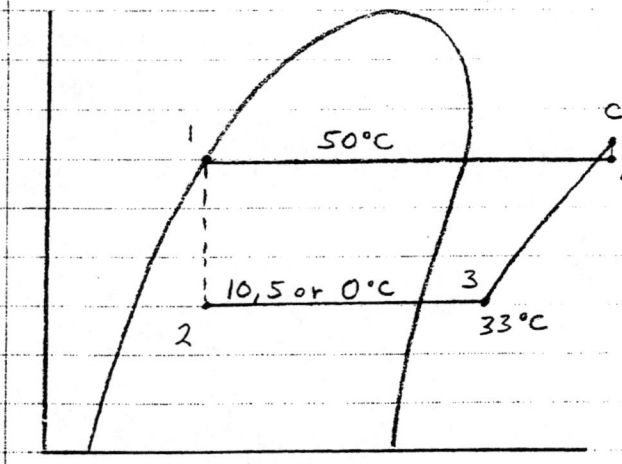

From Table 4.2, n = 1.12
By Table A.3SI, $P_1 = P_4 = 1.9419$ MPa, $h_1 = h_2 = 107.85 \frac{kJ}{kg}$

$P_c = P_4 + \Delta P_{disch} = 1.9419 + 0.01 = 1.9519$ MPa

For $T_{evap} = 10°C$,

By Table A.3SI, $P_2 = P_3 = P_b = 0.6805$ MPa
By Chart C-2SI, $h_3 = 271$ kJ/kg
$v_b = v_3 = 0.039$ m³/kg

By Eq. 4.5

$\dot{m} = \left[1 + 0.01 - 0.01 \left(\frac{1.9519}{0.6805}\right)^{1/1.12}\right] \frac{0.001 \, m^3/s}{0.039 \, m^3/kg} = 0.0252 \frac{kg}{s}$

$\dot{Q} = \dot{m}(h_3 - h_2) = 0.0252 \frac{kg}{s} (271 - 107.85) \frac{kJ}{kg}$

$\dot{Q} = 4.11$ kW

4.13 Cont'd

By Eq. 4.15

$$\dot{W}_{elec} = \frac{1.12(0.6805\,MPa)0.001\,m^3/s}{(1.12-1)\,0.54}\left[1+0.01-0.01\left(\frac{1.9519}{0.6805}\right)^{1/1.12}\right] \times$$

$$\left[\left(\frac{1.9519}{0.6805}\right)^{\frac{1.12-1}{1.12}} - 1\right]$$

$$\dot{W}_{elec} = 1380\,W$$

Using a similar procedure for $T_{evap} = 5°C$

$P_2 = P_3 = P_b = 0.5836\,MPa$
$h_3 = 272\,kJ/kg,\quad v_b = 0.047\,m^3/kg$

$$\dot{m} = \left[1+0.01-0.01\left(\frac{1.9519}{0.5836}\right)^{1/1.12}\right]\frac{0.001}{0.046} = 0.0214\,kg/s$$

$$\dot{Q} = 0.0214(272-107.85) = 3.50\,kW$$

$$\dot{W}_{elec} = \frac{1.12(0.5836)0.001}{(1.12-1)0.54}\left[1+0.01-0.01\left(\frac{1.9519}{0.5836}\right)^{1/1.12}\right] \times$$

$$\left[\left(\frac{1.9519}{0.5836}\right)^{\frac{1.12-1}{1.12}} - 1\right] = 1366\,W$$

For $T_{evap} = 0°C$

$P_2 = P_3 = P_b = 0.4974\,MPa$
$h_3 = 274\,kJ/kg,\quad v_b = 0.054\,m^3/kg$

$$\dot{m} = \left[1+0.01-0.01\left(\frac{1.9519}{0.4974}\right)^{1/1.12}\right]\frac{0.001}{0.054} = 0.0181\,kg/s$$

$$\dot{Q} = 0.0181(274-107.85) = 3.00\,kW$$

4.13 Cont'd

$$\dot{W}_{elec} = \frac{1.12\,(0.4974)\,0.001}{(1.12-1)\,0.54}\left[1+0.01-0.01\left(\frac{1.9519}{0.4974}\right)^{1/1.12}\right] \times$$

$$\left[\left(\frac{1.9519}{0.4974}\right)^{\frac{1.12-1}{1.12}} - 1\right] = 1324\ W$$

Capacity results are slightly below those shown in Fig. 4.10, the power requirements are nearly identical

4.14

[P-V diagram: isobar at P_4 from c^* to c', polytropic curve $PV^n = C$ from c to b, isobar at P_3; volumes V_D and V_S marked on V-axis]

For undercompression

$$\frac{W}{\text{cycle}} = \frac{n}{n-1} P_3 V_S \left[\left(\frac{P_c}{P_b}\right)^{\frac{n-1}{n}} - 1 \right] + (P_4 - P_c) V_D$$

This is identical to Eq.(4.27) for overcompression

4.15 Scroll compressor with R-22
Compute ω when $V_r = 4$, $n = 1.122$ and suction conditions are sat. vapor at 0°F

a) condensing temp. is 90°F

By Eqn. 4.29

$$\omega = \frac{n}{n-1} P_3 v_b \left(V_r^{n-1} - 1\right) - \left(P_3 V_r^n - P_4\right)\frac{v_b}{V_r}$$

we will assume $v_b = v_3$ (suction conditions)
From Table A.3E
$P_3 = P_{sat}(0°F) = 38.64$ psia, $v_3 = v_g(0°F) = 1.373$ ft³/lbm

$P_4 = P_{sat}(90°F) = 183.05$ psia

$$\omega = \frac{1.122}{0.122} \cdot 38.64 \frac{lb_f}{in^2} \cdot 1.373 \frac{ft^3}{lbm}\left(4^{0.122} - 1\right) - \left(38.64 \times 4^{1.122} - 183.05\right)\frac{lb_f}{in^2} \cdot \frac{1.373}{4} \frac{ft^3}{lbm}$$

$$= (89.91 + 0)\frac{lb_f \cdot ft^3}{in^2 \cdot lbm}\left(\frac{144 in^2}{ft^2}\right) = 12,950 \frac{ft \cdot lb_f}{lbm}$$

b) condensing temp. is 80°F
$P_4 = P_{sat}(80°F) = 158.28$ psia
(all other parameters remain as in part a)

$$\omega = \left(89.91 - \left(38.64 \times 4^{1.122} - 158.28\right) 1.373/4\right) 144 = 11,720 \frac{ft \cdot lb_f}{lbm}$$

c) condensing temp. is 70°F
$P_4 = P_{sat}(70°F) = 136.08$ psia

$$\omega = \left(89.91 - \left(38.64 \times 4^{1.122} - 136.08\right) 1.373/4\right) 144 = 10,630 \frac{ft \cdot lb_f}{lbm}$$

4.16

a) $t_{cond} = 80°F$, $t_{evap} = 0°F$, $t_{suction} = 20°F$

From Chart C-2E, $v_b = 1.45$ ft³/lb$_m$

By Eq. 4.29 and values in Prob. 4.15

$$\omega = \left[\frac{1.122}{0.122} \cdot 38.64 \frac{lb_f}{in^2} \cdot 1.45 \frac{ft^3}{lb_m}\left(4^{0.122}-1\right) - \left(38.64 \times 4^{1.122} - 158.28\right)\frac{1.45}{4}\right](1$$

$\omega = 12,380$ ft lb$_f$/lb$_m$

b) $t_{suction} = 40°F$

From Chart C-2E, $v_b = 1.55$ ft³/lb$_m$

By Eq. 4.29

$$\omega = \left[\frac{1.122}{0.122} \cdot 38.64 \frac{lb_f}{in^2} \cdot 1.55 \frac{ft^3}{lb_m}\left(4^{0.122}-1\right) - \left(38.64 \times 4^{1.122} - 158.28\right)\frac{1.55}{4}\right] 144$$

$\omega = 13,230$ ft lb$_f$/lb$_m$

4.17

By Eq. 4.31, $P_r = V_r^n$ or $V_r = P_r^{1/n}$

From Table A.2SI, $P_{cond} = 1.5549$ MPa
$P_{evap} = 0.2909$ MPa

$n = 1.2$

$$V_r = \left(\frac{1.5549}{0.2909}\right)^{1.2} = 7.47$$

4.18
R-22

By Eq. 4.32, $\eta_s = \dfrac{(h_4 - h_3)_s}{h_d - h_3}$

$P_{cond} = 150\ psia$, $n = 1.18$

a) $t_{evap} = 10°F$

From Table A.3E, $h_3 = 105.4\ Btu/lbm$, $s_3 = 0.2259\ \dfrac{Btu}{lbm\ °R}$
$P_3 = 47.45\ psia$, $v_3 = 1.129\ ft^3/lbm$

From Chart C-2E, $h_{4_s} = 117.5\ Btu/lbm$

$v_c = v_3/3 = 1.129/3 = 0.376\ ft^3/lbm$

$P_c = 47.45\ psia\ \cdot 3^{1.18} = 173.5\ psia$

From chart C-2E, $h_c = h_d = 121\ Btu/lbm$

$\eta_s = \dfrac{117.5 - 105.4}{121 - 105.4} = 0.78$

b) $t_{evap} = 20°F$

$h_3 = 106.4\ Btu/lbm$, $s_3 = 0.2238\ Btu/lbm\ °R$
$P_3 = 57.71\ psia$, $v_3 = 0.937\ ft^3/lbm$

$h_{4_s} = 116.0\ Btu/lbm$

$v_c = 0.937/3 = 0.312\ ft^3/lbm$

$P_c = 57.71\ psia\ \cdot 3^{1.18} = 211\ psia$

4.18 Cont'd

$h_c = h_d = 122$ Btu/lbm

$\eta_s = \dfrac{116 - 106.4}{122 - 106.4} = 0.62$

c) $t_{evap} = 30°F$

$h_3 = 107.3$ Btu/lbm, $s_3 = 0.2217$ Btu/lbm·°R
$P_3 = 69.57$ psia, $v_3 = 0.782$ ft³/lbm

$h_{4s} = 115$ Btu/lbm

$v_c = 0.782/3 = 0.261$ ft³/lbm

$P_c = 69.57$ psia $\cdot 3^{1.18} = 254$ psia

$h_c = h_d = 126$ Btu/lbm

$\eta_s = \dfrac{115 - 107.3}{126 - 107.3} = 0.41$

4.19

By Eq. 4.43, $\eta_p = \dfrac{n}{n-1}\left(\dfrac{k-1}{k}\right)$

at state ①, $v_1 = 6.2$ ft³/lbm

at state ②, $v_2 = 2.3$ ft³/lbm

$P_r = V_r^n$ or $n = \dfrac{\log P_r}{\log V_r} = \dfrac{\log(180/48)}{\log(6.2/2.3)} = 1.33$

From Table 4.2, $k = 1.29$ for ammonia

$\eta_p = \dfrac{1.33}{1.33-1}\left(\dfrac{1.29-1}{1.29}\right) = 0.91$

4.20

By Eq. (4.37)

$$\frac{\omega_p}{\omega_s} = \frac{\dfrac{n}{n-1}\left[\left(\dfrac{P_2}{P_1}\right)^{\frac{n-1}{n}} - 1\right]}{\dfrac{k}{k-1}\left[\left(\dfrac{P_2}{P_1}\right)^{\frac{k-1}{k}} - 1\right]}$$

By Eq. (4.43)

$$\eta_p = \frac{n}{n-1}\left(\frac{k-1}{k}\right) \quad \text{or} \quad n = \frac{k\eta_p}{k(\eta_p - 1) + 1}$$

$$n = \frac{1.3(0.76)}{1.3(0.76-1)+1} = 1.44$$

$$\frac{\omega_p}{\omega_s} = \frac{\dfrac{1.44}{0.44}\left[\left(\dfrac{P_2}{P_1}\right)^{\frac{0.44}{1.44}} - 1\right]}{\dfrac{1.3}{0.3}\left[\left(\dfrac{P_2}{P_1}\right)^{\frac{0.3}{1.3}} - 1\right]} = 0.755\left[\frac{P_r^{0.306} - 1}{P_r^{0.231} - 1}\right]$$

Problem 4.20

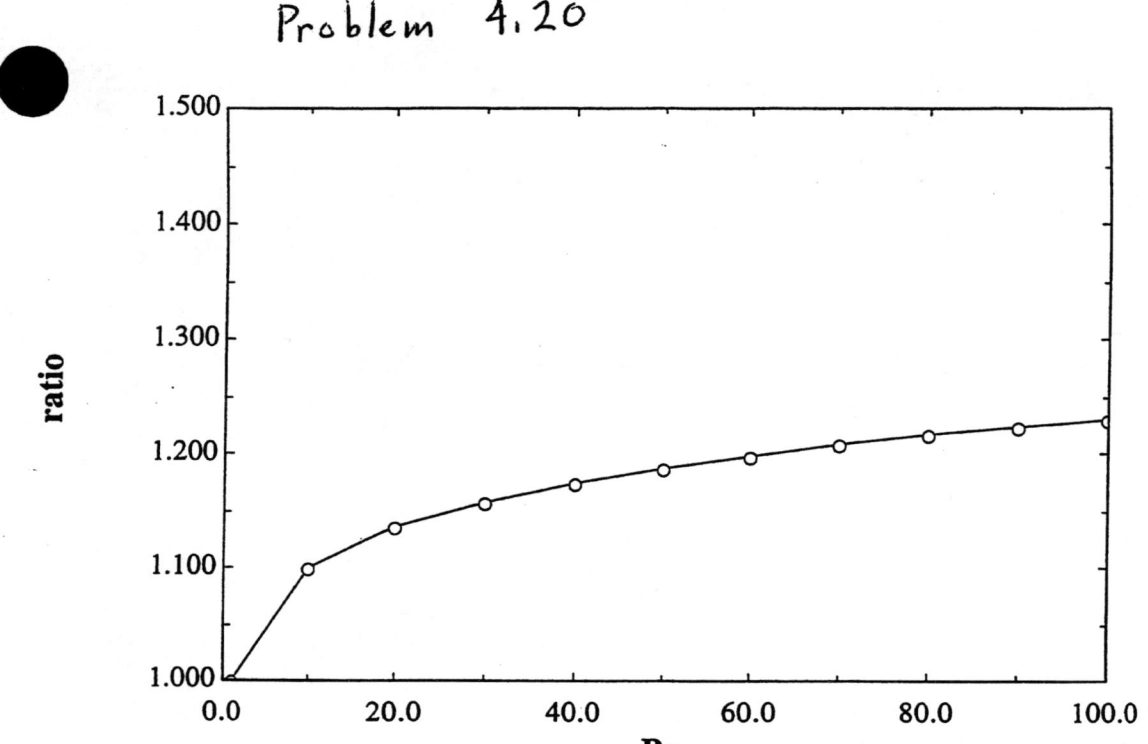

4.21

a) States 6, 8 and 9 are altered

Find state 6 from compressor conditions

By Eq. (4.44)

$$w_{p_{5-6}} = \eta_p \frac{\left[\left(\frac{P_6}{P_5}\right)^{\frac{k-1}{\eta_p k}} - 1\right]}{\left[\left(\frac{P_6}{P_5}\right)^{\frac{k-1}{k}} - 1\right]} (h_6 - h_5)$$

$$w_{p_{5-6}} = 0.8 \frac{\left[\left(\frac{57.71}{38.64}\right)^{\frac{0.3}{0.8(1.3)}} - 1\right]}{\left[\left(\frac{57.71}{38.64}\right)^{\frac{0.3}{1.3}} - 1\right]} (108.3 - 104.5) \text{ Btu/lbm}$$

$$w_{p_{5-6}} = 3.84 \text{ Btu/lbm}$$

$$h_6' = h_5 + w_{p_{5-6}} = 104.5 + 3.84 \cong 108.3 \text{ Btu/lbm}$$

$$h_8' \cong h_8 = 107.9 \text{ Btu/lbm}$$

$$w_{p_{8-9}} = 0.8 \frac{\left[\left(\frac{183.05}{57.71}\right)^{\frac{0.3}{0.8(1.3)}} - 1\right]}{\left[\left(\frac{183.05}{57.71}\right)^{\frac{0.3}{1.3}} - 1\right]} (121.2 - 107.9) \text{ Btu/lbm}$$

$$w_{p_{8-9}} = 13.77 \text{ Btu/lbm}$$

$$h_9' = h_8 + w_{p_{8-9}} = 107.9 + 13.77 = 121.7 \text{ Btu/lbm}$$

4.21 Cont'd

$$C.O.P._1 = \frac{h_5 - h_4}{\omega_{p_{5-6}} + \frac{\dot{m}_8}{\dot{m}_3} \omega_{p_{8-9}}}$$

$$= \frac{104.5 - 15.8}{3.84 + 1.291(13.77)} = 4.10$$

b) with a single compression process

$$\omega_{p_{5-9''}} = \eta_p \frac{\left[\left(\frac{P_9}{P_5}\right)^{\frac{k-1}{\eta_p k}} - 1\right]}{\left[\left(\frac{P_9}{P_5}\right)^{\frac{k-1}{k}} - 1\right]} (h_{9*} - h_5)$$

$$= 0.8 \frac{\left[\left(\frac{183.05}{38.64}\right)^{\frac{0.3}{0.8(1.3)}} - 1\right]}{\left[\left(\frac{183.05}{38.64}\right)^{\frac{0.3}{1.3}} - 1\right]} (121.5 - 104.5) \; Btu/lbm$$

$$\omega_{p_{5-9''}} = 17.83 \; Btu/lbm$$

$$C.O.P. = \frac{h_5 - h_1}{\omega_{p_{5-9''}}} = \frac{104.5 - 36.2}{17.83} = 3.83$$

c) The use of the flash tank increases the system C.O.P. from 3.83 to 4.10, a 7% increase

5.1 Sat. water vapor at 50°F mixes with a sat. LiBr-water solution with $X = 0.6$, $\dot{m}_{liquid} = 5\,\dot{m}_{vapor}$

a) find final concentration

mass balance of LiBr
$$\dot{m}_{vapor} X_{vapor} + \dot{m}_{liquid} X_{liquid} = \dot{m}_{Total} X_{mix}$$

total mass balance
$$\dot{m}_{vapor} + \dot{m}_{liquid} = \dot{m}_{Total}$$

or $\dot{m}_{vapor} + 5\,\dot{m}_{vapor} = 6\,\dot{m}_{vapor} = \dot{m}_{Total}$

substitute in LiBr mass balance eqn.

$$\dot{m}_{vapor} \times 0 + 5\,\dot{m}_{vapor}\, 0.6 = 6\,\dot{m}_{vapor} X_{mix}$$

$$X_{mix} = \frac{5 \times 0.6}{6} = 0.5 \text{ lbm LiBr / lb mix}$$

From Table A.1E
$P = P_{sat}(50°F) = 0.36353 \text{ in Hg}$
$= 9.23 \text{ mm Hg}$
$h_v = h_g(50°F) = 1083.10 \text{ Btu/lbm}$

From Fig. C-4E
$h_L = h_{sat}(9.33 \text{ mm Hg}, 0.6 \frac{\text{lbm LiBr}}{\text{lb mix}})$
$= -66 \text{ Btu/lbm mix}$
$h_3 = -72 \text{ Btu/lbm mix}$

b)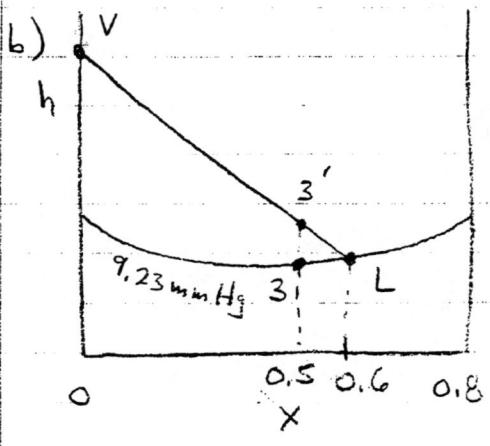

Energy balance on mixing process

$$\dot{Q} = \dot{m}_{mix} h_3 - \dot{m}_{vapor} h_v - \dot{m}_{liquid} h_L$$

Divide through by $\dot{m}_{mix}$

79

5.1 $\dfrac{\dot{Q}}{\dot{m}_{mix}} = h_3 - \dfrac{\dot{m}_{vapor}}{\dot{m}_{mix}} h_v - \dfrac{\dot{m}_{liquid}}{\dot{m}_{mix}} h_L$

$= h_3 - \dfrac{\dot{m}_{vapor}}{6\,\dot{m}_{vapor}} h_v - \dfrac{5\,\dot{m}_{vapor}}{6\,\dot{m}_{vapor}} h_L$

$= -72 - \dfrac{1083.10}{6} - \dfrac{5}{6} \times (-66)$

$\dfrac{\dot{Q}}{\dot{m}_{mix}} = -198 \ Btu/lb_m\,mix$

5.2 Saturated Soln. of $LiBr-H_2O$ at $40°C$ and $X=0.25$ heated to $100°C$

a) Determine mass proportions of liq. and vapor

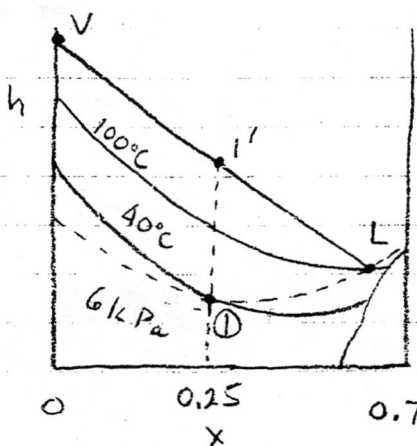

Pressure = $6 kPa$

$X_L = 0.68$

Total Mass Balance
$$m_v + m_L = m_1$$

LiBr Mass Balance
$$m_v X_v + m_L X_L = m_1 X_1$$

or $m_v \times 0 + m_L \times 0.68 = m_1 \times 0.25$

$$\frac{m_L}{m_1} = \frac{0.25}{0.68} = 0.37 \text{ or } 37\% \text{ liquid by mass}$$

$$\frac{m_v}{m_1} = 1 - \frac{m_L}{m_1} = 1 - 0.37 = 0.63 \text{ or } 63\% \text{ vapor}$$

b) h values for liquid and vapor phase
By Fig. C-4 SI,
$h_L = -48 \text{ kJ/kg mix}$

By Eq. 2.36 for low pressure water vapor
$h_v \cong 2501 + 1.86(100) = 2687 \text{ kJ/kg}$

5.3

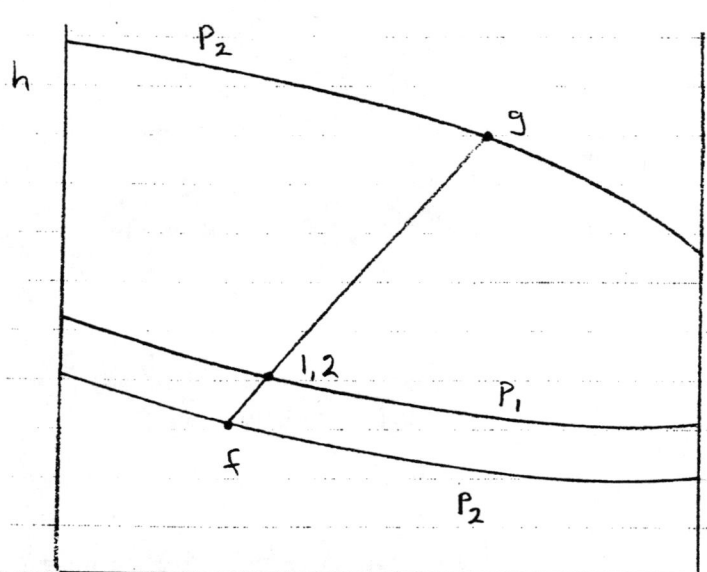

$t_1 = 220°F$
$P_1 = 200 \text{ psia}$
$P_2 = 10 \text{ psia}$
$h_1 = h_2$
$X_1 = X_2 = 0.355 \dfrac{lb_m \; NH_3}{lb_m \; sol}$

$X_f = 0.222 \dfrac{lb_m \; NH_3}{lb_m \; sol}$

$X_g = 0.945 \dfrac{lb_m \; NH_3}{lb_m \; sol}$

a) $t_2 = 92°F$ from chart C-3E

b) Liquid fraction $= \dfrac{X_g - X_2}{X_g - X_f} = \dfrac{0.945 - 0.355}{0.945 - 0.222} = 0.816$

or 81.6%

Vapor fraction $= 1 - 0.816 = 0.184$ or 18.4%

5.4

Rewriting Eqns. (5.22) - (5.24)

$$\dot{m}_1 = \dot{m}_2 + \dot{m}_3 \qquad (1)$$
$$\dot{m}_1 X_1 = \dot{m}_2 X_2 + \dot{m}_3 X_3 \qquad (2)$$
$$\dot{m}_1 h_1 + \dot{Q}_G = \dot{m}_2 h_2 + \dot{m}_3 h_3 + \dot{Q}_D \qquad (3)$$

By Eqns. (1) and (2)

$$\dot{m}_2 X_1 + \dot{m}_3 X_1 = \dot{m}_2 X_2 + \dot{m}_3 X_3$$

$$\dot{m}_2 (X_1 - X_2) = \dot{m}_3 (X_3 - X_1) \qquad (4)$$

By Eqns. (1) and (3)

$$\dot{m}_2 h_1 + \dot{m}_3 h_1 + \dot{Q}_G = \dot{m}_2 h_2 + \dot{m}_3 h_3 + \dot{Q}_D$$

$$\dot{m}_3 (h_3 - h_1) + \dot{Q}_G = \dot{m}_2 (h_1 - h_2) + \dot{Q}_D \qquad (5)$$

Dividing Eq. (5) by Eq. (4) gives

$$\frac{(h_3 + \dot{Q}_D / \dot{m}_3) - h_1}{X_3 - X_1} = \frac{h_1 - (h_2 - \dot{Q}_D / \dot{m}_2)}{X_1 - X_2}$$

5.5

a) From Chart C-3E we find:
$$X_1 = X_3 = 0.435 \text{ lbm NH}_3/\text{lbm sol}$$
$$X_4 = X_6 = 0.380 \text{ lbm NH}_3/\text{lbm sol}$$
$$X_7 = X_{12} = 0.994 \text{ lbm NH}_3/\text{lbm sol}$$

For the absorber
$$\dot{m}_6 + \dot{m}_{12} = \dot{m}_1$$

$$\dot{m}_6 X_6 + \dot{m}_{12} X_{12} = \dot{m}_1 X_1$$

Thus
$$\dot{m}_1 = \dot{m}_{12} \frac{(X_{12} - X_6)}{(X_1 - X_6)} = 100 \frac{\text{lbm}}{\text{min}} \frac{(0.994 - 0.380)}{(0.435 - 0.380)} = 1116 \frac{\text{lbm}}{\text{min}}$$

b) From the construction of the POL on Chart C-3E

$$\dot{Q}_D / \dot{m}_7 = 745 - 657 = 88 \text{ Btu/lbm}$$

Thus

$$\dot{Q}_D = 100 \frac{\text{lbm}}{\text{min}} (88 \text{ Btu/lbm}) = 8800 \text{ Btu/min}$$

$$\dot{m}_w c_{pw} \Delta t_w = \dot{Q}_D, \quad \text{or} \quad \dot{m}_w = \dot{Q}_D / (c_{pw} \Delta t_w)$$

$$\dot{m}_w = \frac{8800 \text{ Btu/min}}{1 \frac{\text{Btu}}{\text{lbm °F}} (15°F)} = 587 \text{ lbm/min of water}$$

5.6

Property values obtained from Chart C-3SI and computed values are summarized below:

State	P MPa	t °C	x kg NH₃/kg sol	h kJ/kg sol	ṁ kg/s
1	0.2	20	0.450	-80	0.0922
2	2.0		0.450	-78	0.0922
3	2.0	100	0.450	300	0.0922
4	2.0	120	0.345	400	0.0773
5	2.0		0.345	-51	0.0773
6	0.2		0.345	-51	0.0773
7	2.0	60	0.995	1535	0.0149
8	2.0	51	0.995	435	0.0149
9	2.0	41	0.995	380	0.0149
10	0.2	-18	0.995	380	0.0149
11	0.2		0.995	1385	0.0149
12	0.2		0.995	1440	0.0149

a & b)

From an energy balance on the combined evaporator, expansion valve, heat exchanger (streams 8 & 12)

$$\dot{Q}_E = \dot{m}_{12}(h_{12} - h_8), \text{ or } \dot{m}_{12} = \frac{\dot{Q}_E}{h_{12} - h_8}$$

$$\dot{m}_{12} = \frac{15 \text{ kW}}{(1440 - 435) \text{ kJ/kg sol}} = 0.0149 \text{ kg sol/s}$$

From an ammonia mass balance on the absorber

$$\dot{m}_{12} x_{12} + \dot{m}_6 x_6 = \dot{m}_1 x_1$$

$$\dot{m}_{12} + \dot{m}_6 = \dot{m}_1$$

Combining these and solving for $\dot{m}_1$

5.6 Cont'd

$$\dot{m}_1 X_1 = \dot{m}_{12} X_{12} + (\dot{m}_1 - \dot{m}_{12}) X_6$$

$$\dot{m}_1 (X_1 - X_6) = \dot{m}_{12}(X_{12} - X_6)$$

$$\dot{m}_1 = \dot{m}_{12} \frac{(X_{12} - X_6)}{(X_1 - X_6)} = \frac{0.0149 \text{ kg}}{s} \frac{(0.995 - 0.345)}{(0.450 - 0.345)} = 0.0922 \frac{\text{kg}}{s}$$

$$\dot{m}_6 = 0.0922 - 0.0149 = 0.0773 \frac{\text{kg sol}}{s}$$

c) Suction line heat exchanger performance not specified, assume 10°C change in liquid temperature from 8 to 9

$$t_9 = t_8 - 10°C = 51°C - 10°C = 41°C, \quad h_9 = 380 \text{ kJ/kg} = h_1$$

$$h_{11} = h_{12} - (h_8 - h_9) = 1440 - (435 - 380) = 1385 \text{ kJ/kg}$$

For the pump, $h_2 = h_1 + (P_2 - P_1) v_1$

$$v_1 = (1-x) v_{H_2O} + (0.85) X_1 \, v_{NH_3}$$

By Tables A.1SI and A.2SI, saturated liquid at 20°

$$v_1 = (1 - 0.450) \, 0.001002 \, \frac{m^3}{kg} + (0.85)(0.450) \frac{1}{610.3} \frac{m^3}{kg}$$

$$v_1 = 0.00118 \, m^3/kg$$

$$h_2 = -80 \frac{kJ}{kg} + (2.0 - 0.2) \text{MPa} \, (0.00118) \, m^3/kg = -7.8 \frac{kJ}{kg}$$

5.6 Cont'd

For the solution heat exchanger
$$\dot{m}_2 h_2 + \dot{m}_4 h_4 = \dot{m}_2 h_3 + \dot{m}_4 h_5$$

$$h_5 = h_4 - \frac{\dot{m}_2}{\dot{m}_4}(h_3 - h_2)$$

$$= 400 - \frac{0.0922}{0.0773}(300 - (-78)) = -51 \frac{kJ}{kg}$$

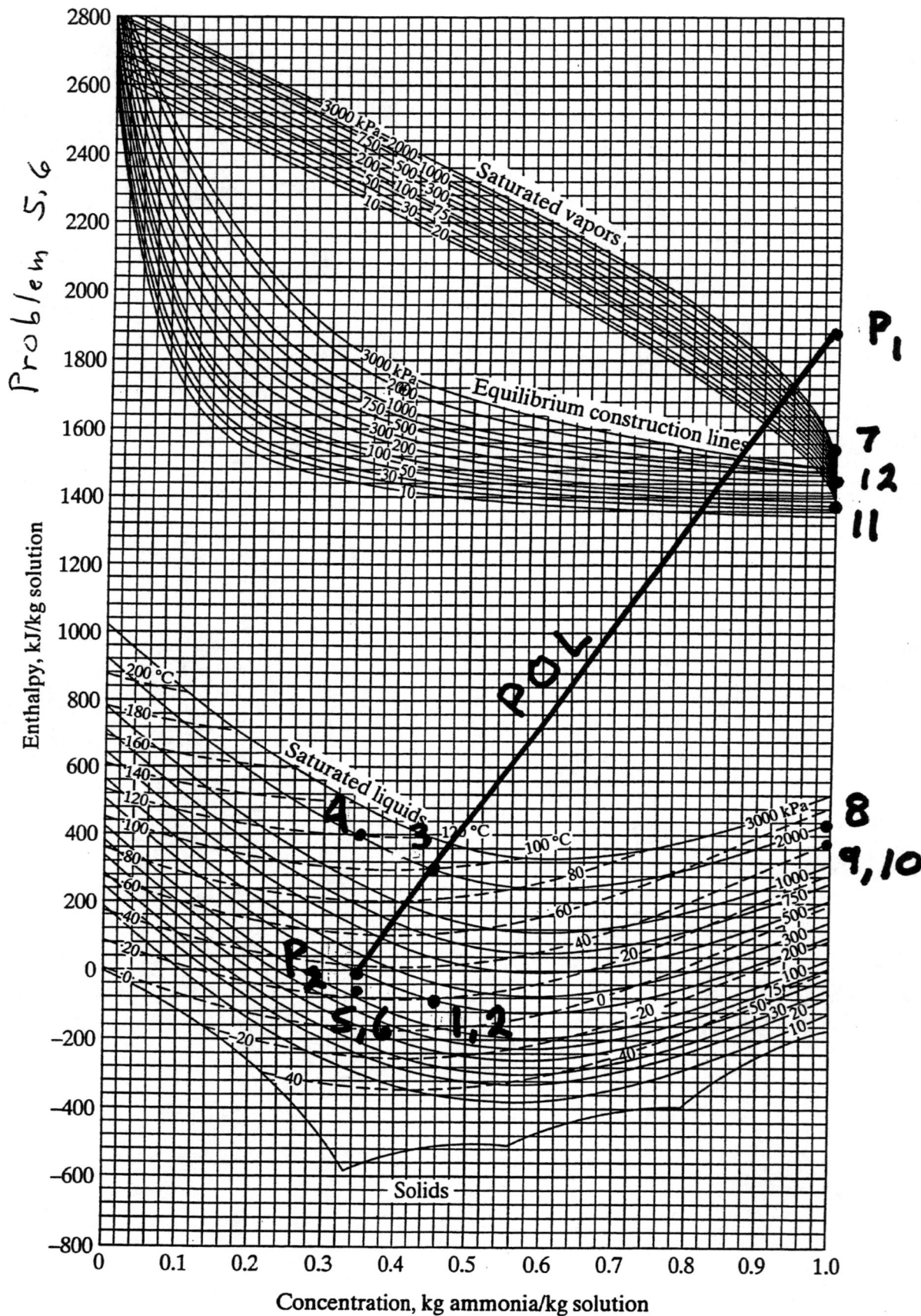

Chart C-3SI

5.7
Procedure follows Prob. 5.6

state	P MPa	t °C	x kg NH$_3$/kg sol	h kJ/kg sol	$\dot{m}$ kg/s
1	0.2	45	0.305	40	0.113
2	1.0	46	0.305	40.1	0.113
3	1.0	100	0.305	310	0.113
4	1.0	120	0.215	425	0.100
5	1.0		0.215	120	0.100
6	0.2		0.215	120	0.100
7	1.0	60	0.985	1590	0.0132
8	1.0	26	0.985	300	0.0132
9	1.0	15	0.985	250	0.0132
10	0.2	-18	0.985	250	0.0132
11	0.2		0.985	1390	0.0132
12	0.2		0.985	1440	0.0132

a & b)

$$\dot{m}_{12} = \frac{15}{1440 - 300} = 0.0132 \text{ kg/s}$$

$$\dot{m}_1 = 0.0132 \frac{(0.985 - 0.215)}{(0.305 - 0.215)} = 0.113 \text{ kg/s}$$

$$\dot{m}_6 = 0.113 - 0.0132 = 0.100 \text{ kg/s}$$

c) Assume $t_8 - t_9 = 10°C$, $h_9 = h_{10} = 250$ kJ/kg

$h_{11} = 1440 - (300 - 250) = 1390$ kJ/kg

By Tables A.1SI and A.2SI, saturated liquid, 45°C

$$v_1 = (1-0.305)\,0.00101 + (0.85)(0.305)\frac{1}{571.3} = 0.001156 \text{ m}^3/\text{kg}$$

5.7 Cont'd

$$h_2 = 40 + (1.0 - 0.2)(0.001156) = 40.1 \text{ kJ/kg}$$

$$h_5 = 425 - \frac{0.113}{0.100}(310 - 40.1) = 120 \text{ kJ/kg}$$

Conclusions:

1) The condenser temp. is reduced from 51°C to 26°C indicating the need for lower cooling water temp.

2) The absorber temp. is much higher ($t_1 = 45°C$ compared to 20°C) so the absorber cooling is less critical

3) All the concentrations change with the refrigerant concentration being less than before.

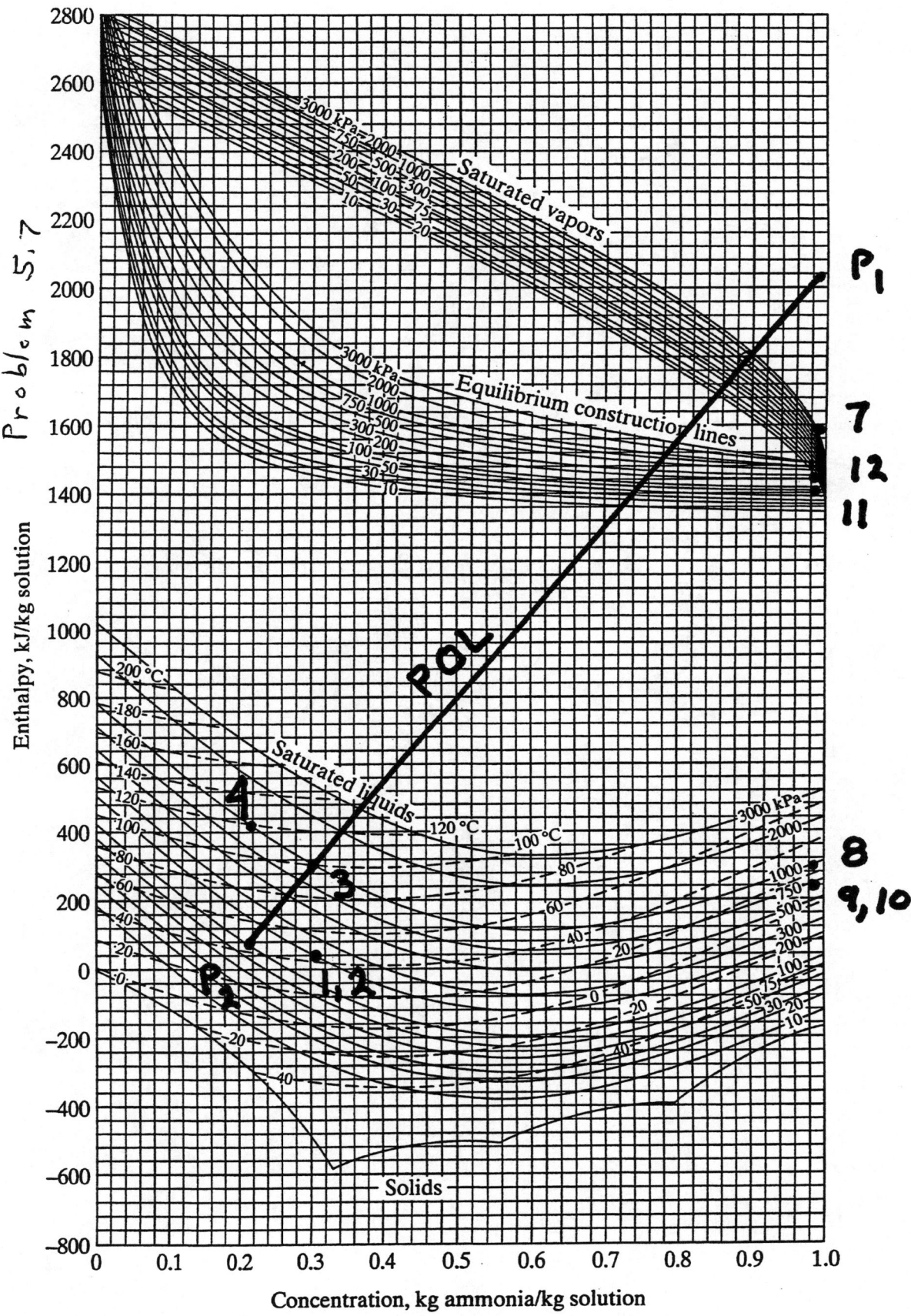

Chart C-3SI

5.8

All statepoints and mass flow rates remain unchanged except for state points 9, 10 & 11. State 9 equals state 8 and state 11 equals 12. State 10 is at the same pressure and temperature as before (L & V mix at 30 psia, $X = 0.997$) although the enthalpy is higher, $h_{10} = h_8 = 149 \, Btu/lb_m$

The energy flows are identical to Example 5.5.

The only difference is that the refrigerant leaving the evaporator is now 61°F rather than 46°F indicating a higher temperature of the load.

Thus the heat exchanger allows a lower load temperature to be maintained (46°F vs 61°F) without affecting the C.O.P. of the system.

5.9

a) $t_3 = 180°F$, all other parameters are as described in Ex. 5.5.

state	P psia	t °F	x lbm NH$_3$/lbm mix	h Btu/lbm mix	$\dot{m}$ lbm/min
1	30	64 *	0.458 *	-40 *	169.8 *
2	200	65 *	0.458 *	-39 *	169.8 *
3	200	180 *	0.458 *	90 *	169.8 *
4	200	240	0.293	158	130.0 *
5	200	85 *	0.293	-11 *	130.0 *
6	30	85 *	0.293	-11 *	130.0 *
7	200	130	0.997	655	39.8
8	200	95	0.997	149	39.8
9	200	85	0.997	137	39.8
10	30	-2	0.997	137	39.8
11	30	46	0.997	640	39.8
12	30	61	0.997	652	39.8

* indicates changed value

Parts a) & b)

$$\dot{m}_6 = \frac{39.8 (0.997 - 0.458)}{(0.458 - 0.293)} = 130.0 \text{ lbm/min}$$

$$\dot{m}_1 = 130.0 + 39.8 = 169.8 \text{ lbm/min}$$

$$v_1 = (1 - 0.458)(0.01604) + 0.85(0.458) \frac{1}{38.30} = 0.0189 \text{ ft}^3/\text{lbm}$$

$$h_2 = -40 + \frac{(200 - 30)(0.0189) 144}{778} = -39.41 \text{ Btu/lbm}$$

$$h_5 = h_6 = 158 - \frac{169.8 (90 - (-39.4))}{130.0} = -11.0 \text{ Btu/lbm}$$

5.9 a) Cont'd

Part c) $\dot{W}_{1-2} = \dfrac{169.8(0.59)}{42.4(0.75)} = 3.15$ hp

Part d) $h_{P_1} = 770$ Btu/lbm, $h_{P_2} = -120$ Btu/lbm

$\dot{Q}_G / \dot{m}_4 = 158 - (-120) = 278$ Btu/lbm

$\dot{Q}_D / \dot{m}_7 = 770 - 655 = 115$ Btu/lbm

$\dot{Q}_G = 278(130.0) = 36{,}140$ Btu/min

$\dot{Q}_D = 115(39.8) = 4577$ Btu/min

C.O.P. $= \dfrac{20{,}000}{36{,}140} = 0.553$

Part e) $\eta_R = \dfrac{0.553}{1.54} = 0.36$

Part f) C.O.P. $= 3.82$ (unchanged)

Part g)

Component	Symbol	Gains Btu/min	Losses Btu/min
Absorber	$\dot{Q}_A$		31,310
Pump	$\dot{W}_P$	134	
Generator	$\dot{Q}_G$	36,140	
Dephlegmator	$\dot{Q}_D$		4577
Condenser	$\dot{Q}_C$		20,140
Evaporator	$\dot{Q}_E$	20,000	
Totals		56,270	56,030

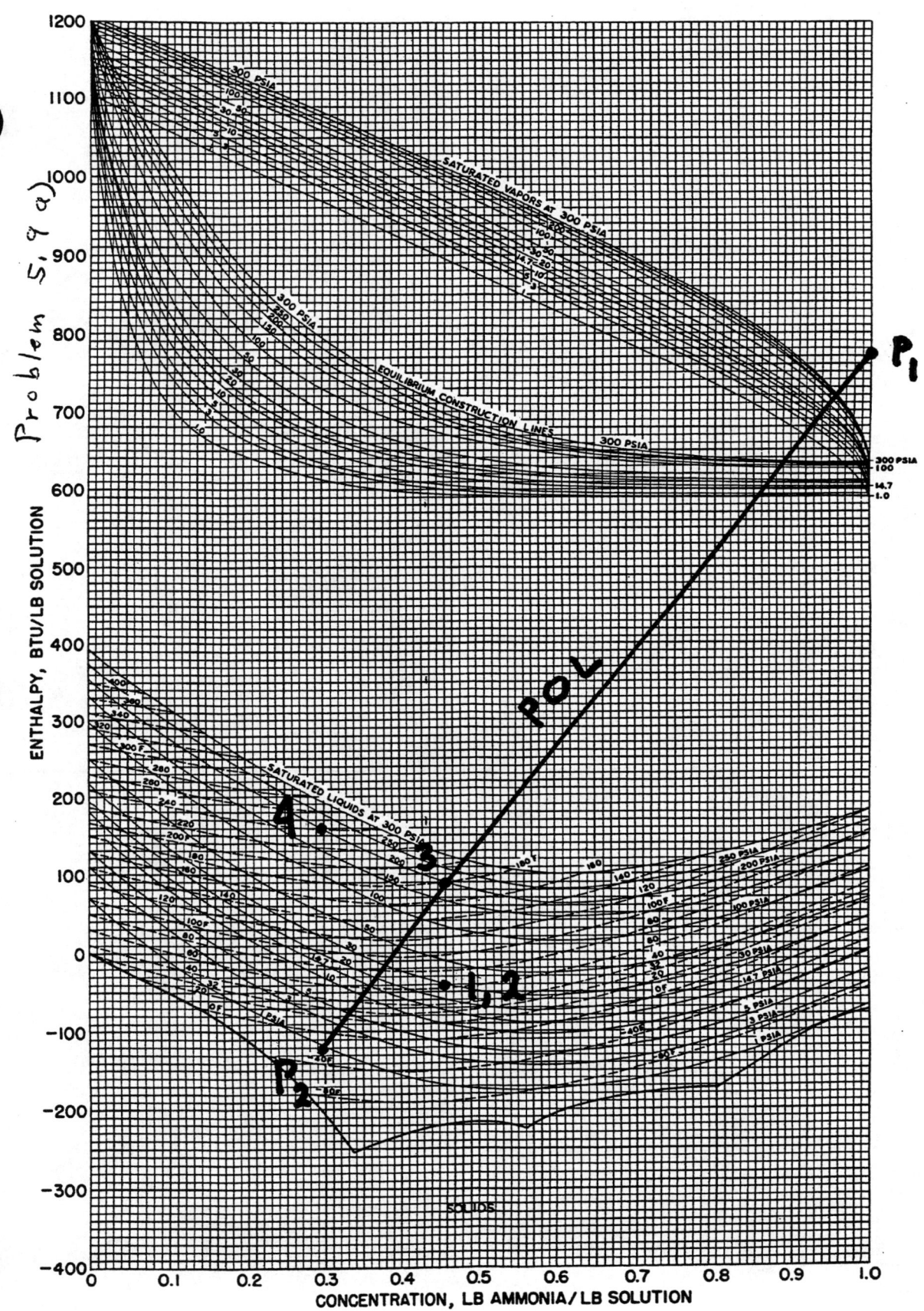

Chart C-3E

5.9 cont'd
b) $t_3 = 160°F$, all other parameters are as described in Ex. 5.5

state	P psia	t °F	x $lb_m NH_3/lb_m$ mix	h Btu/lb_m mix	$\dot{m}$ lb_m/min
1	30	48 *	0.528 *	-52 *	119.2 *
2	200	49 *	0.528 *	-51 *	119.2 *
3	200	160 *	0.528 *	72 *	119.2 *
4	200	240	0.293	158	79.4 *
5	200	70 *	0.293	-27 *	79.4 *
6	30	70 *	0.293	-27 *	79.4 *
7	200	130	0.997	655	39.8
8	200	95	0.997	149	39.8
9	200	85	0.997	137	39.8
10	30	-2	0.997	137	39.8
11	30	46	0.997	640	39.8
12	30	61	0.997	652	39.8

* indicates changed value

Parts a) & b)

$$\dot{m}_6 = \frac{39.8(0.997-0.528)}{(0.528-0.293)} = 79.4 \ lb_m/min$$

$$\dot{m}_1 = 79.4 + 39.8 = 119.2 \ lb_m/min$$

$$v_1 = (1-0.528)0.01602 + (0.85)0.528(1/39.10) = 0.0190 \ ft^3/lb_m$$

$$h_2 = -52 + \frac{(200-30)(0.0190)144}{778} = -51.4 \ Btu/lb_m$$

$$h_5 = h_6 = 158 - \frac{119.2}{79.4}(72-(-51.4)) = -27.3 \ Btu/lb_m$$

5.9 b) Cont'd

Part c) $\dot{W}_{1-2} = \dfrac{119.2\,(0.59)}{42.4\,(0.75)} = 2.21\ hp$

Part d) $h_{P_1} = 702\ Btu/lbm$, $h_{P_2} = -240\ Btu/lbm$

$\dot{Q}_G = 79.4\,(158-(-240)) = 31{,}600\ Btu/min$

$\dot{Q}_D = 39.8\,(702-655) = 1870\ Btu/min$

$C.O.P. = \dfrac{20{,}000}{31{,}600} = 0.633$

Part e) $\eta_R = \dfrac{0.633}{1.54} = 0.41$

Part f) $C.O.P. = 3.82$ (unchanged)

Component	Symbol	Gains Btu/min	Losses Btu/min
Absorber	$\dot{Q}_A$		30,000
Pump	$\dot{W}_P$	71	
Generator	$\dot{Q}_G$	31,600	
Dephlegmator	$\dot{Q}_D$		1870
Condenser	$\dot{Q}_C$		20,140
Evaporator	$\dot{Q}_E$	20,000	
Totals		51,670	52,014

Although the C.O.P. improves as t_3 is reduced, the temperature of state ① leaving the absorber becomes unrealistically low so the cycle may not operate at reduced values of t_3.

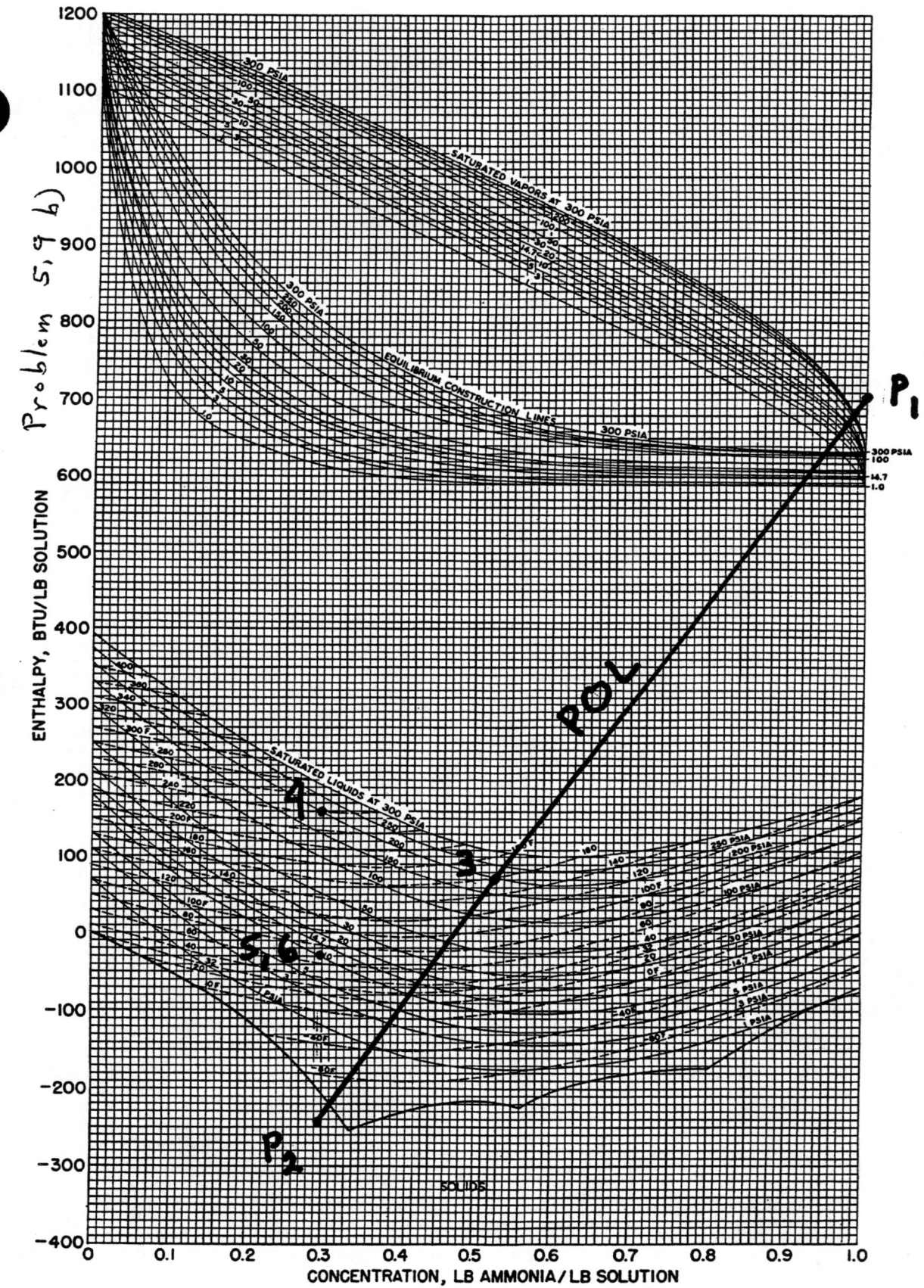

Chart C-3E

5.10

Property values and rates of flow obtained from the given data are shown below.

State	P psia	t F	x lb NH$_3$/lb sol	h Btu/lb	m lb/min
1	20				1100
2	220				1100
3	220				1100
4	220	210	0.398	120	1000
5	220		0.398		1000
6	20		0.398		1000
7	220		0.998		100
8	220	80	0.998	131	100
9	220		0.998		100
10	20		0.998		100
11	20		0.998		100
12	20	40	0.998	627	100

(a) For the absorber

$$m_6 + m_{12} = m_1$$

$$m_6 x_6 + m_{12} x_{12} = m_1 x_1$$

Thus

$$x_3 = x_1 = x_6 + \frac{m_{12}}{m_1}(x_{12} - x_6)$$

$$= 0.398 + \frac{100}{1100}(0.600) = 0.398 + 0.0545$$

$$= 0.453 \text{ lb NH}_3/\text{lb sol}$$

Thus States 1 and 3 may be located on Fig. E-3.

(b) $v_1 = (1 - x_1) v_{H_2O} + 0.85\, x_1 v_{NH_3}$

$$= (0.547)(0.01603) + \frac{(0.85)(0.453)}{39.1}$$

$$= 0.0186 \text{ cu ft/lb}$$

5.10

and
$$h_2 = h_1 + \frac{(P_2 - P_1)v_1}{J} = -58 + \frac{(200)(144)(0.0186)}{778}$$

$$= -57.3 \text{ Btu/lb}$$

For the combination of absorber and heat exchanger

$$m_{12}h_{12} + m_4 h_4 + m_2 h_2 = m_1 h_1 + m_3 h_3 + q_A$$

$$\begin{aligned} q_A &= m_{12}h_{12} + m_4 h_4 + m_2(h_2 - h_1 - h_3) \\ &= (100)(627) + (1000)(120) + (1100)(-57.3 + 58 - 99) \\ &= 62{,}700 + 120{,}000 - 108{,}130 \\ &= 74{,}570 \text{ Btu/min} \end{aligned}$$

(c) $q_{E^i} = m_{12}(h_{12} - h_8) = (100)(627 - 131)$

$$= 49{,}600 \text{ Btu/min}$$

and

$$\text{Tons} = \frac{49{,}600}{200} = 248$$

5.11

Procedure follows Example 5.6

From Table A.1SI, $P_{cond} = 0.007384$ MPa
$P_{Evap} = 0.0008721$ MPa

From Chart C-4SI, $h_3 = -83$ kJ/kg, $x_3 = 0.575$ kg LiBr/kg sol
$h_4 = -49$ kJ/kg, $x_4 = 0.665$ kg LiBr/kg sol

By Eq. 2.35
$$h_7 = 2501 + 1.86(100) = 2687 \text{ kJ/kg}$$

From Table A.1SI, $h_8 = h_9 = 167.26$ kJ/kg
$h_{10} = 2510.10$ kJ/kg
$h_{fg,st} = 2200$ kJ/kg st

$\dot{m}_{10} = 1 \text{ kW}/(2510.10 - 167.26) = 1.537$ kg/hr

$\dot{m}_6 = 1.537 \text{ kg/hr} (0.575)/(0.665 - 0.575) = 9.820$ kg/hr

$\dot{m}_1 = 1.537 + 9.820 = 11.357$ kg/hr

$\dot{Q}_G = (9.820(-49) + 1.537(2687) - 11.357(-83))/3600 = 1.275$ kW

$\dot{m}_{st} = \dfrac{1.275 \text{ kW} (3600)}{2200 \text{ kJ/kg st}} = 2.09$ kg/hr

5.12

Procedure follows Prob. 5.11

a) $t_3 = 60°C$

From Chart C-4 SI, $h_3' = -90 \, kJ/kg$, $x_3' = 0.465 \, kg \, LiBr/kg$

$\dot{m}_6' = 1.537(0.465)/(0.665 - 0.465) = 3.574 \, kg/hr$

$\dot{m}_1' = 1.537 + 3.574 = 5.111 \, kg/hr$

$\dot{Q}_G' = (3.574(-49) + 1.537(2687) - 5.111(-91))/3600 = 1.228 \, kW$

$\dot{m}_{st}' = \dfrac{1.228 \,(3600)}{2200} = 2.01 \, kg/hr$

b) $t_3 = 50°C$

From Chart C-4 SI, $h_3'' = -68 \, kJ/kg$, $x_3'' = 0.382 \, kg \, LiBr/kg \, sol$

$\dot{m}_6'' = 1.537(0.382)/(0.665 - 0.382) = 2.075 \, kg/hr$

$\dot{m}_1'' = 1.537 + 2.075 = 3.612 \, kg/hr$

$\dot{Q}_G'' = (2.075(-49) + 1.537(2687) - 3.612(-68))/3600 = 1.187 \, kW$

$\dot{m}_{st}'' = \dfrac{1.187 \,(3600)}{2200} = 1.94 \, kg/hr$

These conditions will not actually exist because state 5 will lie in the solid region of the chart. The heat exchanger is necessary to ensure proper operation of the system.

Problems 5.11 and 5.12

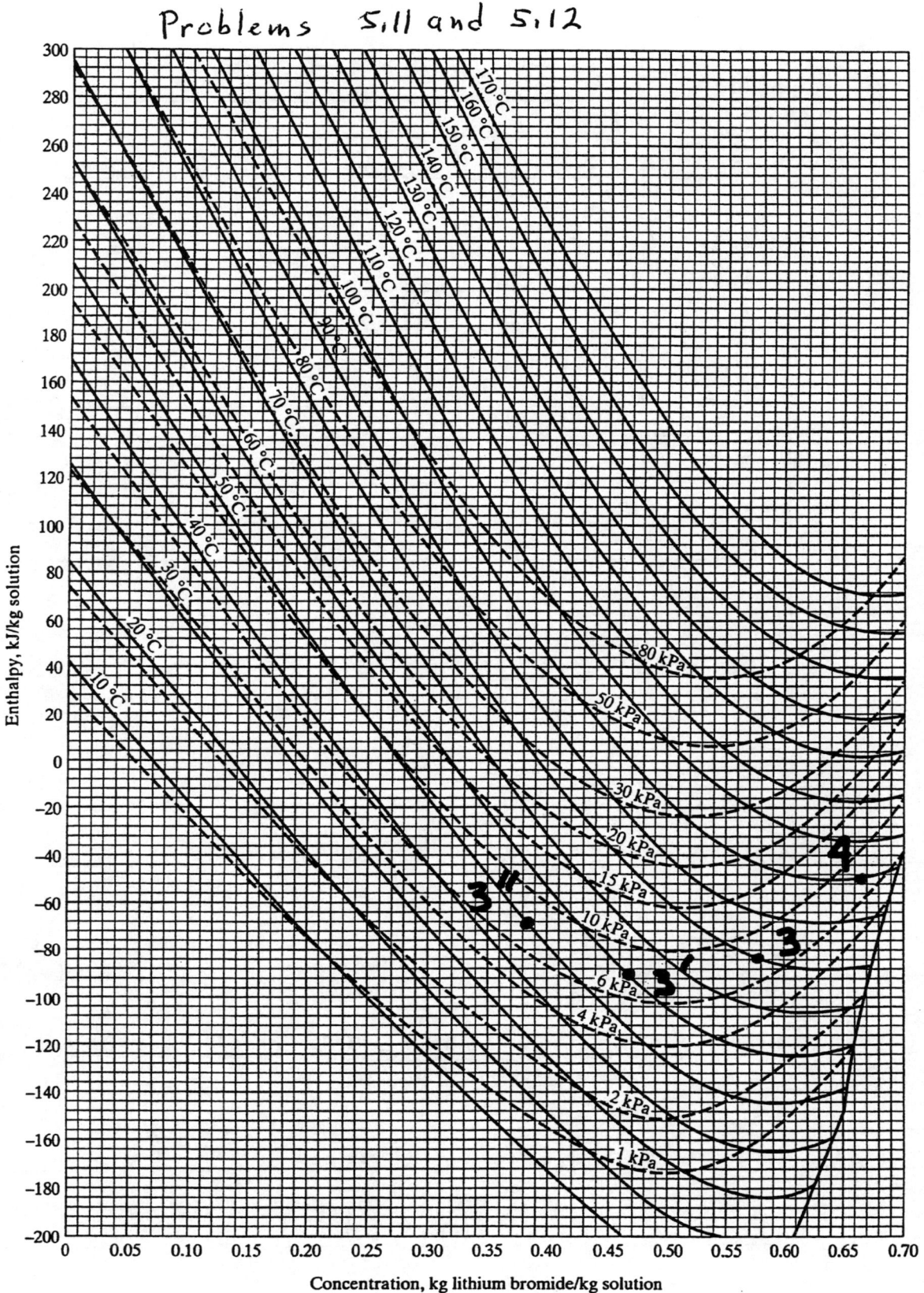

Chart C-4SI

5.13

State	P mmHg	t °F	x lbm LiBr/lbm mix	h Btu/lbm mix
1	5.8	100	0.58	
2	66.2		0.58	
3	66.2	180	0.58	-34
4	66.2	210	0.65	-23
5	66.2		0.65	
6	5.8		0.65	
7	66.2	210	0.00	1153
8	66.2	110	0.00	78
9	5.8	38	0.00	78
10	5.8	38	0.00	1078

Mass flow rate of returning water: (52°F)

$$\dot{m}_w = \frac{\dot{V}}{v} = \frac{600 \text{ gal/min}}{0.01603 \text{ ft}^3/\text{lbm}} \left(\frac{35.32 \text{ ft}^3}{264.2 \text{ gal}} \right) = 5004 \text{ lbm/min}$$

For the evaporator energy balance

$$\dot{m}_q = \frac{\dot{m}_w c_{pw} \Delta t_w}{h_{10} - h_9}$$

From Table A.6E, $c_{pw}(48°F) = 1.0019$ Btu/lbm °F

$$\dot{m}_q = \frac{5004 \text{ lbm/min} (1.0019 \text{ Btu/lbm °F}) 8°F}{(1078 - 78) \text{ Btu/lbm}} = 40.1 \text{ lbm/min}$$

LiBr mass balance on absorber

$$\dot{m}_6 = \dot{m}_{10} (x_1 - x_{10})/(x_6 - x_1)$$

$$= \frac{40.1 \text{ lbm}}{\text{min}} \frac{(0.58)}{(0.65 - 0.58)} = 332.3 \text{ lbm/min}$$

5.13 cont'd

$\dot{m}_1 = \dot{m}_{10} + \dot{m}_6 = 40.1 + 332.3 = 372.4 \text{ lbm/min}$

Energy balance on the generator

$\dot{Q}_G = \dot{m}_4 h_4 + \dot{m}_7 h_7 - \dot{m}_3 h_3$

$= 332.3(-23) + 40.1(1153) - 372.4(-34) = 51,254 \text{ Btu/min}$

$\dot{m}_{st} = \dfrac{\dot{Q}_G}{h_{fg,st}}$

From Table A.1E, $h_{fg,st} = 952 \text{ Btu/lbm}$

$\dot{m}_{st} = \dfrac{51,254 \text{ Btu/min}}{952 \text{ Btu/lbm}} \left(\dfrac{60 \text{ min}}{\text{hr}}\right) = 3230 \text{ lbm/hr}$

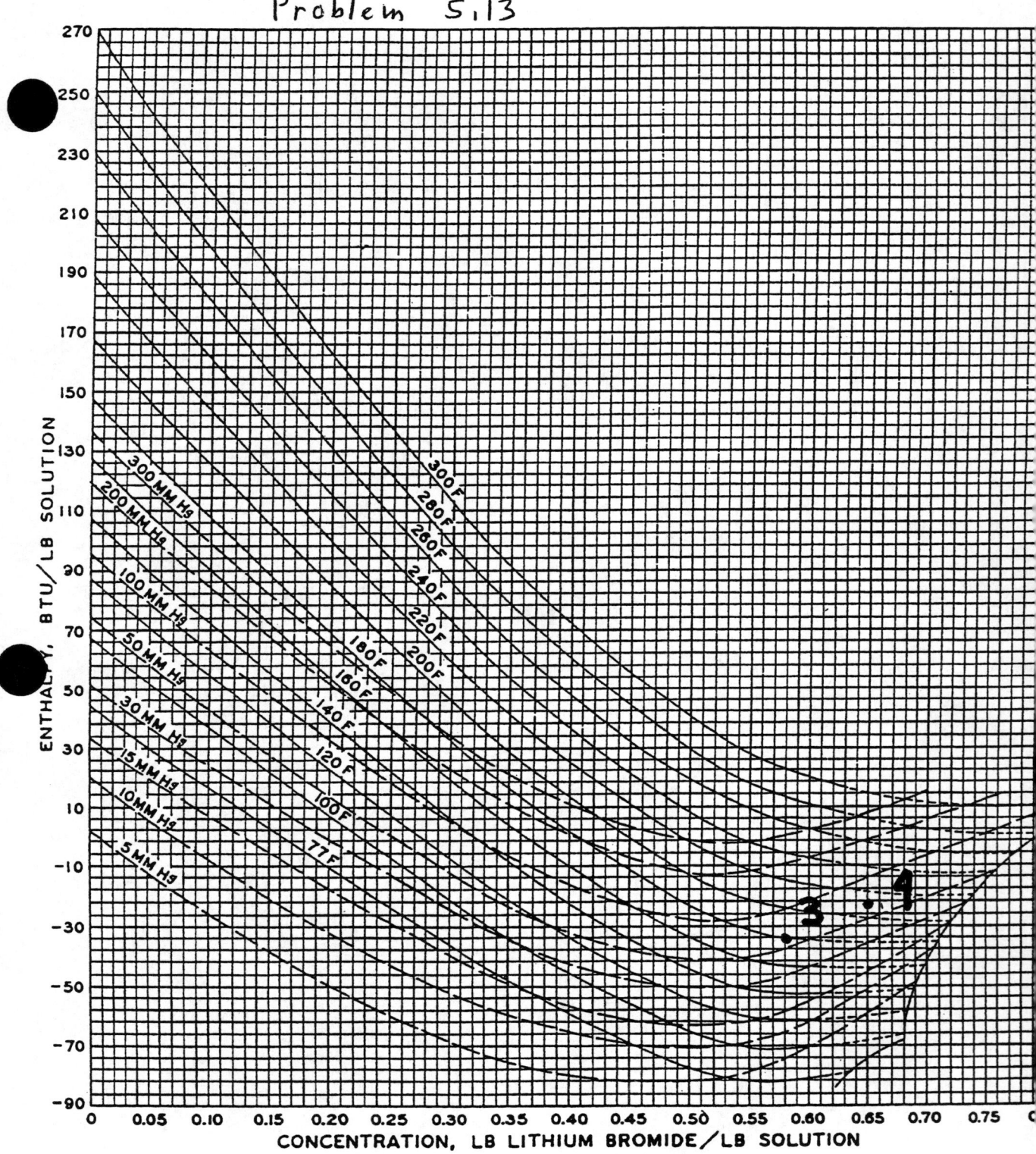

Chart C-4E

5.14

a) Repeat Ex. 5.6 with $T_{evap} = 50°F$

By Table A.1E, the evaporating pressure changes to
0.37729 in Hg × (25.4 mm/in) = 9.58 mm Hg

Parts (a) and (b)

State	P mmHg	t °F	x $lb_{m\,LiBr}/lb_{m\,mix}$	h $Btu/lb_{m\,mix}$	$\dot{m}$ $lb_{m\,mix}/min$
1	9.58	100	0.598	-70.3	2.70
2	49.1	100	0.598	-70.3	2.70
3	49.1	180	0.598	-34.5	2.70
4	49.1	200	0.645	-27.5	2.51
5	49.1	108	0.645	-66.0	2.51
6	9.58	108	0.645	-66.0	2.51
7	49.1	200	0	1150	0.197
8	49.1	100	0	67.9	0.197
9	9.58	50	0	67.9	0.197
10	9.58	50	0	1083.1	0.197

$$\dot{m}_{10} = \dot{m}_9 = \frac{\dot{Q}_{9-10}}{h_{10}-h_9} = \frac{200\,Btu/min}{(1083.1-67.9)\,Btu/lb_m} = 0.197\,lb_m/min$$

LiBr mass balance on absorber

$$\dot{m}_6 = \frac{\dot{m}_{10}(x_1)}{x_6 - x_1} = \frac{0.197 \times 0.598}{0.645 - 0.598} = 2.51\,lb_m/min$$

$$\dot{m}_1 = \dot{m}_{10} + \dot{m}_6 = 0.197 + 2.51 = 2.70\,lb_m/min$$

Energy balance on heat exchanger to determine h_5

$$\dot{m}_1(h_2 - h_3) = \dot{m}_6(h_5 - h_4)$$

5.14 cont'd

$$h_5 = \frac{\dot{m}_7}{\dot{m}_6}(h_2 - h_3) + h_4$$

$$= \frac{2.70}{2.51}(-70.3 + 34.5) - 27.5 = -66.0 \text{ Btu/lbm} = h_6$$

Locating state 5 on Fig. C-4E, $t_5 = 108°F$
State 6 is also subcooled liquid so $t_6 = 108°F$

Part (c)

$$\dot{Q}_G = \dot{m}_4 h_4 + \dot{m}_7 h_7 - \dot{m}_3 h_3$$

$$= \left[2.51(-27.5) + 0.197(1150) - 2.70(-34.5)\right] \frac{lbm}{min} \frac{Btu}{lbm}$$

$$\dot{Q}_G = 250.7 \text{ Btu/min}$$

$$C.O.P. = \frac{200 \text{ Btu/min}}{250.7 \text{ Btu/min}} = 0.798$$

Part (d) $C.O.P._{max} = \frac{510(100)}{50(660)} = 1.55$

$$\eta_R = \frac{0.798}{1.55} = 0.515$$

Part (e)

$$\dot{m}_s = \frac{(60 \text{ min/hr}) \, 250.7 \text{ Btu/min}}{(964.9) \text{ Btu/lbm}} = 15.59 \text{ lbm/hr}$$

Problem 5.14a

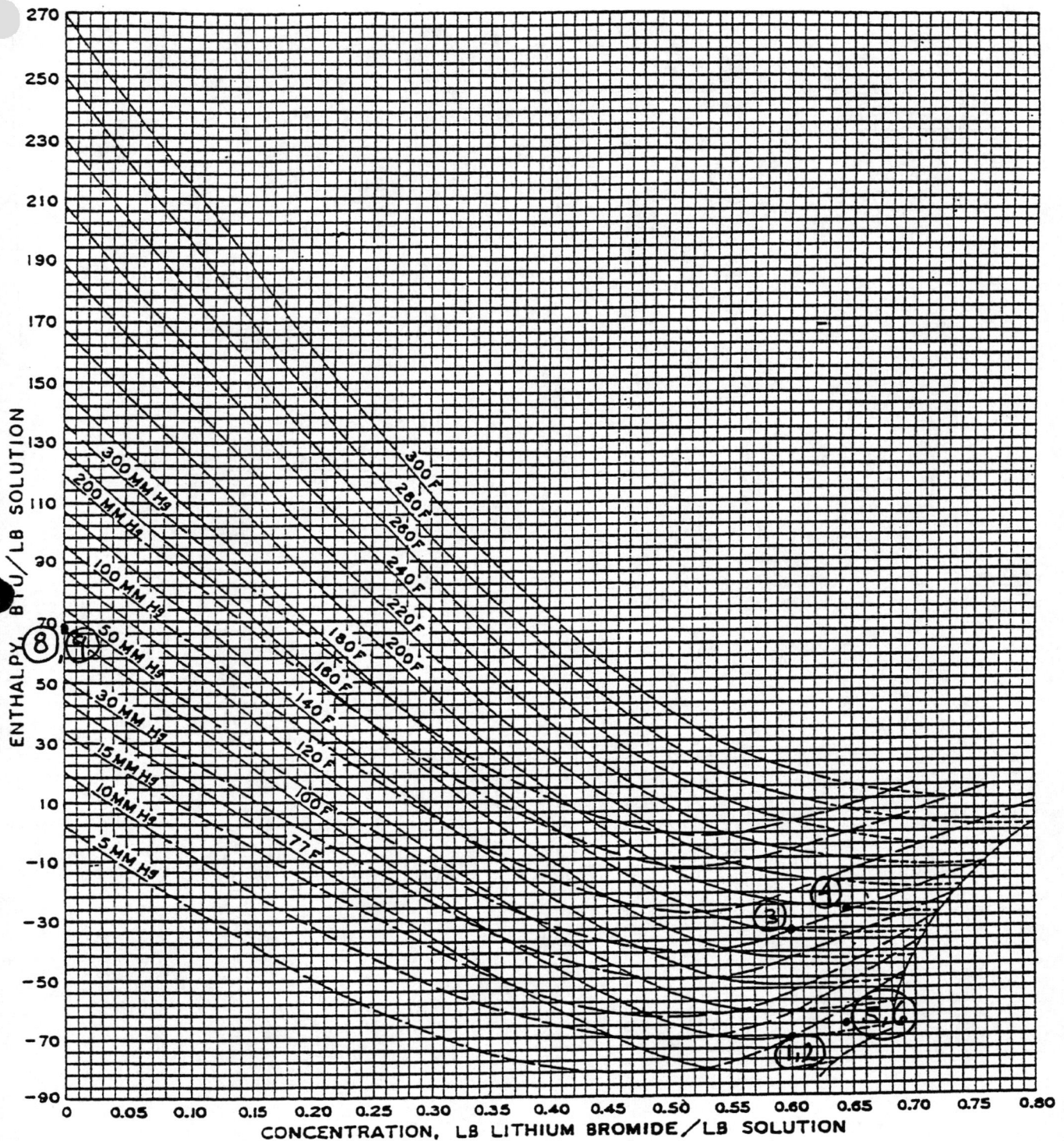

FIGURE C-4E

5.15

State	P kPa	t °C	x kg LiBr/kg mix	h kJ/kg mix	$\dot{m}$ kg mix/sec
1	1	42	0.575	-160	4.92
2	6	42	0.575	-160	4.92
3	6	75	0.575	-93	4.92
4	6	86	0.629	-76	4.50
5	6		0.629		4.50
6	1		0.629		4.50
7	6	86	0	2661	0.423
8	6	36	0	153	0.423
9	1	7	0	153	0.423
10	1	7	0	2514	0.423

$$\dot{m}_6 = \frac{\dot{m}_{10}(x_1 - x_{10})}{x_6 - x_1} = \frac{0.423(0.575 - 0)}{0.629 - 0.575} = 4.50 \text{ kg mix/sec}$$

$$\dot{m}_1 = \dot{m}_6 + \dot{m}_{10} = 4.50 + 0.423 = 4.92 \text{ kg mix/sec}$$

By Eqn. 2.35, $h_7 = 2501 + 1.86(86) = 2661$ kJ/kg

$$\dot{Q}_G = \dot{m}_4 h_4 + \dot{m}_7 h_7 - \dot{m}_3 h_3$$

$$= 4.50(-76) + 0.423(2661) - 4.92(-93) = 1.241 \text{ MW}$$

Neglecting Pump Work

$$\text{C.O.P.} = \frac{\dot{Q}_E}{\dot{Q}_G} = \frac{1 \text{ MW}}{1.241 \text{ MW}} = 0.806$$

This is lower than the result of Ex. 5.7, 1.06, by about 24%.

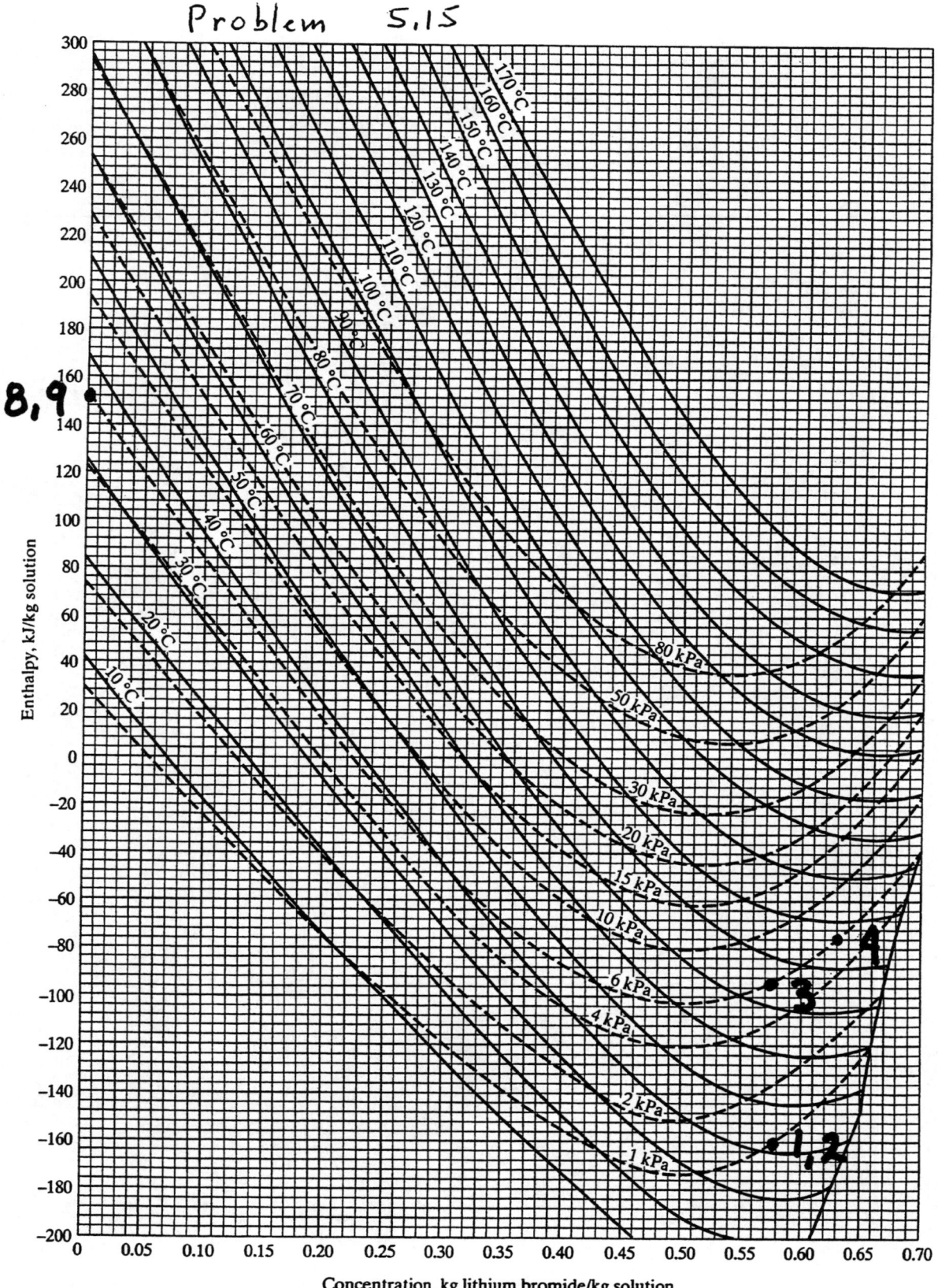

Chart C-4SI

5.16
Procedure follows Ex. 5.7

state	P kPa	t °C	x kg LiBr/kg mix	h kJ/kg mix	ṁ kg mix/sec
1	1	42	0.575	-160	7.876
2	80	42	0.575	-160	7.876
3	80	80	0.575	-84	7.876
4	80	80	0.575	-84	3.072
5	80	152	0.62	48	2.849
6	6	96	0.62	48	2.849
7	80	80	0.575	-84	4.804
8	6	76	0.575	-84	4.804
9	80	152	0	2784	0.223
10	6	80	0.60	-87	4.604
11	6				
12	1				
13	6	80	0	2650	0.200
14	80	93.5	0	392	0.223
15	6	36	0	392	0.223
16	6	36	0	150	0.423
17	1	7	0	150	0.423
18	1	7	0	2514	0.423

$h_9 = 2501 + 1.86(152) = 2784 \text{ kJ/kg}$

$h_{13} = 2501 + 1.86(80) = 2650 \text{ kJ/kg}$

$\dot{m}_{18} = \dfrac{\dot{Q}_E}{h_{18} - h_{17}} = \dfrac{1 \text{ MW}}{(2514 - 150) \text{ kJ/kg}} = 0.423 \text{ kg/sec}$

$\dfrac{\dot{m}_q}{\dot{m}_g} = \dfrac{[2650 - (-84)] + \dfrac{0.575}{0.16}(-87 - 2650)}{2784 - 392} = 0.0464$

5.16 Cont'd

$$\frac{\dot{m}_4}{\dot{m}_8} = \frac{0.0464(0.62)}{0.62 - 0.575} = 0.6396$$

$$\frac{\dot{m}_3}{\dot{m}_8} = 0.6396 + 1 = 1.6396$$

$$\frac{\dot{m}_4}{\dot{m}_3} = 0.3901$$

$$\frac{\dot{m}_9}{\dot{m}_3} = \frac{0.3901(0.62 - 0.575)}{0.62} = 0.0283$$

$$\frac{\dot{m}_5}{\dot{m}_3} = 0.3901 - 0.0283 = 0.3618$$

$$\frac{\dot{m}_{10}}{\dot{m}_3} = \frac{0.6099(0.575)}{0.60} = 0.5845$$

$$\frac{\dot{m}_{13}}{\dot{m}_3} = 0.6099 - 0.5845 = 0.0254$$

$$\frac{\dot{m}_{16}}{\dot{m}_3} = 0.0254 + 0.0283 = 0.0537$$

$$\dot{m}_3 = 0.423 \text{ kg/sec} / 0.0537 = 7.876 \text{ kg/sec}$$

$$\dot{m}_{13} = \dot{m}_8 - \dot{m}_{10} = 4.804 - 4.604 = 0.200 \text{ kg/sec}$$

$$\dot{m}_5 = \dot{m}_4 - \dot{m}_9 = 3.072 - 0.223 = 2.849 \text{ kg/sec}$$

5.16 cont'd

$$\dot{Q}_G = 0.223(2784) + 2.849(48) - 3.072(-84) = 1016 \text{ kW}$$

$$\text{C.O.P.} = \frac{1 \text{ MW}}{1016 \text{ kW}} = 0.984$$

The conditions specified in Ex. 5.7 provide a slightly larger value for C.O.P. (1.06 vs 0.984)

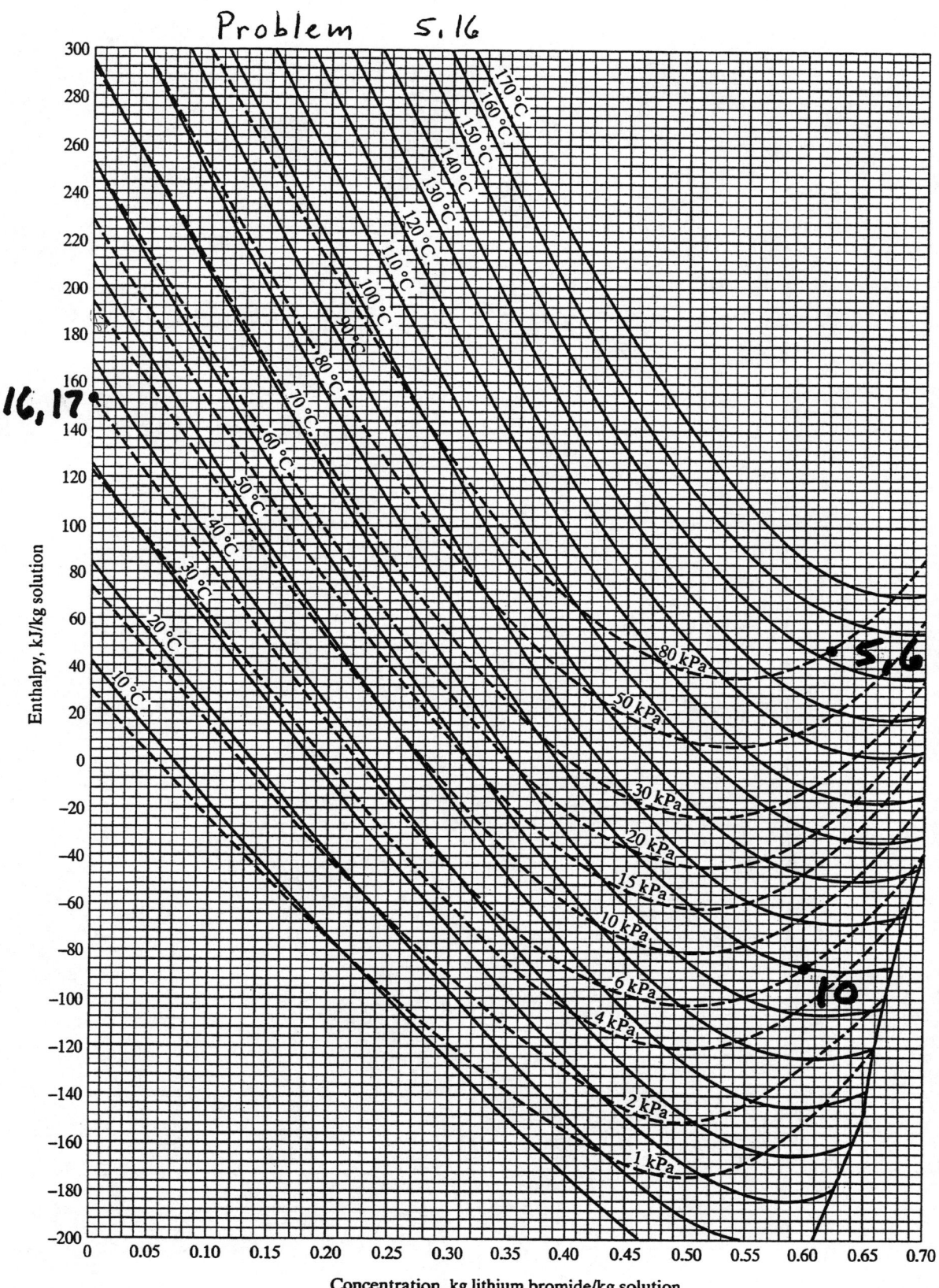

Chart C-4SI

6.1

By Eq. (6.2), $\omega_{z,min} = T_o(s_o - s_f) - (h_o - h_f)$

From chart C-5E, $h_f = 46$ Btu/lb$_m$, $s_f = 0.00$ Btu/lb$_m$°R
$h_o = 227.3$ Btu/lb$_m$, $s_o = 0.922$ Btu/lb$_m$°R

$\omega_{z,min} = (530°R)(0.922 \text{ Btu/lb}_m°R) - (227.3 - 46) \text{ Btu/lb}_m$

$\omega_{z,min} = 307.4$ Btu/lb$_m$

6.2

By Eq. (6.2), $w_{z,min} = T_0(s_0 - s_f) - (h_0 - h_f)$

From chart C-5 SI, $\bar{h}_f = 0$, $\bar{s}_f = 0$
$\bar{h}_0 = 12,400$ J/mole, $\bar{s}_0 = 112.5$ J/mole·K

$w_{z,min} = [300K(112.5 \text{ J/mole·K}) - (12,400 \text{ J/mole})]\left(\dfrac{1 \text{ mole}}{28.96 \text{ gm}}\right)$

$w_{z,min} = 737.2$ kJ/kg

6.3

We will use the nomenclature of Fig. 6.3
a)
By Eq. (6.5), $Z_L = \dfrac{h_7 - h_2}{h_7 - h_5}$

From Chart C-SE, $h_7 = h_1 = 227.3 \text{ Btu/lb}_m$
$s_1 = s_7 = 0.922 \text{ Btu/lb}_m\text{°R}$
$h_2 = 214 \text{ Btu/lb}_m$, $s_2 = 0.542 \text{ Btu/lb}_m\text{°R}$
$h_5 = 46 \text{ Btu/lb}_m$

$Z_L = \dfrac{227.3 - 214}{227.3 - 46} = 0.0734 \text{ lb liq/lb}_m \text{ compressed}$

b) By Eq. (6.6)

$\omega_{z,L} = \left[T_1(s_1 - s_2) - (h_1 - h_2) \right] \left(\dfrac{h_7 - h_5}{h_7 - h_2} \right)$

$= \left[530°R(0.922 - 0.542)\dfrac{\text{Btu}}{\text{lb}_m\text{°R}} - (227.3 - 214)\dfrac{\text{Btu}}{\text{lb}_m} \right] \left(\dfrac{227.3 - 46}{227.3 - 214} \right)$

$\omega_{z,L} = 2563 \text{ Btu/lb}_m \text{ liq.}$

Problem 6.3

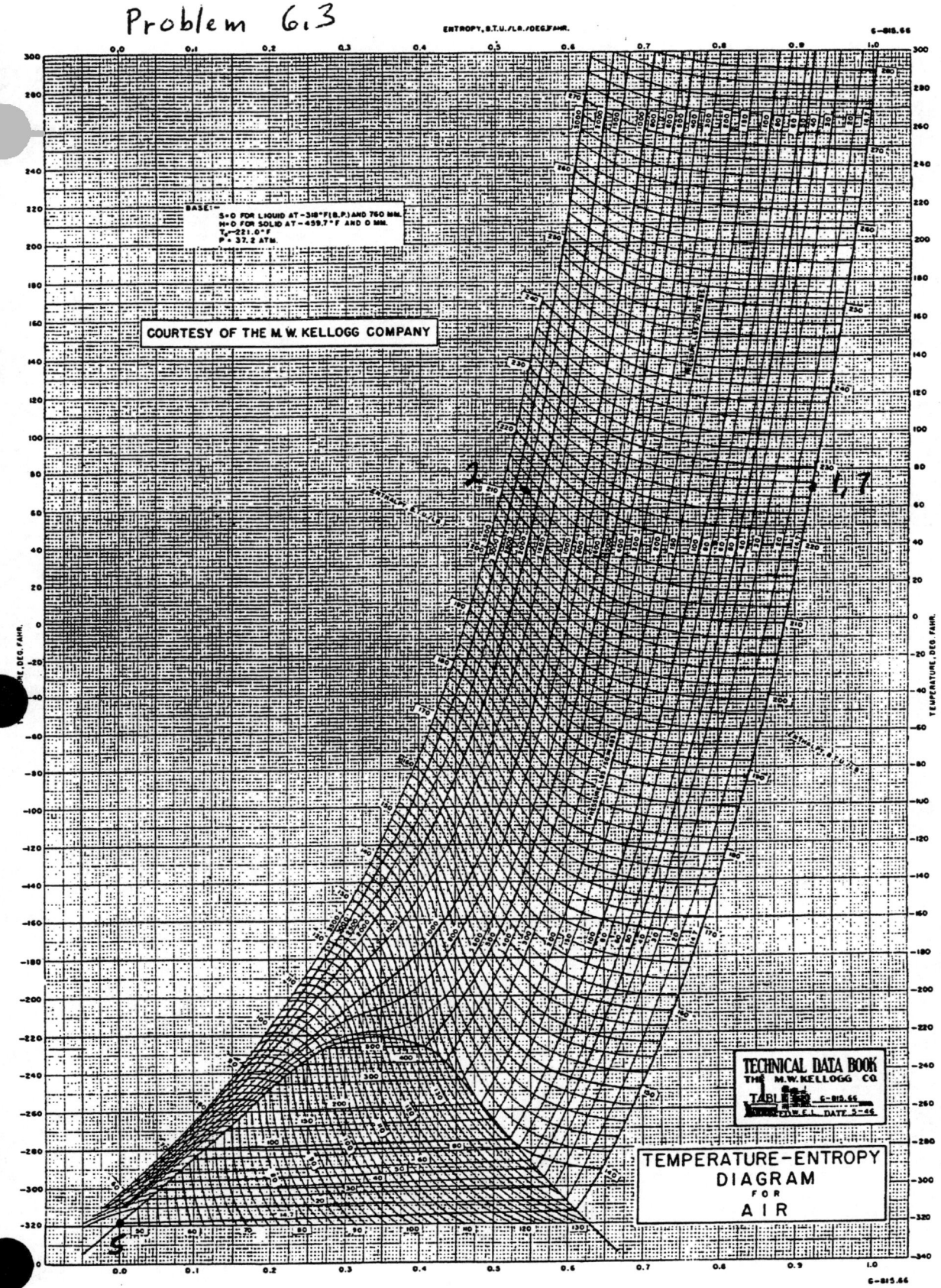

Chart C-5E

6.4

We will use the nomenclature of Fig. 6.3

a) By Eq. (6.5), $Z_L = \dfrac{h_7 - h_2}{h_7 - h_5}$

From chart C-5 SI, $\bar{h}_1 = \bar{h}_7 = 12{,}400$ J/mole
$\bar{s}_1 = \bar{s}_7 = 112.5$ J/mole K
$\bar{h}_2 = 11{,}400$ J/mole, $\bar{s}_2 = 65.5$ J/mole K
$\bar{h}_5 = 0$

$Z_L = \dfrac{12{,}400 - 11{,}400}{12{,}400 - 0} = 0.0806$ kg liq / kg compressed

b) By Eq. (6.6)

$w_{z,L} = \left[T_1(\bar{s}_1 - \bar{s}_2) - (\bar{h}_1 - \bar{h}_2) \right] / Z_L$

$= \dfrac{\left[300K \,(112.5 - 65.5)\,\text{J/mole K} - (12{,}400 - 11{,}400)\,\text{J/mole} \right]}{0.0806} \left(\dfrac{1 \text{ mole}}{28.96 \text{ gm}} \right)$

$w_{z,L} = 5612$ kJ / kg liq.

Problem 6.4

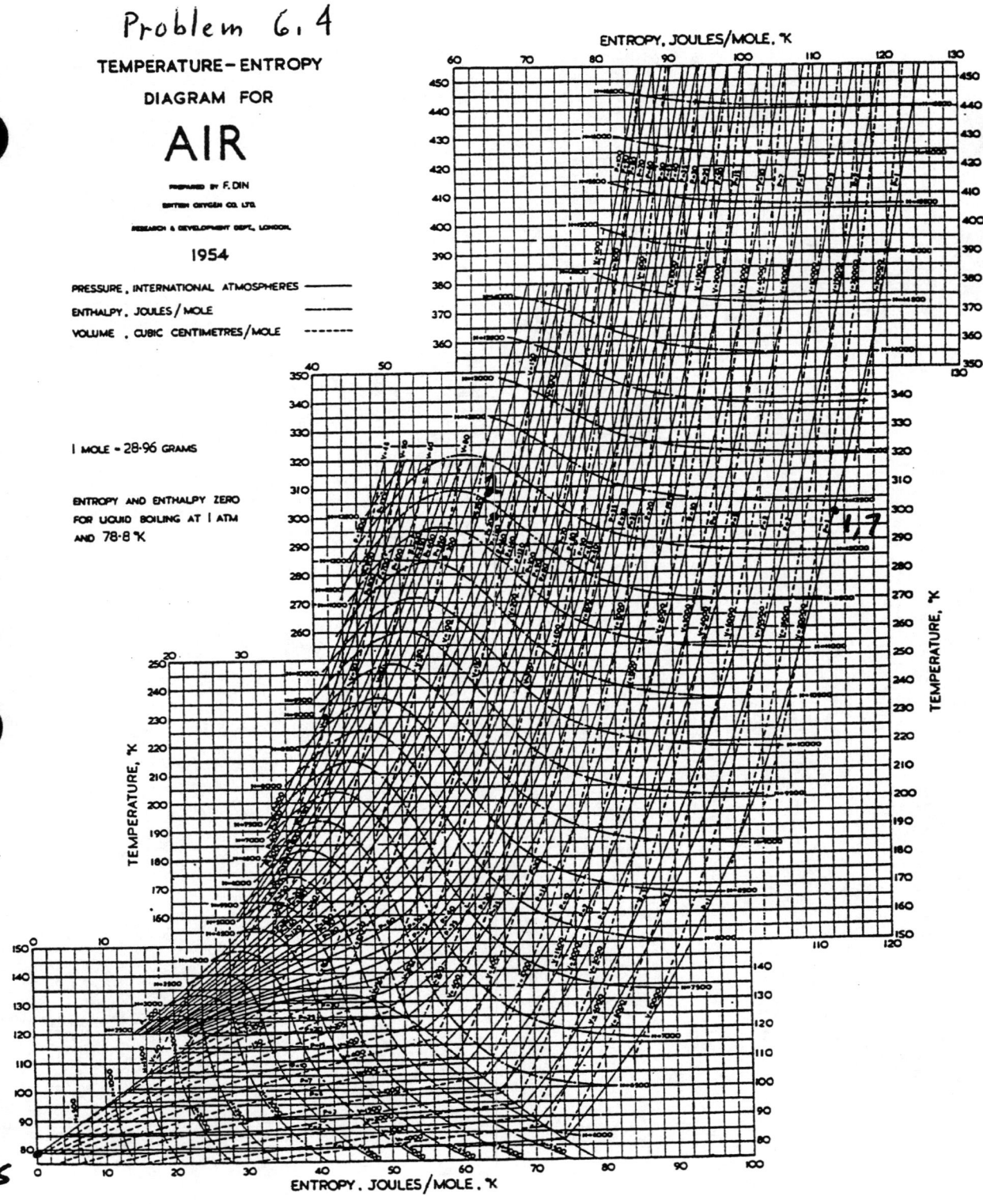

Chart C-5SI

6.5

Using the nomenclature of Fig. 6.3 and assuming states 1 and 7 are equal.

a)
From Chart C-S E

cycle	t_3 °F	t_4 °F	liquid produced
1	70	14	no
2	14	-57	no
3	-57	-160	no
4	-160	-314	yes

liquid is produced during the 4th cycle

b) At steady state, from Prob. 6.3 we know that $z_L = 0.0734$ lbm liq / lbm compressed

Assuming an ideal heat exchanger between states 2-3 and 6-7

$$\dot{m}_2 (h_2 - h_3) = \dot{m}_6 (h_7 - h_6)$$

$$h_3 = h_2 - \frac{\dot{m}_6}{\dot{m}_2}(h_7 - h_6)$$

$$\dot{m}_6 / \dot{m}_2 = 1 - z_L$$

$$h_3 = h_2 - (1-z_L)(h_7 - h_6)$$

From Chart C-SE, $h_2 = 214$ Btu/lbm
$h_7 = 227.3$ Btu/lbm, $h_6 = 134$ Btu/lbm

6.5 cont'd

$h_3 = 214 - (1-0.0734)(227.3 - 134) = 127.5 \; Btu/lbm$

From Chart C-5E, $t_3 \approx -161°F$

This is nearly identical to the result for the 4th cycle obtained in part a).

Problem 6.5

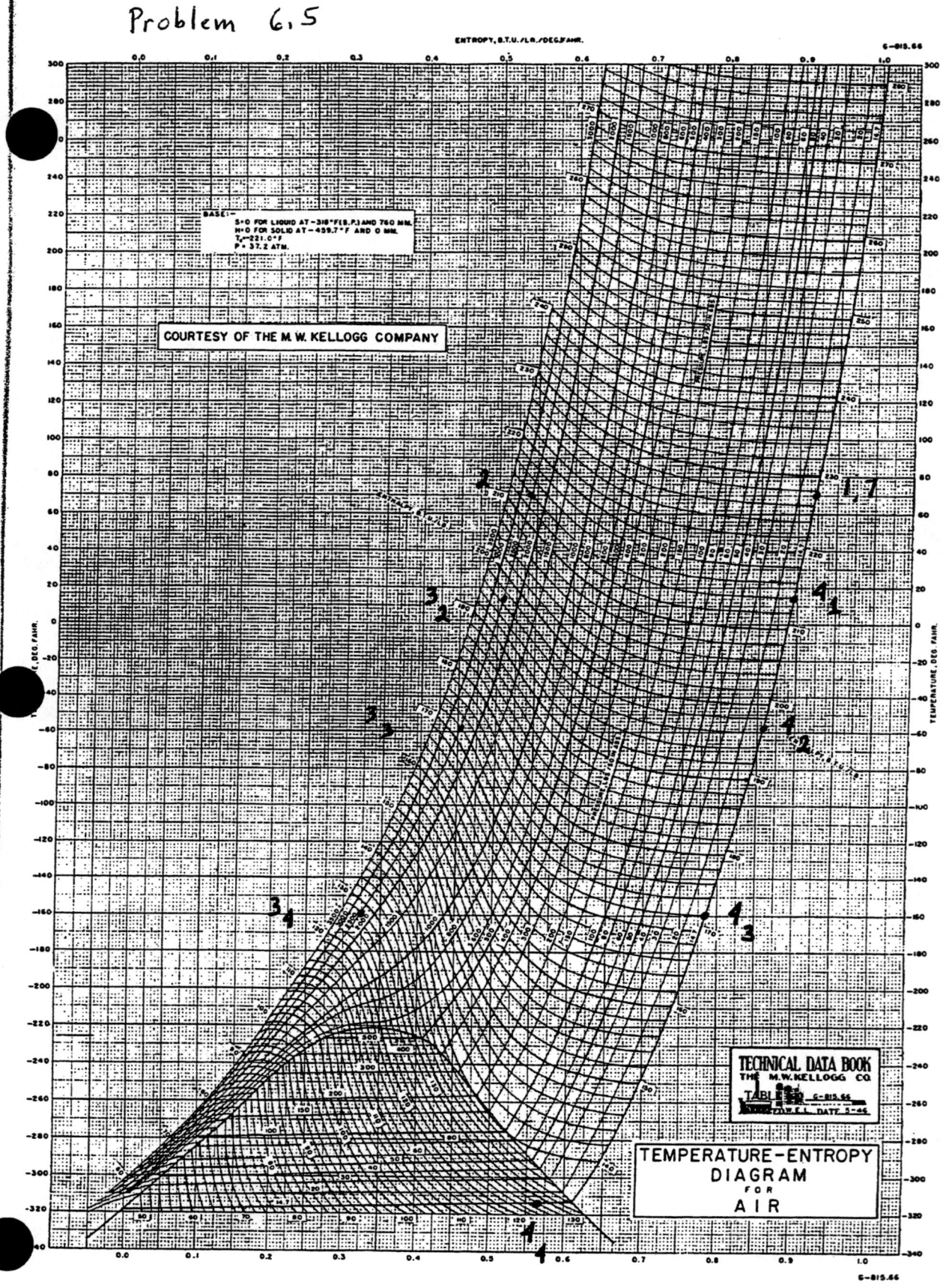

Chart C-5E

6.6

a)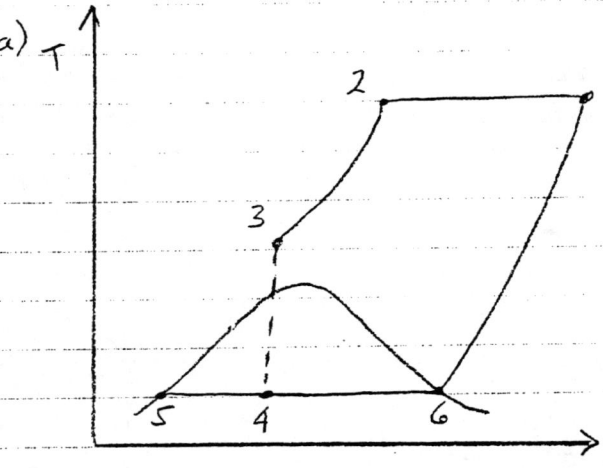

assuming states 1 & 7 are identical

From a mass and energy balance around a combination of the heat exchanger and the separator,

$$\dot{m}_2 = \dot{m}_5 + \dot{m}_6$$

$$\dot{m}_2 h_2 + \dot{m}_2 h_4 = \dot{m}_2 h_3 + \dot{m}_5 h_5 + \dot{m}_6 h_7$$

Eliminating $\dot{m}_6$ and solving for $\dot{m}_5/\dot{m}_2$

$$\frac{\dot{m}_5}{\dot{m}_2} = Z = \frac{(h_2 - h_3) - (h_4 - h_7)}{h_5 - h_7} \qquad (A)$$

From an energy balance on the heat exchanger alone

$$\frac{\dot{m}_5}{\dot{m}_2} = Z = 1 - \frac{h_2 - h_3}{h_7 - h_6} \qquad (B)$$

6.6 Cont'd

From Chart C-SSI, $\bar{h}_1 = \bar{h}_7 = 12{,}400$ J/mole
$\bar{h}_2 = 11{,}400$ J/mole, $\bar{h}_5 = 0$, $\bar{h}_6 = 5900$ J/mole

Setting Eqns. (A) and (B) equal and solving for $\bar{h}_3$ and $\bar{h}_4$ by trial

$\bar{h}_3 = 7200$ J/mole, $\bar{h}_4 = 3800$ J/mole, $t_3 = 202$ K

Then from Eq. (A), $Z = 0.355$ kg liq/kg compresse[d]

b) The net power requirement is

$$\dot{W}_{1-2} - \dot{W}_{3-4} = \dot{m}_2 \left[T_1 (\bar{s}_1 - \bar{s}_2) - (\bar{h}_1 - \bar{h}_2) \right] - \dot{m}_2 (\bar{h}_3 - \bar{h}_4)$$

Converting to specific work requirement

$$w_z = \left[T_1 (\bar{s}_1 - \bar{s}_2) - (\bar{h}_1 - \bar{h}_2) - (\bar{h}_3 - \bar{h}_4) \right] / Z$$

From Chart C-SSI, $\bar{s}_1 = 112.5$ kJ/mole K
$\bar{s}_2 = 65.5$ J/mole K

$$w_z = \left[300\text{K}(112.5 - 65.5)\text{J/mole K} - (12{,}400 - 11{,}400)\text{J/mole} - (7200 - 38\ldots) \right.$$
$$\left. \text{J/mole} \right] \left(\frac{1 \text{ mole}}{28.96 \text{ gm}} \right) / 0.355$$

$w_z = 944$ kJ/kg liq.

Problem 6.6

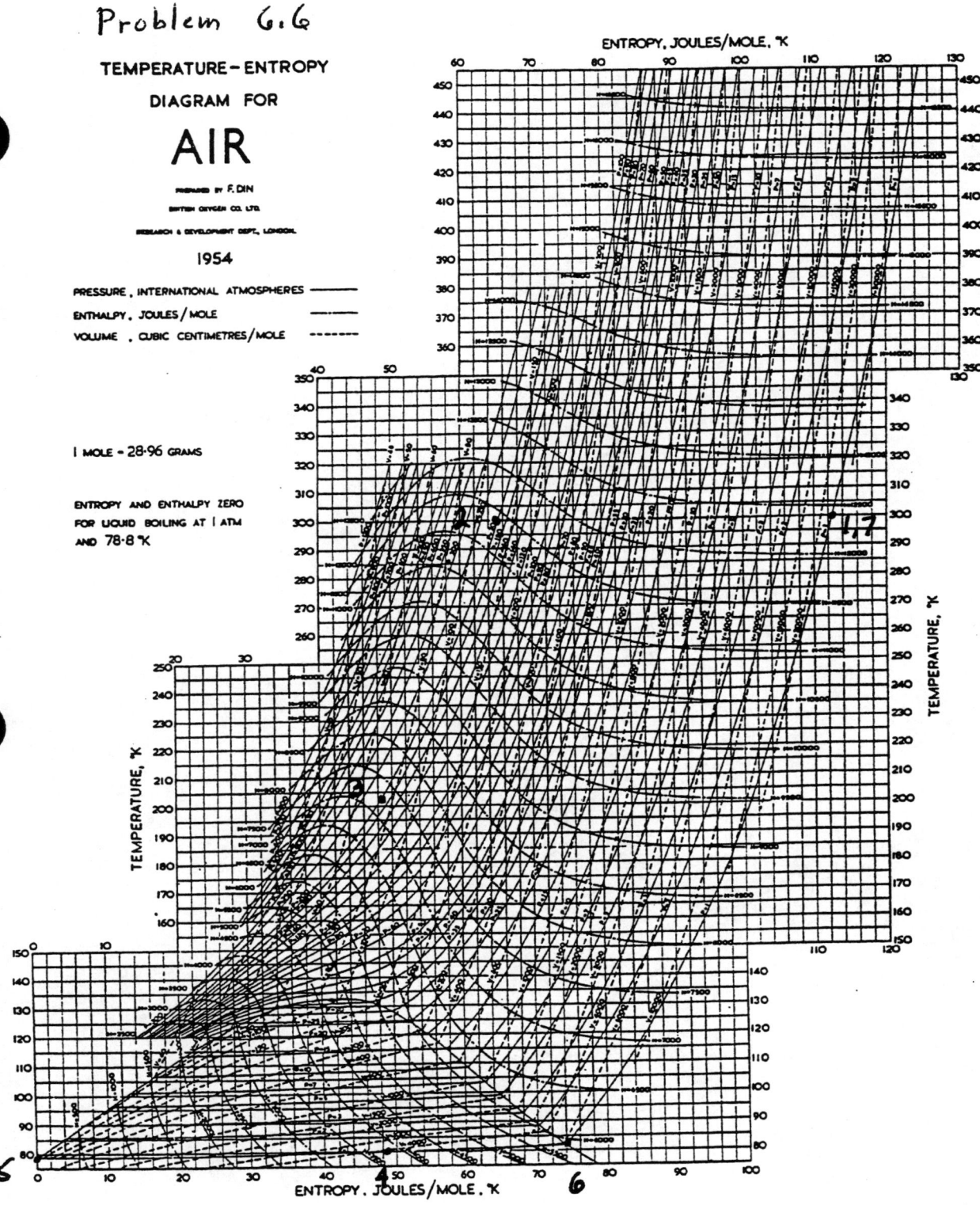

Chart C-5SI

6.7

For the heat exchanger and two separators

$$m_4 = m_9 + m_{11} + m_{13} \tag{1}$$

$$m_4 h_4 = m_9 h_9 + m_{11} h_{11} + m_{13} h_{13} \tag{2}$$

For the low pressure separator

$$m_8 = m_9 + m_{11}$$

$$m_8 h_8 = m_9 h_9 + m_{11} h_{10}$$

and

$$\frac{m_{11}}{m_9} = \frac{h_8 - h_9}{h_{10} - h_8} \tag{3}$$

By Eq. (1)

$$m_{13} = m_4 - m_9 - m_{11} \tag{4}$$

6.7

By Eqs. (2) and (4)

$$\frac{m_4}{m_9} = \frac{h_{13} - h_9}{h_{13} - h_4} = \frac{m_{11}(h_{11} - h_{13})}{m_9(h_{13} - h_4)} \qquad (5)$$

We may also write

$$m_2 = m_9 + m_{11} \qquad (6)$$

$$\frac{m_2}{m_9} = 1 + \frac{m_{11}}{m_9}$$

By Eqs. (3) and (6)

$$\frac{m_2}{m_9} = \frac{h_{10} - h_9}{h_{10} - h_8} \qquad (7)$$

We have

$$W_{Z,2L} = \frac{m_2}{m_9} {}_1W_2 + \frac{m_4}{m_9} {}_3W_4 \qquad (8)$$

Be Eqs. (3), (5), (7) and (8)

$$W_{Z,2L} = \frac{(h_{10} - h_9)_1W_2}{(h_{10} - h_8)} + \left[\frac{h_{13} - h_9}{h_{13} - h_4} - \frac{(h_8 - h_9)(h_{11} - h_{13})}{(h_{10} - h_8)(h_{13} - h_4)} \right] {}_3W_4$$

where

$${}_1W_2 = T_1(s_1 - s_2) - (h_1 - h_2)$$
$${}_3W_4 = T_3(s_3 - s_4) - (h_3 - h_4)$$

6.8

We will use the nomenclature of Fig. 6.4

a)
By Eq. (7) of the solution to Prob. 6.7

$$\frac{\dot{m}_9}{\dot{m}_1} = \frac{h_{10} - h_8}{h_{10} - h_9} = Z_{2L}$$

From Chart C-5E, $h_9 = 46$ Btu/lbm, $h_{10} = 134$ Btu/lbm, $h_7 = h_8 = 76$ Btu/lbm

Thus $Z_{2L} = \frac{134 - 76}{134 - 46} = 0.659$ lbm liq/lbm compressed

b) From Chart C-5E, $h_1 = h_{11} = 227.3$ Btu/lbm
$s_1 = 0.922$ Btu/lbm °R, $h_2 = h_3 = h_{13} = 225.8$ Btu/lbm
$s_2 = s_3 = 0.740$ Btu/lbm °R, $h_4 = 214$ Btu/lbm
$s_4 = 0.542$ Btu/lbm °R

By the equations developed in Prob. 6.7

$$w_{1-2} = 530°R\,(0.922 - 0.740)\frac{Btu}{lbm\,°R} - (227.3 - 225.8)\frac{Btu}{lbm}$$

$$w_{1-2} = 95.0 \text{ Btu/lbm}$$

$$w_{3-4} = 530(0.740 - 0.542) - (225.8 - 214) = 93.1 \text{ Btu/lbm}$$

$$w_{2,2L} = \frac{95.0}{0.659} + \left[\frac{225.8 - 46}{225.8 - 214} - \frac{(76 - 46)(227.3 - 225.8)}{(134 - 76)(225.8 - 214)}\right]$$

$$w_{2,2L} = 1556 \text{ Btu/lbm liq}$$

6.9
Using the nomenclature shown in Fig. 6.5

a)
A mass and energy balance around the combined system of the two air-to-air heat exchangers, the throttle, and the separator:

$$\dot{m}_2 = \dot{m}_7 + \dot{m}_{10}$$

$$\dot{m}_2 h_2 + \dot{m}_2 h_4 = \dot{m}_2 h_3 + \dot{m}_7 h_7 + \dot{m}_{10} h_{10}$$

eliminating $\dot{m}_{10}$ and solving for $\dot{m}_7/\dot{m}_2$

$$\frac{\dot{m}_7}{\dot{m}_2} = z = \frac{h_{10} - h_2}{h_{10} - h_7} + \frac{h_3 - h_4}{h_{10} - h_7} = z_L + \frac{h_3 - h_4}{h_{10} - h_7}$$

From chart C-5 SI, $\bar{h}_{10} = 12,400$ J/mole $= \bar{h}_1$
$\bar{h}_2 = 11,400$ J/mole, $\bar{h}_7 = 0$
$\bar{h}_3 = 10,200$ J/mole, $\bar{h}_4 = 9,000$ J/mole

$$z = \frac{(12,400 - 11,400) + (10,200 - 9,000)}{12,400 - 0} = 0.177 \text{ kg liq/kg comp.}$$

b) For the air compressor, By Eq. (6.6)

$$w_{z,air} = [T_1(s_1 - s_2) - (h_1 - h_2)]/z$$

From chart C-5 SI, $\bar{s}_1 = 112.5$ J/mole K
$\bar{s}_2 = 65.5$ J/mole K

$$w_{z,air} = \left[300K \frac{(112.5 - 65.5)J}{\text{mole K}} - \frac{(12,400 - 11,400)J}{\text{mole}} \right] \left(\frac{1 \text{ mole}}{28.96 \text{ gm}} \right) / 0.177$$

$$= 2556 \text{ kJ/kg liq.}$$

6.9 Con't.

For the refrigeration compressor, using Fig. 3.3 for the nomenclature,

From Table A.3SI, $h_{1'} = h_{2'} = 77.43$ kJ/kg$_{ref}$
$h_{3'} = 236.39$ kJ/kg$_{ref}$

From Chart C-2SI, $h_{4'} = 290.0$ kJ/kg$_{ref}$

Mass Balance on evaporator:
$$\dot{m}_{ref}(h_{3'} - h_{2'}) = \dot{m}_2(h_3 - h_4)$$

or $\dfrac{\dot{m}_{ref}}{\dot{m}_7} = \dfrac{(h_3 - h_4)}{Z(h_{3'} - h_{2'})}$

$$= \dfrac{(10.2 - 9.0)\text{ kJ/mole}\left(\dfrac{1\text{ mole}}{0.02896\text{ kg}}\right)}{0.177} / (236.39 - 77.43)\text{ kJ/kg}_{ref}$$

$\dfrac{\dot{m}_{ref}}{\dot{m}_7} = 1.473$ kg$_{ref}$/kg liq

$w_{z,ref} = \dfrac{\dot{m}_{ref}(h_{4'} - h_{3'})}{\dot{m}_7}$

$= 1.473 \dfrac{\text{kg}_{ref}}{\text{kg liq}}(290 - 236.39)\text{ kJ/kg}_{ref} = 79.0 \dfrac{\text{kJ}}{\text{kg liq}}$

$w_z = w_{z,air} + w_{z,ref} = 2556 + 79 = 2635 \dfrac{\text{kJ}}{\text{kg liq}}$

Problem 6.9

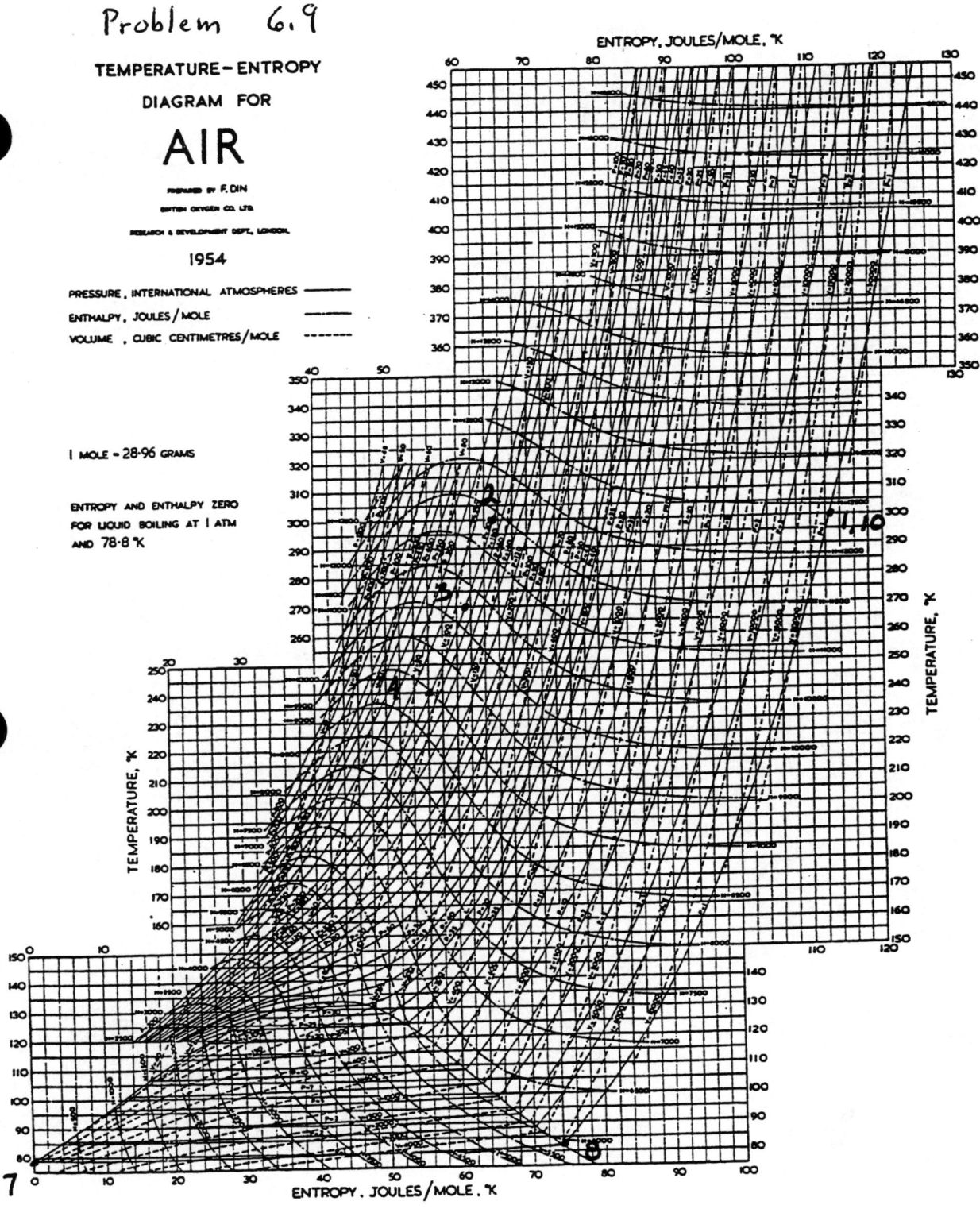

Chart C-5SI

6.10

We will use the nomenclature of Fig. 6.6

a) By Eq. (6.8)

$$z_c = z_L + \frac{\dot{m}_8}{\dot{m}_2} \frac{(h_3 - h_8)}{(h_{11} - h_6)}$$

z_L for these conditions from Prob. 6.2 is 0.0734.

From Chart C-5E, $h_1 = h_{11} = 227.3$ Btu/lbm,
$s_1 = 0.922$ Btu/lbm °R, $h_2 = 214$ Btu/lbm,
$s_2 = 0.542$ Btu/lbm °R, $h_3 = 150.5$ Btu/lbm,
$h_A = 103$ Btu/lbm, $h_6 = 46$ Btu/lbm

$$z_c = 0.0734 + \frac{(0.8)(0.5)(150.5 - 103)}{227.3 - 46} = 0.1782 \text{ lbm liq/lbm comp.}$$

b) $\omega_{z,c} = \frac{\omega_{1-2}}{z_c} - \frac{\dot{m}_8}{\dot{m}_6} \omega_{3-8}$

From Prob. 6.2, $\omega_{1-2} = 188.1$ Btu/lbm

$\omega_{3-8} = (0.5)(h_3 - h_A) = 0.5(150.5 - 103) = 23.8$ Btu/lbm

$\frac{\dot{m}_8}{\dot{m}_6} = 0.8/z_c$

$\omega_{z,c} = \frac{188.1}{0.1782} - \frac{0.8(23.8)}{0.1782} = 949$ Btu/lbm liq

c) $\omega_{z,c} = \frac{188.1}{0.1782} = 1056$ Btu/lbm liq

6.11

The state points will remain equal to the values given in Prob. 6.10

a) $\dot{m}_8 / \dot{m}_2 = 0.6$

$$z_c = 0.0734 + \frac{(0.6)(0.5)(150.5-103)}{227.3 - 46} = 0.1520 \; \frac{lb_m \; liq}{lb_m \; comp}$$

$$w_{z,c} = \frac{188.1 - 0.6(23.8)}{0.1520} = 1144 \; Btu/lb_m \; liq$$

if the expander output is wasted, $w_{z,c} = 1238 \; \frac{Btu}{lb_m \; liq}$

b) $\dot{m}_8 / \dot{m}_2 = 0.4$

$$z_c = 0.0734 + \frac{(0.4)(0.5)(150.5-103)}{227.3 - 46} = 0.1258 \; \frac{lb_m \; liq}{lb_m \; comp}$$

$$w_{z,c} = \frac{188.1 - 0.4(23.8)}{0.1258} = 1420 \; Btu/lb_m \; liq$$

if the expander output is wasted, $w_{z,c} = 1495 \; \frac{Btu}{lb_m \; liq}$

As $\dot{m}_8/\dot{m}_2 \to 0$, the results approach those of the Linde cycle. The best performance for the values considered here is when $\dot{m}_8/\dot{m}_2 = 0.8$. The optimum performance should be near this value.

6.12

For a combination of the two heat exchangers and the separator

$$\dot{m}_2 = \dot{m}_8 + \dot{m}_5$$

$$\dot{m}_2 = \dot{m}_6 + \dot{m}_{11}$$

$$\dot{m}_2 h_2 + \dot{m}_5 h_5 + \dot{m}_8 h_8 = \dot{m}_8 h_3 + \dot{m}_5 h_4 + \dot{m}_6 h_6 + \dot{m}_{11} h_{11}$$

setting $\dot{m}_5 = \dot{m}_2 - \dot{m}_8$ and $\dot{m}_{11} = \dot{m}_2 - \dot{m}_6$ and eliminating $\dot{m}_5$ & $\dot{m}_{11}$

$$\dot{m}_2 h_2 + (\dot{m}_2 - \dot{m}_8) h_5 + \dot{m}_8 h_8 = \dot{m}_8 h_3 + (\dot{m}_2 - \dot{m}_8) h_4$$
$$+ \dot{m}_6 h_6 + (\dot{m}_2 - \dot{m}_6) h_{11}$$

dividing by $\dot{m}_2$ and rearranging

$$\frac{\dot{m}_6}{\dot{m}_2}(h_{11} - h_6) = (h_{11} - h_2) + \frac{\dot{m}_8}{\dot{m}_2}(h_3 - h_8) + \frac{\dot{m}_8}{\dot{m}_2}(h_5 - h_4) + (h_4 - h_5)$$

solving for $\dot{m}_6 / \dot{m}_2 = Z_c^*$

$$Z_c^* = \frac{h_{11} - h_2}{h_{11} - h_6} + \frac{\dot{m}_8}{\dot{m}_2} \frac{(h_3 - h_8)}{(h_{11} - h_6)} + \left(1 - \frac{\dot{m}_8}{\dot{m}_2}\right) \frac{(h_4 - h_5)}{(h_{11} - h_6)}$$

6.13

We will follow the nomenclature used in Figs. 6.7 and 6.8

a) The ideal gas equation of state written on a molar basis (see Eqns. (2.13 and 2.14) for states 1 and 7 is

$$P_1 \dot{V}_1 = \dot{m}_1 R_u T_1, \quad P_7 \dot{V}_7 = \dot{m}_7 R_u T_7$$

where $\dot{m}_1$ and $\dot{m}_7$ represent the number of moles at states 1 and 7 respectively and R_u is the universal gas constant

$$\dot{V}_1 = \dot{V}_7 \frac{\dot{m}_1}{\dot{m}_7} \frac{T_1}{T_7} \qquad (1)$$

also $\dot{m}_1 = \dot{m}_7 + \dot{m}_q$

$$\dot{m}_1 x_1 = \dot{m}_7 x_7 + \dot{m}_q x_q$$

eliminating n_q

$$\frac{\dot{m}_1}{\dot{m}_7} = \frac{x_q - x_7}{x_q - x_1} \qquad (2)$$

combining Eqns. (1) and (2)

$$\dot{V}_1 = \dot{V}_7 \frac{(x_q - x_7)}{(x_q - x_1)} \frac{T_1}{T_7} = 350 \frac{ft^3}{hr} \frac{(0.91 - 0.01)}{(0.91 - 0.79)} \frac{(530°R)}{(520°R)}$$

$$\dot{V}_1 = 2675 \; ft^3/hr$$

6.13 Cont'd

b) Yield $= \dfrac{\dot{m}_7 (1-X_7)}{\dot{m}_1 (1-X_1)} = \dfrac{(X_9 - X_1)(1-X_7)}{(X_9 - X_7)(1-X_1)}$

$\text{Yield} = \dfrac{(0.91 - 0.79)(1-0.01)}{(0.91 - 0.01)(1 - 0.79)} = 0.629$

c) State 3 is a liquid-vapor mixture at a pressure of 10 atm. By trial using Chart C-6E, $T_3 = T_f = T_g = 194°R$ or $t_3 = -266°F$

6.14

We will follow the nomenclature given in Figs. 6.9 and 6.10.

a) Using an energy balance on the air passing through the generator

$$\dot{Q}_A = \dot{m}_1 (h_1 - h_2)$$

Using the ideal gas equation of state for the incoming air (see Eqs. (2.13) and (2.14))

$$\dot{m}_1 = P_{atm} V_{atm} / R_u T_{atm}$$

$$= \frac{(14.696 \, lb_f/in^2)(144 \, in^2/ft^2) \, 200,000 \, ft^3/hr}{(1545 \, ft \, lb_f / lb_{mole} \, °R)(530 \, °R)} = 517 \frac{lb \, moles}{hr}$$

From Chart C-6E, $h_1 = 4130$ Btu/lb mole, $h_2 = 2200$ Btu/lb mole

$$\dot{Q}_A = \left(517 \frac{lb \, moles}{hr}\right)(4130 - 2200) \frac{Btu}{lb \, mole} = 998,000 \, Btu/hr$$

b) For the entire double column

$$\dot{m}_1 = \dot{m}_g + \dot{m}_q$$
$$\dot{m}_1 X_1 = \dot{m}_g X_g + \dot{m}_q X_q$$

combining these and eliminating $\dot{m}_g$

$$\frac{\dot{m}_q}{\dot{m}_1} = \frac{X_g - X_1}{X_g - X_q} = \frac{0.95 - 0.79}{0.95 - 0.01} = 0.170 \text{ or } 17\%$$

Problem 6.14

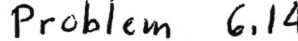

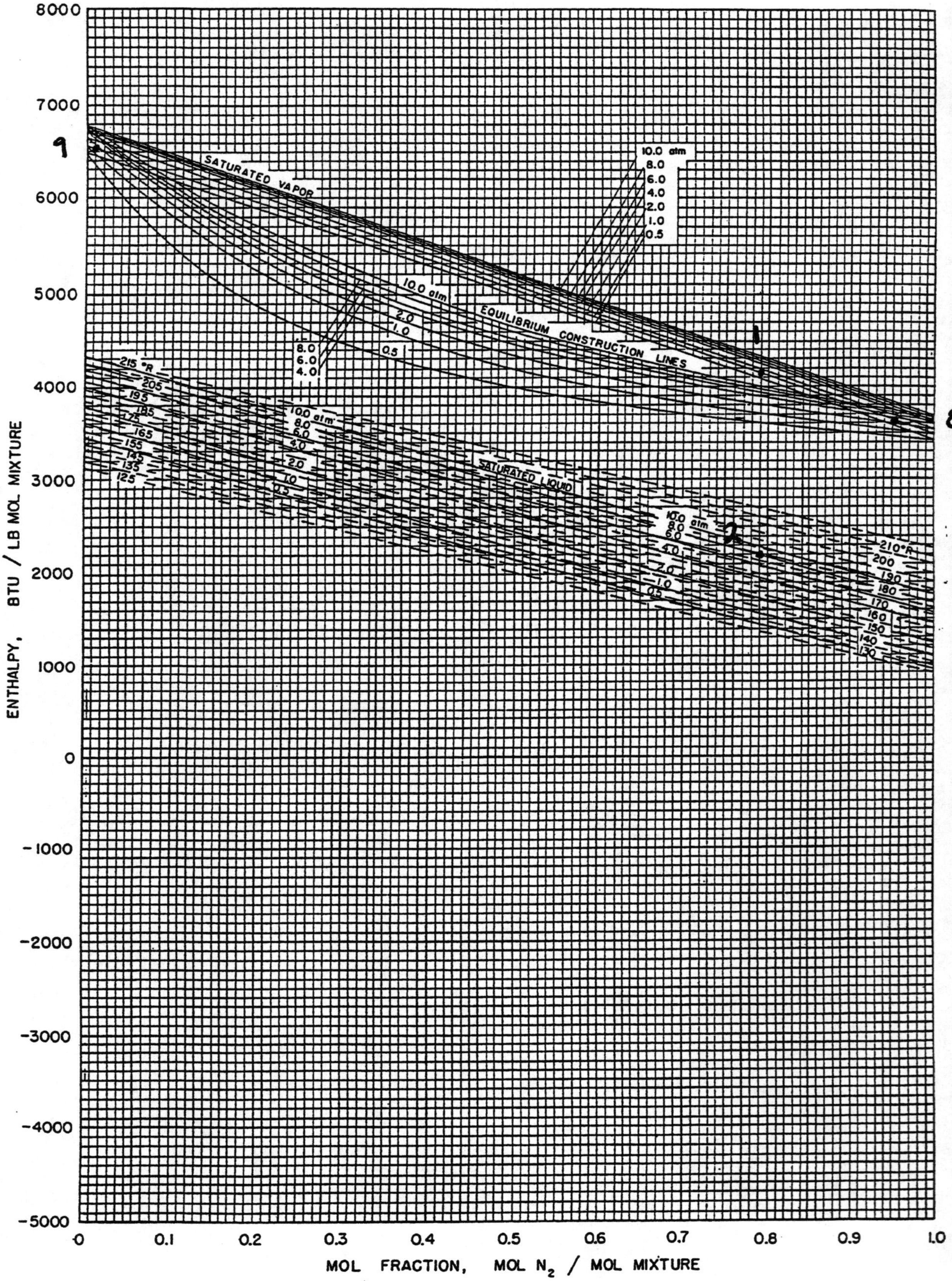

Chart C-6E

6.15

The nomenclature follows that shown in Figs. 6.9 & 6.10.

a)
1. Locate states 8 and 9 ($P = 1$ atm, $X_8 = 0.99$, $X_9 = 0.005$)
2. Locate state 1 on the line between 8 & 9 at $X_1 = 0.79$
3. Locate states 2 & 3 at $P = 6$ atm, $X = 0.79$, sat. liquid
4. Locate states 4 & 5 at $P = 4$ atm, $X = 0.59$, sat. liquid
5. Draw isotherm t_8, locate states 6 & 7 on t_8, sat. liquid at $P = 4$ atm
6. Locate P_1 at the intersection of $X = X_6$ and the line connecting states 1 and 4
7. Draw the P.O.L. for the lower column by drawing the line that passes through P_1 and state 2
8. Locate P_2 at the intersection of the lower column P.O.L. and $X = 0.59$
9. Locate state 11 at the intersection of the line connecting states 4 and 6 and $X = 0.79$
10. Locate P_2' at the intersection of $X = X_9$ and the line connecting 8 and 11
11. Locate P_1' at the intersection of extensions of the lines through P_2' and 5, and 7_5 and 8

State	P atm	T °R	t °F	h Btu/lbmole	X lbmoles N_2/lbmoles mix
1	6	180	-280	4120	0.79
2	6	175	-285	2090	0.79
3	4	167	-293	2090	0.79
4	4	170	-290	2440	0.59
5	1	144	-316	2440	0.59
6	4	165	-295	1520	0.965
7	1	139	-321	1520	0.965
8	1	139	-321	3520	0.99
9	1	162	-298	6550	0.005
10	1	162	-298	141 3600	0.00

6.15 cont'd.

b)

Pole	P_1	P_2	P_1'	P_2'
h, Btu/lbmole	5650	-1950	7580	-4280
x, lbmole N_2/lbmole mix	0.965	0.59	1.04	0.005

c) $\dfrac{\dot{Q}_A}{\dot{m}_4} = h_4 - h_{P_2} = 2440 - (-1950) = 4390 \text{ Btu/lbmole}$

d) $\dfrac{\dot{Q}_B}{\dot{m}_q} = h_q - h_{P_2} = 6550 - (-4280) = 10,830 \text{ Btu/lbmole}$

e) Oxygen yield is determined as in Prob. 6.14

$$\dfrac{\dot{m}_q}{\dot{m}_1} = \dfrac{x_8 - x_1}{x_8 - x_q}$$

$$\text{Yield} = \dfrac{\dot{m}_q (1-x_q)}{\dot{m}_1 (1-x_1)} = \dfrac{(x_8 - x_1)(1-x_q)}{(x_8 - x_q)(1-x_1)}$$

$$\text{Yield} = \dfrac{(0.99 - 0.79)(1 - 0.005)}{(0.99 - 0.005)(1 - 0.79)} = 0.962$$

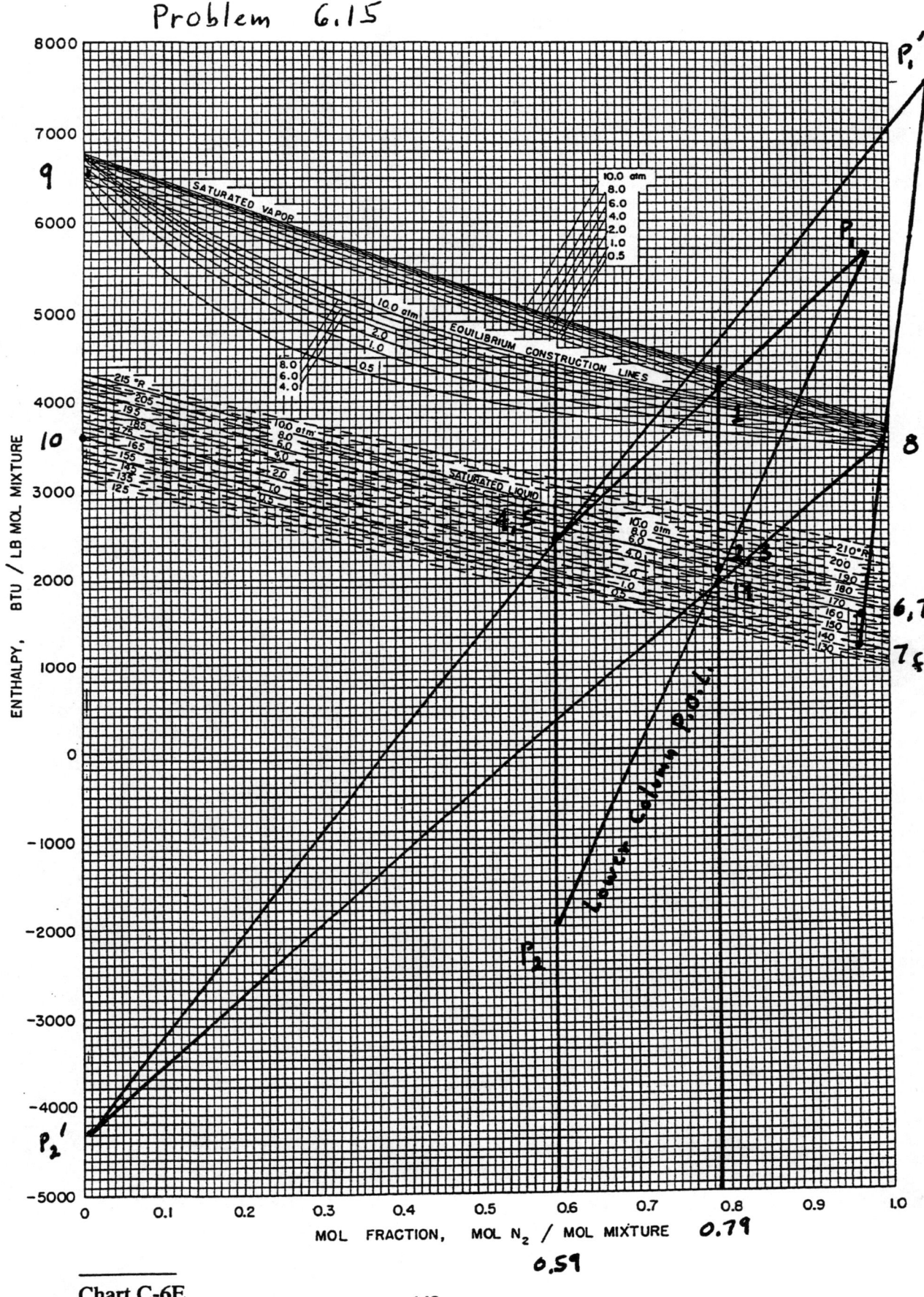

Chart C-6E

6.17

Using property values shown in Fig. 6.13

$$\dot{m}_{16} = \frac{(10 \text{ Tons/hr})(2000 \text{ lbm/ton})}{(32.0 \text{ lbm/lbmole } O_2)} = 625 \text{ lbmole } O_2/\text{hr}$$

For the entire system

$$\dot{m}_1 = \dot{m}_{16} + \dot{m}_{20}$$
$$\dot{m}_1 X_1 = \dot{m}_{16} X_{16} + \dot{m}_{20} X_{20}$$

Rearranging and eliminating $\dot{m}_{20}$

$$\dot{m}_1 = \frac{\dot{m}_{16}(X_{20} - X_{16})}{X_{20} - X_1} = \frac{625 \text{ lbmoles}}{\text{hr}} \frac{(0.990 - 0.005)}{(0.990 - 0.790)} = 3078 \frac{\text{lbmoles}}{\text{hr}}$$

or $\dot{m}_1 = 3078 \frac{\text{lbmoles}}{\text{hr}} \left(\frac{28.966 \text{ lbm}}{\text{lbmole}} \right) = 89,157 \text{ lbm/hr}$

The compressor specific work assuming no valve pressure losses (see Eq. (4.7))

$$\omega_{1-2} = \frac{n}{n-1} P_1 v_1 \left[\left(\frac{P_2}{P_1} \right)^{\frac{n-1}{n}} - 1 \right]$$

By Table A.4E, dry air at 70°F, $v_1 = 13.349 \text{ ft}^3/\text{lbm}$

$$\omega_{1-2} = \frac{1.35}{0.35} \frac{14.696 \text{ lbf}}{\text{in}^2} \left(\frac{144 \text{ in}^2}{\text{ft}^2} \right) \frac{13.349 \text{ ft}^3/\text{lbm}}{(778 \text{ ft lbf/Btu})} \left[\left(\frac{89.696}{14.696} \right)^{\frac{0.35}{1.35}} - 1 \right]$$

$$\omega_{1-2} = 83.8 \text{ Btu/lbm}$$

$$\dot{W}_{1-2} = \frac{\dot{m}_1 \omega_{1-2}}{\eta_m} = \frac{(89,157 \text{ lbm/hr}) 83.8 \text{ Btu/lbm}}{0.7 (2545 \text{ Btu/hr h.p.})} = 4194 \text{ h.p.}$$

Chapter 7

7.1
(a.) By Table A4E, with $W = 0.0150$ lbm$_w$/lbm$_a$, $t_d = 68.4$ °F
(b.) By Table A4E at 80 °F, $W_s = 0.022340$ lbm$_w$/lbm$_a$ By Eq. (7.6)

$$\phi = \frac{1 + \frac{0.622}{0.02234}}{1 + \frac{0.622}{0.0150}} = 0.679$$

(c.) By Table A4E at 80 °F, $v_a = 13.602$ ft^3/lbm$_a$, $v_{as} = 0.487$ ft^3/lbm$_a$, By Eq. (7.15)

$$v = 13.602 + \frac{0.0150}{0.02234}(0.487) = 13.929 \text{ ft}^3/\text{lbm}_a$$

(d.) By Table A4E at 80 °F, $h_a = 19.222$ Btu/lbm$_a$, $h_{as} = 24.479$ Btu/lbm$_a$ By Eq. (7.16)

$$h = 19.222 + \frac{0.0150}{0.02234}(24.479) = 35.658 \text{ Btu/lbm}_a$$

7.2
(a.) <u>Saturated air at 70 °F, 14.696 psia</u>
By Table A1E, $P_{w,s} = 0.36324$ psi, $h_g = 1091.85$ Btu/lbm$_w$, By Eq. (7.18)

$$W_s = \frac{0.622(0.36324)}{14.696 - 0.36324} = 0.01576 \text{ lbm}_w/\text{lbm}_a \quad (0.01583 \text{ by Table A4E})$$

By Eqn. (7.21)
$h_s = 0.240(70) + 0.01576(1091.85) = 34.01$ Btu/lbm$_a$, (34.10 by Table A4E)
By Eqn. (7.25)

$$v_s = \frac{(53.35)(530)}{(14.696 - 0.36324)(144)} = 13.70 \text{ ft}^3/\text{lbm}_a \quad (13.69 \text{ by Table A4E})$$

(b.) <u>Saturated air at -20 °F, 14.696 psia</u>
By Table A1E, $P_{w,s} = 6.186 \times 10^{-3}$ psi, $h_g = 1052.23$ Btu/lbm$_w$, By Eq. (7.18)

$$W_s = \frac{0.622(0.006186)}{14.696 - 0.006186} = 0.000262 \text{ lbm}_w/\text{lbm}_a \quad (0.000263 \text{ by Table A4E})$$

By Eqn. (7.21)
$h_s = 0.240(-20) + 0.000262(1052.23) = -4.524$ Btu/lbm$_a$, (-4.527 by Table A4E)
By Eqn. (7.25)

$$v_s = \frac{(53.35)(440)}{(14.696 - 0.006186)(144)} = 11.097 \text{ ft}^3/\text{lbm}_a \quad (11.078 \text{ by Table A4E})$$

7.3
By Table A4E at 70 °F, $W_s = 0.015832$ lbm$_w$/lbm$_a$, By Eq. (7.4)
$W = 0.5(0.015832) = 0.007916$ lbm$_w$/lbm$_a$
Using this W value in Table A4E, the dew-point temp. in the room is $t_d = 51$ °F. Therefore, since the glass is colder than the dew-point temperature, <u>condensation will occur</u> on the window.

7.4
By Eq. (7.18)
$$P_w = \frac{PW}{0.622+W} = \frac{(14.696)(0.007916)}{0.622+0.007916} = 0.185 \text{ psia} = 26.6 \text{ lbf}/\text{ft}^2,$$
Thus
$$m_w = \frac{P_w V}{R_w T} = \frac{26.6(30\times15\times8)}{85.78(530)} = 2.11 \text{ lbm}_w$$

7.5
(a.) Saturated air at 20 °C, 101.325 kPa
By Table A1SI, $P_{w,s} = 2.339$ kPa, $h_g = 2537.61$ kJ/kg$_w$, By Eq. (7.18)
$$W_s = \frac{0.622(2.339)}{101.325 - 2.339} = 0.01470 \text{ kg}_w/\text{kg}_a \quad (0.01476 \text{ by Table A4SI})$$
By Eqn. (7.21)
$$h_s = 1.00(20) + 0.01470(2537.61) = 57.30 \text{ kJ/kg}_a, \quad (57.56 \text{ by Table A4SI})$$
By Eqn. (7.25)
$$v_s = \frac{(287)(293)}{(101.325 - 2.339)(1000)} = 0.850 \text{ m}^3/\text{kg}_a \quad (0.830 \text{ by Table A4SI})$$

(b.) Saturated air at -25 °C, 101.325 kPa
By Table A1SI, $P_{w,s} = 63.29 \times 10^{-3}$ kPa, $h_g = 2454.69$ kJ/kg$_w$, By Eq. (7.18)
$$W_s = \frac{0.622(0.06329)}{101.325 - 0.06329} = 0.000389 \text{ kg}_w/\text{kg}_a \quad (0.000392 \text{ by Table A4SI})$$
By Eqn. (7.21)
$$h_s = 1.00(-25) + 0.000389(2454.69) = -24.05 \text{ kJ/kg}_a, \quad (-24.18 \text{ by Table A4SI})$$
By Eqn. (7.25)
$$v_s = \frac{(287)(248)}{(101.325 - 0.06329)(1000)} = 0.703 \text{ m}^3/\text{kg}_a \quad (0.703 \text{ by Table A4SI})$$

7.6
By Table A4SI, $W_s = 0.049141$ kg$_w$/kg$_a$, $h_a = 40.253$ kJ/kg$_a$, $h_{as} = 126.430$ kJ/kg$_a$
Eq. (7.6) can be rewritten as
$$W = \frac{0.622}{\frac{1}{\phi}\left[1+\frac{0.622}{W_s}\right]-1} = \frac{0.622}{\frac{1}{0.2}\left[1+\frac{0.622}{0.049141}\right]-1} = 0.00924 \text{ kg}_w/\text{kg}_a$$
By Eqn. 7.16)
$$h = 40.253 + \frac{0.00924}{0.049141}(126.430) = 64.026 \text{ kJ/kg}_a$$

7.7
By Table A4E, $W_s = 0.043219$ lbm$_w$/lbm$_a$, $h_a = 24.031$ Btu/lbm$_a$, $h_{as} = 47.730$ Btu/lbm$_a$
Eq. (7.6) can be rewritten as
$$W = \frac{0.622}{\frac{1}{\phi}\left[1+\frac{0.622}{W_s}\right]-1} = \frac{0.622}{\frac{1}{0.2}\left[1+\frac{0.622}{0.043219}\right]-1} = 0.00819 \text{ lbm}_w/\text{lbm}_a$$

By Eqn. 7.16)
$$h = 24.031 + \frac{0.00819}{0.043219}(47.730) = 33.076 \text{ Btu/lbm}_a$$

7.8
By Table A1E, $P_w = P_{w,s}$ at the dew-point temp. $= 0.30569$ psia
(a.) By Eqn. (7.18)
$$W = \frac{0.622(0.30569)}{14.00 - 0.30569} = 0.01389 \text{ lbm}_w/\text{lbm}_a$$

(b.) By Eqn. (7.27)
$$P_{w,s} = \frac{P_w}{\phi} = \frac{0.30569}{.603} = 0.50695 \text{ psia}$$

By Table A1E, $t = 80$ °F
By Eqn. (7.25)
$$v = \frac{(53.35)(540)}{(14.00 - 0.30569)(144)} = 14.61 \text{ ft}^3/\text{lbm}_a$$

7.9
By Table A1SI, $P_w = P_{w,s}$ at the dew-point temp. $= 2.339$ kPa
(a.) By Eqn. (7.18)
$$W = \frac{0.622(2.339)}{96.5 - 2.339} = 0.01545 \text{ kg}_w/\text{kg}_a$$

(b.) By Eqn. (7.27)
$$P_{w,s} = \frac{P_w}{\phi} = \frac{2.339}{.603} = 3.879 \text{ kPa}$$

By Table A1SI, $t = 28.4$ °C
By Eqn. (7.25)
$$v = \frac{(287)(301.4)}{(96.5 - 2.339)(1000)} = 0.919 \text{ m}^3/\text{kg}_a$$

7.10
To avoid condensation the dew-point temperature of the air must be less than the window temperature of 5 °C. Therefore find the relative humidity at $t_d = 5$ °C, $t = 20$ °C, $P = 101.325$ kPa. Since this is at standard atmospheric pressure we can use Table A4SI or use perfect gas approach.

Approach 1: Using values from Table A4SI, $W = 0.005436$ kg$_w$/kg$_a$, $W_s = 0.014758$ kg$_w$/kg$_a$
By Eqn. (7.6)
$$\phi = \left(\frac{0.005436}{0.014758}\right)\frac{0.622 + 0.014758}{0.622 + 0.005436} = 0.37$$

Approach 2: By Table A1SI, $P_w = 0.8721$ kPa, $P_{w,s} = 2.339$ kPa
By Eqn. (7.27)
$$\phi = \frac{0.8721}{2.339} = 0.37$$
For humidity less than 37% no condensation will occur.

7.11
(a.) Approach is to find $P_{w,s}$ and P_w then use Eqn. (7.27) to find ϕ. By Table A1E $P_{w,s} = 0.81627$ psia. To find P_w, use Eqn. (7.29) to get W and then Eqn. (7.18) to determine P_w.
By Table A1E, $h^*_{fg} = 1048.50$ Btu/lbm$_w$, $h^*_f = 47.70$ Btu/lbm$_w$, $P^*_{w,s} = 0.50734$ psia, $h_g = 1102.68$ Btu/lbm$_w$. By Eqn. (7.18)
$$W^*_s = \frac{0.622(0.50734)}{13.20 - 0.50734} = 0.02486 \text{ lbm}_w/\text{lbm}_a$$
By Eqn. (7.29)
$$W = \frac{(0.02486)(1048.5) - 0.240(95 - 80)}{1102.68 - 47.70} = 0.02129$$
By Eqn. (7.18)
$$P_w = \frac{PW}{0.622 + W} = \frac{(13.20)(0.02129)}{0.622 + 0.02129} = 0.4369 \text{ psia}$$
By Eqn. (7.27)
$$\phi = \frac{0.4369}{0.81627} = 0.53$$
(b.) By Table A1E at $P_w = 0.4369$ psia, $t_d = 75.4$ °F

7.12
(a.) Approach is to find $P_{w,s}$ and P_w then use Eqn. (7.27) to find ϕ. By Table A1SI $P_{w,s} = 5.628$ kPa. To find P_w, use Eqn. (7.29) to get W and then Eqn. (7.18) to determine P_w.
By Table A1SI, $h^*_{fg} = 2442.80$ kJ/kg$_w$, $h^*_f = 103.93$ kJ/kg$_w$, $P^*_{w,s} = 3.169$ kPa, $h_g = 2564.83$ kJ/kg$_w$.
By Eqn. (7.18)
$$W^*_s = \frac{0.622(3.169)}{91.0 - 3.169} = 0.02244 \text{ kg}_w/\text{kg}_a$$
By Eqn. (7.29)
$$W = \frac{(0.02244)(2442.80) - 1.00(35 - 25)}{2564.83 - 103.93} = 0.01821 \text{ kg}_w/\text{kg}_a$$
By Eqn. (7.18)
$$P_w = \frac{PW}{0.622 + W} = \frac{(91.0)(0.01821)}{0.622 + 0.01821} = 2.588 \text{ kPa}$$
By Eqn. (7.27)
$$\phi = \frac{2.588}{5.628} = 0.46$$

(b.) By Table A1SI at $P_w = 2.588$ kPa, $t_d = 21.6$ °C

7.13
Eqn. (7.28) can be rewritten as
$$t = t^* + \frac{(W_s^* - W)h_{fg}^*}{c_{pa} + Wc_{pw}}$$
By Table A1SI, $h_{fg}^* = 2430.81$ kJ/kg$_w$, $P_{w,s}^* = 4.246$ kPa, By Eqn. (7.18)
$$W_s^* = \frac{0.622(4.246)}{(90.0 - 4.246)} = 0.0308 \text{ kg}_w/\text{kg}_a,$$
Therefore:
$$t = t^* + \frac{(W_s^* - W)h_{fg}^*}{c_{pa} + Wc_{pw}} = 30 + \frac{(0.0308 - 0.020)(2430.81)}{1.00 + (0.020)(1.86)} = 55.3 \text{ °C}$$

7.14
By Table A4E, $W = 0.004107$ lbm$_w$/lbm$_a$, $W_s^* = 0.015832$ lbm$_w$/lbm$_a$, $h_s^* = 34.097$ Btu/lbm$_a$
By Table A1E, $h_f^* = 37.65$
By Eqn. (7.14)
$$h = 34.097 - (0.015832 - 0.004107)37.65 = 33.66 \text{ Btu/lbm}_a$$

7.15
Eqn. (7.28) can be rewritten as
$$t = t^* + \frac{(W_s^* - W)h_{fg}^*}{c_{pa} + Wc_{pw}}$$
By Table A1E, $h_{fg}^* = 1048.50$ Btu/lbm$_w$, $P_{w,s}^* = 0.50734$ Psia, By Eqn. (7.18)
$$W_s^* = \frac{0.622(0.50734)}{(13.00 - 0.50734)} = 0.02526 \text{ lbm}_w/\text{lbm}_a,$$
Therefore:
$$t = t^* + \frac{(W_s^* - W)h_{fg}^*}{c_{pa} + Wc_{pw}} = 80 + \frac{(0.02526 - 0.0125)(1048.50)}{0.240 + (0.0125)(0.444)} = 134.5 \text{ °F}$$

7.16
For an element of the atmosphere of thickness dz and at an altitude of z, the differential pressure is
$$dP = -g\rho dz,$$ Substituting for the density using Eqn. 7.1 gives $$dp = -\frac{gP}{RT}dz$$
By Eqn. (7.2) the absolute temperature T can be written as
$T = T_o + Az$, where To is in °R or K. Integrating with $P = P_o$ at $z = 0$ and $P = P$ at $z = z$
$$\int_{P_o}^{P} \frac{dP}{P} = -\frac{g}{R}\int_0^z \frac{dz}{T_o - Az} = \frac{g}{AR}\int_0^z \frac{d\left(1 - \frac{Az}{T_o}\right)}{\left(1 - \frac{Az}{T_o}\right)}$$

$$\therefore \frac{P}{P_o} = \left(1 - \frac{Az}{T_o}\right)^{g/AR}$$ Evaluating the exponent in IP units,

$$\frac{g}{AR} = \frac{32.2 \text{ ft/s}^2}{\left(0.00356 \frac{°R}{\text{ft}}\right)\left(53.35 \frac{\text{ft}-\text{lbf}}{\text{lbm}-°R}\right)\left(32.2 \frac{\text{lbm}-\text{ft}}{\text{s}^2-\text{lbf}}\right)} = 5.27$$

Problem 7.17

a) By Eqn 7.25
$$v = \frac{R_a T}{P - P_w}$$

But
$$P_w = \frac{1.608 \, PW}{1 + 1.608 \, W} \quad \text{and}$$

$$P - P_w = \frac{P + 1.608 \, PW - 1.608 \, PW}{1 + 1.608 \, W} = \frac{P}{1 + 1.608 \, W}$$

Thus
$$v = \frac{R_a T}{P}(1 + 1.608 \, W)$$

b.) By Eqn 7.27
$$\phi = P_w / P_{ws}$$

But
$$P_w = \frac{1.608 \, PW}{1 + 1.608 \, W}$$

Thus
$$\phi = 1.608 \frac{P_w}{P_{ws}}\left[\frac{W}{1 + 1.608 \, W}\right]$$

Problem 7.18

By Eqns 7.4 and 7.25

$$v_a + \mu v_{as} = \frac{R_a T}{P} + \frac{W}{W_s}\left(\frac{R_a T}{P - P_{ws}} - \frac{R_a T}{P}\right)$$

$$= \frac{R_a T}{P}\left[1 + \frac{W \cdot P_{ws}}{W_s (P - P_{ws})}\right]$$

By Eqn 7.18 for saturated conditions

$$\frac{P_{ws}}{W_s (P - P_{ws})} = 1.608$$

$$\therefore \quad v_a + \mu v_{as} = \frac{R_a T}{P}(1 + 1.608 W)$$

From Problem 7.17

$$P - P_w = \frac{P}{1 + 1.608 W}$$

Thus:

$$v_a + \mu v_{as} = \frac{R_a T}{P - P_w}$$

Problem 7.19

For a line of constant dry-bulb temperature, we have by Eqn (7.16)

$$q' = \frac{dh}{dW} = \frac{h_{as}}{W_s}$$

By Table A.4E for $t = 80°F$, $h_{as} = 24.479 \frac{Btu}{lbm_a}$
and $W_s = 0.02234$ lbmw/lbma

Thus:
$$q' = \frac{24.479}{0.02234} = 1095.7 \text{ Btu/lbm}_w$$

For a vertical line $\theta = 90°$ By Eqn. (7.52)
$$\cot \beta = q'/S = 1.0957$$

For a line of constant t^*, we have by Eqn. (7.14)
$$dh/dW = h_f^* = q'$$

By Table A.1E for $t^* = 60°F$, $h_f^* = 27.63$ Btu/lbmw

By Eqn. (7.52)
$$\cot \theta = 0.02763 - 1.0957 = -1.0681$$

$$\tan \theta = \frac{1}{-1.0681} = -0.936$$

Problem 7.20

By Eqns 7.23 and 7.24

$$h = (C_{pa} + W C_{pw})t + W h_{go} \qquad (t \text{ is °C or °F})$$

From Problem 7.17(a)

$$T = \frac{Pv}{R_a(1+1.608W)} \qquad (T \text{ is K or °R})$$

$$\therefore t = \frac{Pv}{R_a(1+1.608W)} - T_0 \qquad (T_0 = 273 \text{ K or } 460\text{°R})$$

Thus

$$h = \frac{(C_{pa} + W C_{pw})Pv}{R_a(1+1.608W)} - (C_{pa} + W C_{pw})T_0 + W h_{go}$$

for a constant volume line

$$\left(\frac{dh}{dW}\right)_v = \frac{C_{pw} Pv}{R_a(1+1.608W)} - \frac{1.608(C_{pa} + W C_{pw})Pv}{R_a(1+1.608W)^2}$$

$$- C_{pw} T_0 + h_{go}$$

Combining the first two terms on the right hand side gives for constant volume

$$\boxed{q' = \left(\frac{dh}{dW}\right)_v = \left(\frac{C_{pw} - 1.608 C_{pa}}{R_a}\right)\left[\frac{Pv}{(1+1.608W)^2}\right] - C_{pw} T_0 + h_{go}}$$

Problem 7.21
Use Chart C-8E
At the intersection of $t = 80°F$ and $W = 0.0150$ lbmw/lbma find the point

a.) Follow a line of const. W to saturation and read $t_d = 68.5°F$
b.) Interpolating between ϕ-lines read $\phi = 0.68$
c.) Interpolating between v-lines $v = 13.93$ ft³/lbma
d.) Read: $h = 35.6$ Btu/lbma

These values are in good agreement with problem 7.1

Problem 7.22
Use Chart C-8SI
At the intersection of $t = 40°C$ and $\phi = 0.20$ locate the point.
Read: $h = 64$ kJ/kga which is good agreement with prob. 7.6

Problem 7.23
Use Chart C-8SI (atmospheric pressure = 101.325 kPa)
At the intersection of $t = 35°C$ and $t^* = 25°C$ locate the point

a.) Read: $\phi = 0.45$, compared with $\phi = 0.46$ in prob. 7.12
b.) Read: $t_d = 21.2°C$, compared with $t_d = 21.6°C$ in prob 7.12

Problem 7.24
Use Chart C-8E
At the intersection of $t = 85°F$ and $\phi = 0.4$ locate the point.
Read:
a.) $t_d = 58°F$
b.) $t^* = 67.2°F$
c.) $W = 0.0103$ lbmw/lbma
d.) $h = 31.6$ Btu/lbma
e.) $v = 13.95$ ft³/lbma

Problem 7.25

Use Chart C-8SI

At the intersection of $t = 25°C$ and $\phi = 0.4$ locate the point

Read:
- a.) $t_d = 10.5°C$
- b.) $t^* = 16.3°C$
- c.) $W = 0.0079$ kgw/kga
- d.) $h = 45.3$ kJ/kga
- e.) $v = 0.855$ m³/kga

Problem 7.26

Use Chart C-8E

Draw a line connecting state 1 ($t_1 = 55°F$, $t_{d,1} = 50°F$) with state 2 ($t_2 = 100°F$, $\phi_2 = 0.3$). Draw a parallel line on the protractor and read:

- a.) $q' = \frac{\Delta h}{\Delta W} \cong 3400$ Btu/lbmw

- b.) SHR ≅ 0.68

- c.) The water vapor added will be $(W_2 - W_1)$ lbmw/lbma

  Read: $W_2 = 0.0124 \frac{lbmw}{lbma}$; $W_1 = 0.0076$ lbmw/lbma

  ∴ $\Delta W = 0.0048$ lbmw/lbma

Problem 7.27

Use Chart C-8SI

Draw a line connecting state 1 ($t_1 = 15°C$, $t_1^* = 10°C$) with state 2 ($t_2 = 35°C$, $t_{d,2} = 10°C$). Draw a parallel line on the protractor and read:

- a.) $q' = \frac{\Delta h}{\Delta W} \cong 14$ kJ/gw [In this region it is difficult to get an accurate estimate from the protractor & f.! May need to use $q' = (\frac{h_2 - h_1}{W_2 - W_1})$]

- b.) SHR = 0.8

Problem 7.28

Use Chart C-8 SI

Locate the state 1, $t_1 = 15°C$ and saturated. Establish the slope of the process line with $SHR = 0.70$. Draw a line with this slope starting at state 1 and where this line intersects $t = 40°C$ establish state 2.

Read:
a.) $W_2 = 0.0159 \; kg_w/kg_a$

b.) $t_{d,2} = 21.1 °C$

c.) $\phi_2 = 0.73$

d) $t_2^* = 27.5 °C$

Problem 8.1

```
        1        2        3
  → t₁=55°F  ▭▭▭  ▭▭▭  ⇉ t₃ = 90°F
  → φ₁=0.4                ⇉ t₃* = 65°F
   ṁₐ =      Heating   Air
   2200 lbma  Coil     Washer
         hr
```

Use Chart C-8E: Locate States 1 & 3.
  Process from 1→2 is constant W
  Process from 2→3 is constant t^*

∴ Draw constant W line through state 1 and constant t^* line through state 3. The intersection of the process lines establishes state 2.

a.) State 2 Read: $t_2 = 107.1°F$ also $W_2 = W_1 = 0.0036 \frac{lbmw}{lbma}$

b.) By Eqn (8.12)

$$\dot{Q}_2 = 2200 \frac{lbma}{hr} [0.24 + 0.444(0.0036)](107.1 - 55) \frac{Btu}{lbma}$$

$$\dot{Q}_2 = 27,700 \; Btu/hr$$

c.) The rate of water added is:
   $\dot{m}_w = \dot{m}_a (W_3 - W_2)$
   from the chart $W_3 = 0.0075 \; lbmw/lbma$
   ∴ $\dot{m}_w = 2200(0.0075 - 0.0036) = 8.6 \; lbmw/hr$

Problem 8.2

By Eqn. (8.30) $SHR = \frac{80,000}{80,000 + 34,000} = 0.70$

Using the chart C-8E, on the protractor locate the process line with SHR = 0.70 and draw a parallel line through state 2 ($t_2 = 70°F$, $t_{d,2} = 44°F$)

Problem 8.2 continued

Extend the line so it intersects the supply air temperature, $t_1 = 95°F$

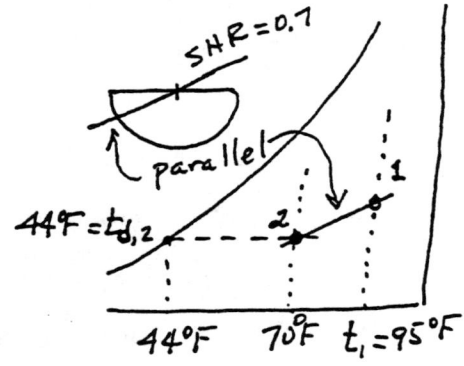

a.) Read: $\phi_1 = 0.24$

b.) By Eqn. (8.25)
$$\dot{m}_a = \frac{\dot{Q}_s + \dot{Q}_L}{h_2 - h_1}$$

Read $h_2 = 23.4$ Btu/lbma ; $h_1 = 32$ Btu/lbma

$$\dot{m}_a = \frac{-(89,000 + 34,000)}{23.4 - 32} = 13,300 \text{ lbma/hr}$$

(Note the $\dot{Q}_s$ and $\dot{Q}_L$ values are negative when using Eqn (8.25) since they are heat <u>losses</u>)

Could also find $\dot{m}_a$ using approximation, Eqn (8.28).

Problem 8.3

See psychrometric chart.

In part b.) the process line has slope given by Eqn 8.23. From Table A.1E
h_g @ 200°F is 1145.75 Btu/lbm_w

∴ $q' \cong 1146$ Btu/lbm_w for the process of part b.)

Problem 8.3

Problem 8.4

By Eqns. (8.14) and (8.15) and neglecting the condensate term

$$\dot{Q}_R = \dot{m}_a(h_1 - h_2)$$

By Eqn (7.21)

$$h_1 = C_{pa}t_1 + W_1 h_{g1} \quad \text{and} \quad h_2 = C_{pa}t_2 + W_2 h_{g2}$$

By Eqn (7.27)

$$P_{w,1} = \phi P_{ws,1}$$

From Table A.1E at $88°F$; $P_{ws,1} = 0.65623$ psia, $h_{g,1} = 1099.66$ Btu/lb

$$\therefore P_{w,1} = 0.45(0.65623) = 0.2953$$

By Eqn (7.18)

$$W_1 = 0.622\left(\frac{0.2953}{12.75 - 0.2953}\right) = 0.01475 \text{ lbmw/lbma}$$

$$h_1 = 0.24(88) + 0.01475(1099.66) = 37.34 \text{ Btu/lbma}$$

From Table A.1E. $P_{w,2} = P_{ws}(t_d) = 0.16521$ psia

and $h_g @ 52° = h_{g,2} = 1083.98$ Btu/lbmw

$$W_2 = 0.622\left(\frac{0.16521}{12.75 - 0.16521}\right) = 0.0082 \text{ lbmw/lbma}$$

$$h_2 = 0.24(52) + 0.0082(1083.98) = 21.37 \text{ Btu/lbma}$$

$\dot{m}_a = \dot{V}/v_1$; From Eqn (7.25)

$$v_1 = \frac{(53.35 \frac{ft \cdot lbf}{lbm \cdot °R})(460 + 88)°R}{(12.75 - 0.2953 \frac{lbf}{in^2})(144 \frac{in^2}{ft^2})} = 16.30 \text{ ft}^3/\text{lbm}$$

$$\dot{m}_a = \left(\frac{4500 \text{ ft}^3/\text{min}}{16.30 \text{ ft}^3/\text{lbma}}\right)\left(60 \frac{min}{hr}\right) = 15,560 \text{ lbma/hr}$$

$$\dot{Q}_R = (15,560)(37.34 - 21.37) = \underline{248,500 \text{ Btu/hr}}$$

Problem 8.5

a.) From the chart: $SHZ = 0.635$

b.) By Eqns (8.14) and (8.15) and neglecting condensate
$$\dot{Q}_R = \dot{m}_a(h_1 - h_2)$$
From the chart: $h_1 = 58.6 \text{ kJ/kg}_a$
$h_2 = 34.2 \text{ kJ/kg}_a$
$\dot{Q}_R = 1.50(58.6 - 34.2) = 36.6 \text{ kW}$

c.) From the chart, $t_d = 10.2°C$

d.) By Eqn (8.18)
$$b = \frac{13 - 10.2}{28 - 10.2} = 0.16$$

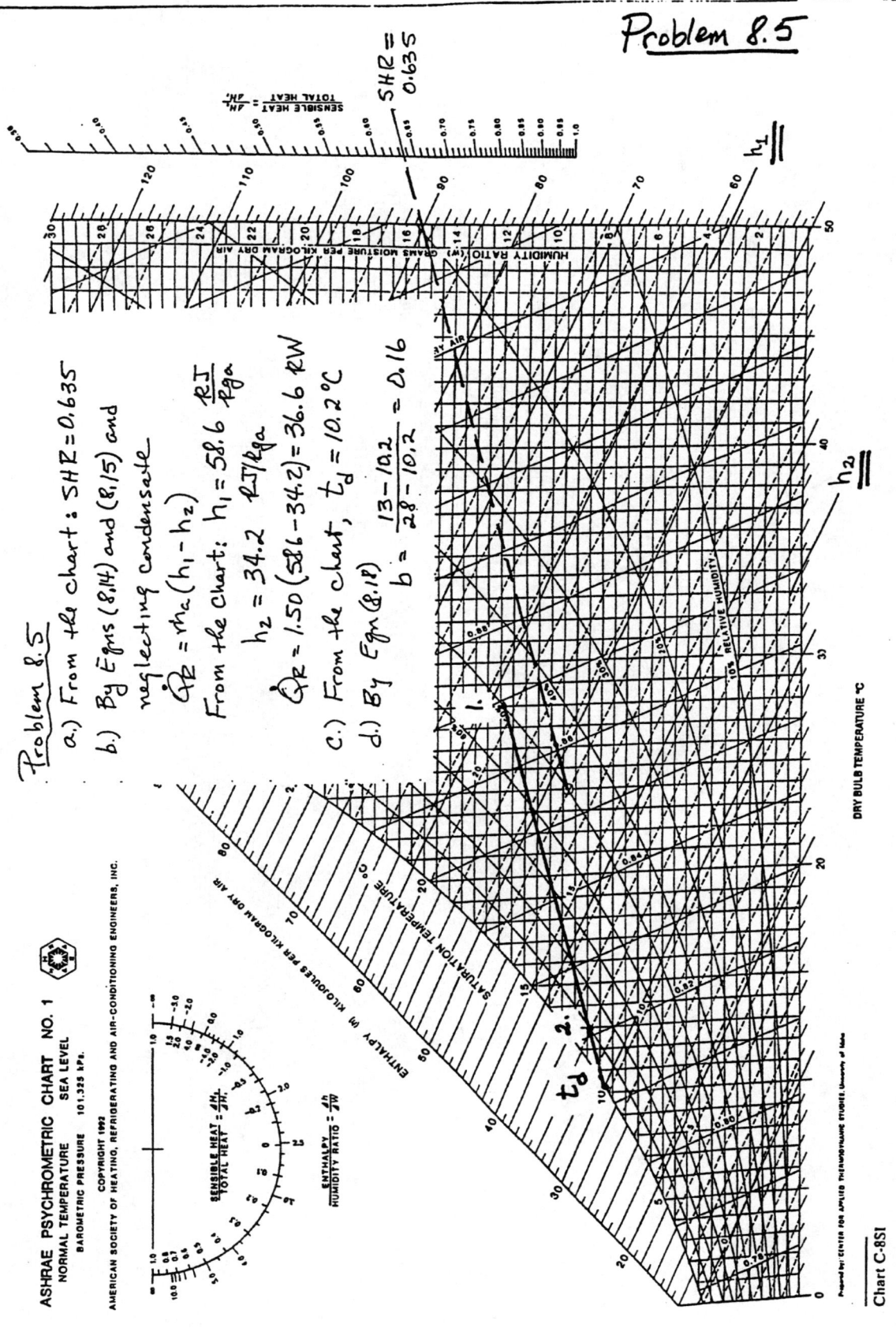

Prob 8.6

Problem 8.6

See idealized process line on the chart.

Read: $h_1 = 34.0$ Btu/lbma
$h_2 = 22.6$ Btu/lbma
$v_1 = 13.97$ ft³/lbma

$\dot{m}_a = \dot{V}_1/v_1 = \dfrac{3500\ \text{ft}^3/\text{min}}{13.97\ \text{ft}^3/\text{lbma}}$

$= 250.5$ lbma/min
$= 15,030$ lbma/hr

By Eqns. (8.14) & (8.15) and neglecting the condensate term.

$\dot{Q}_R = 15,030(34.0 - 22.6)$
$\dot{Q}_R = 171,000$ Btu/hr

or $\dot{Q}_R = \dfrac{171,000\ \text{Btu/hr}}{12,000\ \text{Btu/hr-ton}}$

$\dot{Q}_R = 14.3$ tons

Idealized Process for "Perfect Contact"

ASHRAE PSYCHROMETRIC CHART NO. 1
NORMAL TEMPERATURE
BAROMETRIC PRESSURE: 29.921 INCHES OF MERCURY
COPYRIGHT 1992
AMERICAN SOCIETY OF HEATING, REFRIGERATING AND AIR-CONDITIONING ENGINEERS, INC.

SEA LEVEL

Problem 8.7 -- Use Chart C-8E

State 1 may be located by the intersection of $t_1 = 89°F$ and $t_1^* = 65°F$. We obtain $h_1 = 29.80$ Btu/lbma, $v_1 = 14.0$ ft³/lbma and $t_{d,1} = 50.2°F$. Since the coil surface temperature is $55°F$, no dehumidification occurs and $W_1 = W_2$.

∴ By Eqns. 8.14 and 8.15

$$\dot{Q}_R = \dot{m}_a (h_1 - h_2)$$

Thus $h_2 = 29.80 - \dfrac{(3.5 \text{ tons})(12{,}000 \text{ Btu/hr-ton})}{(1400 \text{ ft}^3/\text{min})(60 \frac{\text{min}}{\text{hr}})(\frac{1}{14.0} \frac{\text{lbma}}{\text{ft}^3})}$

$h_2 = 22.80$ Btu/lbma

At the intersection of h_2 and the line $W_1 = W_2$ we read $t_2 = 60.0°F$ and $t_2^* = 54.5°F$.

Problem 8.8

By Eqn (8.23) $q' = h_w$

For saturated steam at 25 psia from Table A.1E $h_w = h_g = 1160.51$ Btu/lbm$_w$

a) End state (state 2) is saturated. See construction on chart.

$\dot{m}_w = \dot{m}_a(W_2 - W_1) = 2000$
$= 2000 \frac{lbm_a}{min}(0.010 - 0.0043) \frac{lbm_w}{lbm_a}$

$\dot{m}_w = 11.4 \, lbm_w/min$

b.) Read $t_2 = 57°F$

$\frac{q'}{\Delta W} \approx 1161$

parallel

Problem 8.9

Use Chart C-8E

Let: State 1 = recirculated air
State 2 = Outdoor air

$$\therefore \dot{V}_1 = 3\dot{V}_2$$

$$\dot{m}_{a,1} = \dot{V}_1/v_1 \quad ; \quad \dot{m}_{a,2} = \frac{\dot{V}_2}{v_2} = \frac{\dot{V}_1}{3v_2}$$

Let state 3 = mixed air condition

$$\therefore \dot{m}_{a,3} = \dot{m}_{a,1} + \dot{m}_{a,2} = \dot{V}_1\left(\frac{1}{v_1} + \frac{1}{3v_2}\right)$$

By Eqn (8.9)

$$t_3 = \frac{\frac{1}{v_1}}{\left(\frac{1}{v_1} + \frac{1}{3v_2}\right)} \cdot t_1 + \frac{\frac{1}{3v_2}}{\left(\frac{1}{v_1} + \frac{1}{3v_2}\right)} t_2$$

From the chart: $v_1 = 13.5 \frac{ft^3}{lb_{ma}}$; $v_2 = 14.51 \frac{ft^3}{lb_{ma}}$

a.) $\therefore t_3 = 0.763(70) + 0.237(97) = 76.4°F$

b.) Locate state 3 at $t_3 = 76.4°F$ on the line connecting states 1 & 2

Read $t_3^* = 64.5°F$

Problem 8.9

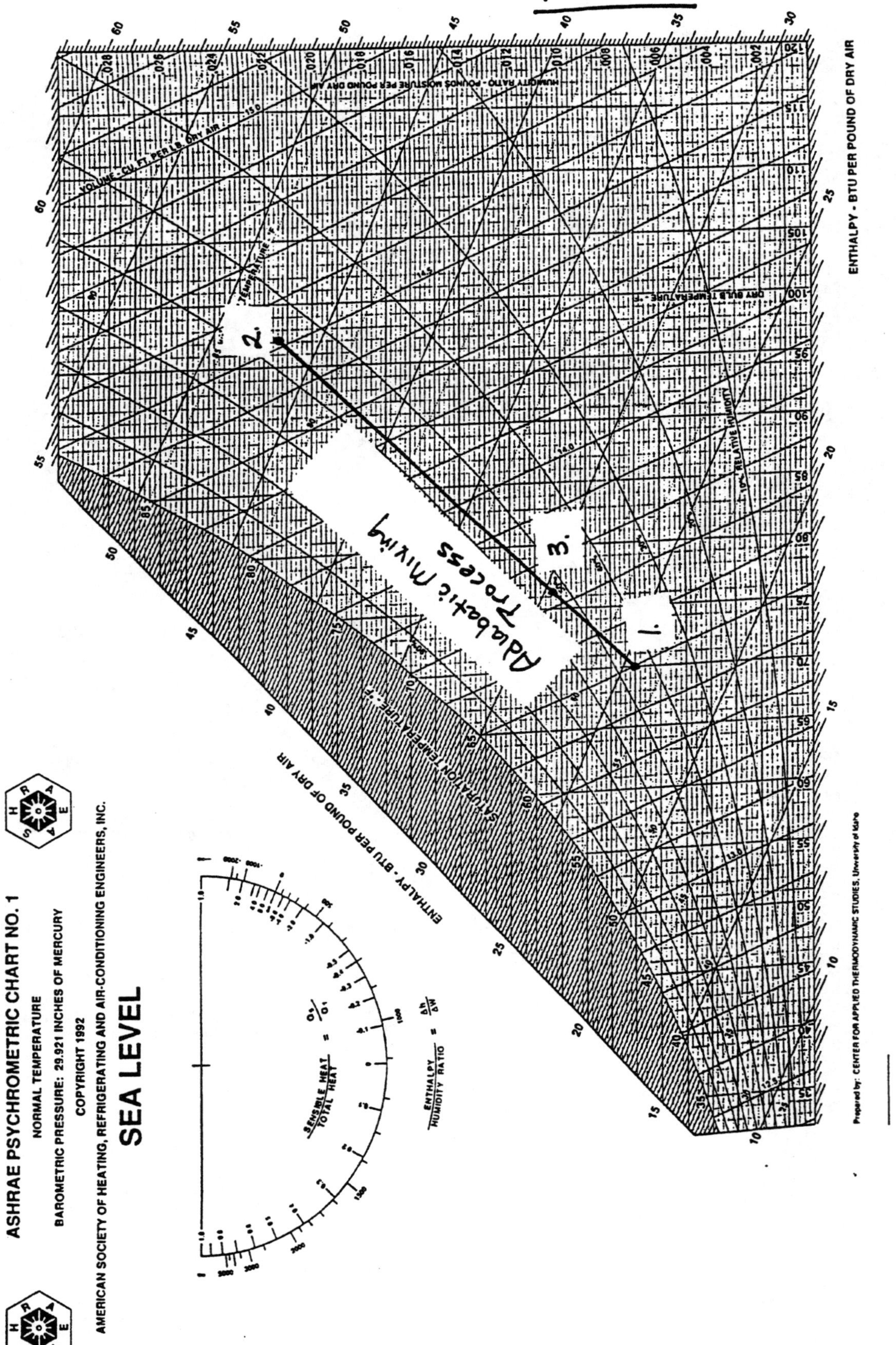
ASHRAE PSYCHROMETRIC CHART NO. 1
NORMAL TEMPERATURE
BAROMETRIC PRESSURE: 29.921 INCHES OF MERCURY
COPYRIGHT 1992
AMERICAN SOCIETY OF HEATING, REFRIGERATING AND AIR-CONDITIONING ENGINEERS, INC.
SEA LEVEL
SENSIBLE HEAT / TOTAL HEAT
ENTHALPY / HUMIDITY RATIO
ENTHALPY - BTU PER POUND OF DRY AIR
SATURATION TEMPERATURE °F
VOLUME - CU. FT. PER LB. DRY AIR
HUMIDITY RATIO - POUNDS MOISTURE PER POUND DRY AIR
DRY BULB TEMPERATURE °F
RELATIVE HUMIDITY
Adiabatic Mixing Process
1.
2.
3.
Prepared by: CENTER FOR APPLIED THERMODYNAMIC STUDIES, University of Idaho

Problem 8.10

Use Chart C-8E.

Determine the necessary apparatus dew-point temperature by drawing the line connecting $t_1 = 80°F$, $t_1^* = 67°F$ and $t_2 = 60°F$, $t_2^* = 54°F$

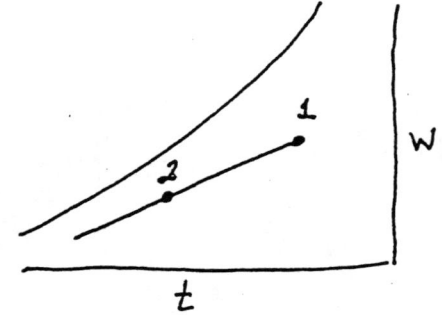

a) Notice that the line does not intersect the saturation curve. Therefore, no apparatus dew-point exists that satisfies the desired process and it can not be achieved.

b) The desired result can be achieved by two processes, a cooling process from state "1" to state "a" and a sensible heating process (reheat) from state "a" to state "2"

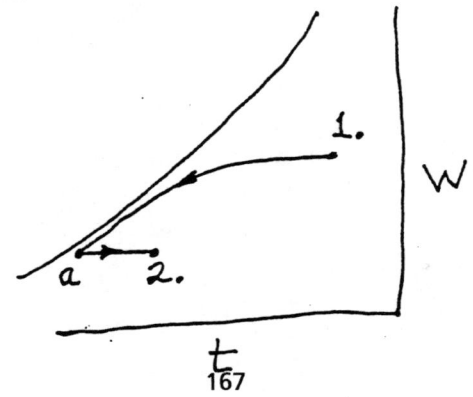

Problem 8.11

ASHRAE PSYCHROMETRIC CHART — Problem 8.11 -- Use Chart C-8E

- Note Process is constant W, $\mu = 100\%$ ∴ State 1 = Saturated.
- Locate State 1 & draw Const W process & where it crosses $\phi_2 = 20\%$ locate state 2; Read $t_2 = 86.5°F$
- Locate State 3 at $t_3 = 75°F$ on the mixing line.

a.) Let $\dot{m}_{a,b}$ = bypass mass flow rate. Combine Eqn. (8.2) with either Eqn (8.4) or (8.9) to find $\dot{m}_{a,b}$

Using Eqns. (8.2) & (8.9)

$$\frac{\dot{m}_{a,b}}{\dot{m}_{a,3}} = \frac{t_2 - t_3}{t_2 - t_1} = \frac{86.5 - 75}{86.5 - 41} = 0.25$$

Since $\dot{m}_{a,3} = \dot{m}_{a,1}$; $\dot{m}_{a,b} = 0.25(120) = 30$ lbma/min

b.) By Eqn (8.11): $\dot{Q}_{coil} = \dot{m}_{a,2}(h_2 - h_1)$

$\dot{m}_{a,2} = 120 - 30$ lbma/min ; Read $h_2 = 26.6$ Btu/lbma
and $h_1 = 15.6$ Btu/lbma

$\dot{Q}_{coil} = (90 \frac{lbma}{min})(26.6 - 15.6)\frac{Btu}{lbma}(60 \frac{min}{hr}) = \underline{59,400\ Btu/hr}$

Problem 8.12

Using Chart C-8E Locate States 1, 2 and 4
(Note State 1 is in the fog region and is found by extending the temperature along $t^* = 50°F$ into the fog region.)

State 3 is located at the intersection of the mixing line connecting states 1 & 2 and the constant W line drawn through state 4.
We find state 3 is also in the fog region.
Read $t_3 = 55.5 °F$

By Eqns 8.2 and 8.9 we get

$$\frac{\dot{m}_{a,1}}{\dot{m}_{a,3}} = \frac{t_2 - t_3}{t_2 - t_1}$$

$$\therefore \dot{m}_{a,3} = \dot{m}_{a,1}\left[\frac{80-50}{80-55.5}\right] = 1.22\, \dot{m}_{a,1}$$

From the chart read $v_1 = 13\ ft^3/lbma$

$$\dot{m}_{a,1} = \frac{(1300\ ft^3/min)(60\ min/hr)}{13.\ ft^3/lbma} = 6,000\ \frac{lbma}{hr}$$

$$\dot{m}_{a,3} = 1.22(6,000) = 7,320\ lbma/hr$$

From the chart read, $h_3 \cong 23.3\ Btu/lbma$
$h_4 \cong 33.9\ Btu/lbma$

By Eqn 8.11

$$_3\dot{Q}_4 = \dot{m}_{a,3}(h_4 - h_3) = 7,320(33.9 - 23.3)$$

$$_3\dot{Q}_4 = 77,600\ Btu/hr.$$

Prob 8.12

ASHRAE PSYCHROMETRIC CHART NO. 1
NORMAL TEMPERATURE
BAROMETRIC PRESSURE: 29.921 INCHES OF MERCURY
COPYRIGHT 1992
SOCIETY OF HEATING, REFRIGERATING AND AIR-CONDITIONING ENGINEERS, INC.
SEA LEVEL

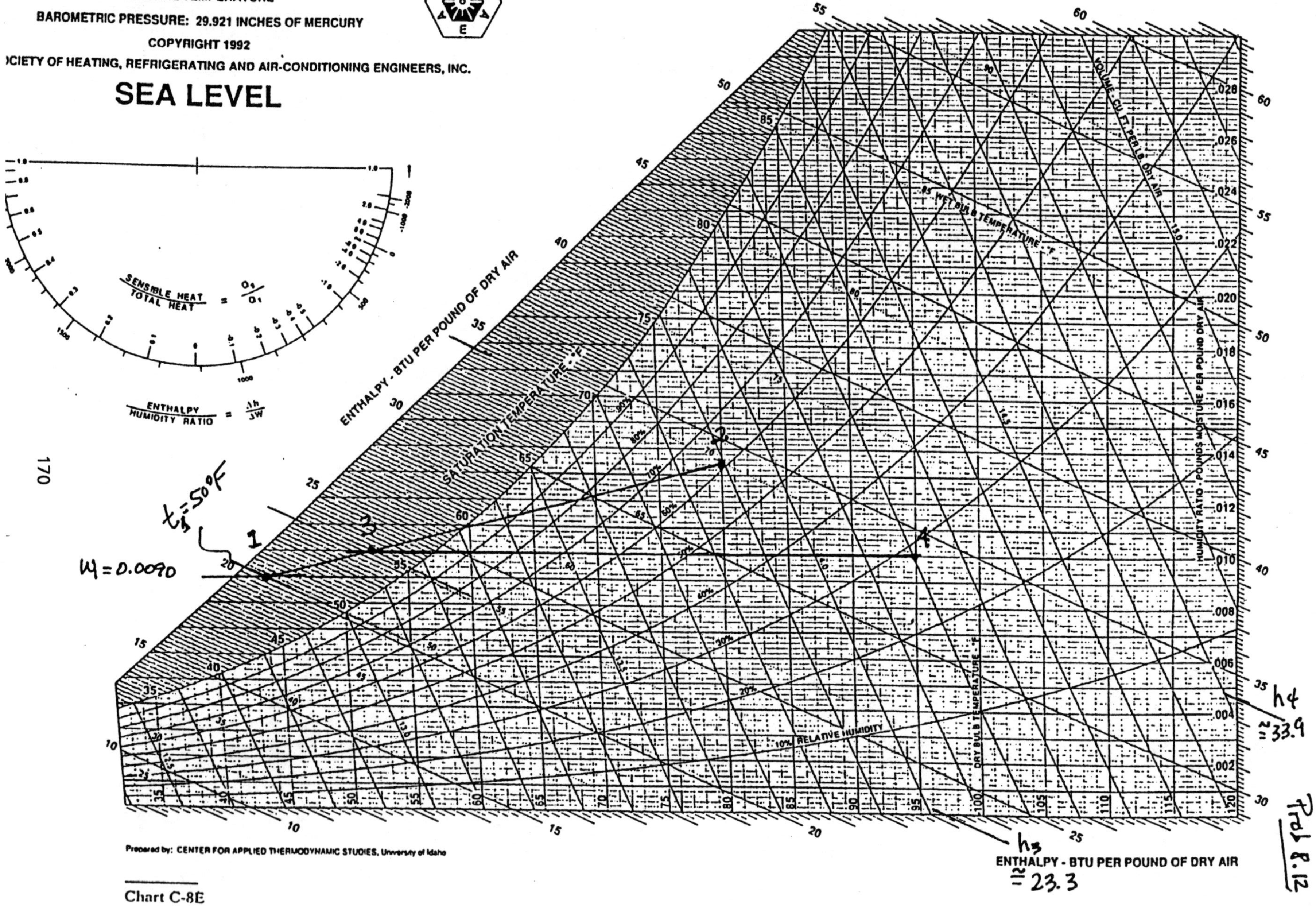

$h_4 \approx 33.9$
$h_3 \approx 23.3$
$W_1 = 0.0090$
$h_1 \approx 20.5$

Chart C-8E

Problem 8.13

Heat Recovery Unit

$$\dot{m}_{a,4}(h_4 - h_5) \times \underbrace{0.8}_{\text{80\% of heat transferred as per problem statement.}} = \dot{m}_{a,0}(h_6 - h_0)$$

$\dot{m}_{a,4} = \dot{m}_{a,0}$

$\therefore h_6 = 0.8(h_4 - h_5) + h_0$

$h_4 \cong 22.8 \frac{Btu}{lbma}$, $h_5 \cong 17.8 \frac{Btu}{lbma}$, $h_0 \cong 11.4 \frac{Btu}{lbma}$

$\therefore \boxed{h_6 = 15.4 \ Btu/lbma}$ & $W_6 = W_0$ to find state 6

State 7

$$h_7 = \frac{\dot{m}_{a,6}}{\dot{m}_{a,7}} h_6 + \frac{\dot{m}_{a,3}}{\dot{m}_{a,7}} h_3$$

$h_7 = 0.3(15.4) + 0.7(22.8) = 20.6 \ Btu/lbma$

$\therefore$ @ h_7 along the mixing line we find state 7

Humidifier uses sat vapor @ 240°F $\therefore h_w = 1160.5 \frac{Btu}{lbmw}$

$\therefore$ State 8 = intersection of line drawn through state 1 with $q' = 1160.5$ and const W line from state 7.

$_7\dot{Q}_8 = \dot{m}_{a,7}(h_8 - h_7)$ $h_8 \cong 27.6 \frac{Btu}{lbma}$

$\dot{m}_{a,7} = \dot{m}_{a,1} = \frac{|\Sigma \dot{Q}_s + \Sigma \dot{Q}_L|}{h_1 - h_2}$; $h_1 \cong 30.8 \frac{Btu}{lbma}$

$\therefore \dot{m}_{a,1} = \frac{100,000 \ Btu/hr}{30.8 - 22.8 \ Btu/lbma} = 12,500 \ lbma/hr$

$_7\dot{Q}_8 = (12,500 \frac{lbma}{hr})(27.6 - 20.6) \frac{Btu}{lbma} = 87,500 \ Btu/hr$

$\dot{m}_w = \dot{m}_{a,1}(W_1 - W_8) = 12,500 \frac{lbma}{hr}(0.0073 - 0.0046) \frac{lbmw}{lbma}$

$\dot{m}_w = 33.8 \ lbmw/hr$

Prob 8.13

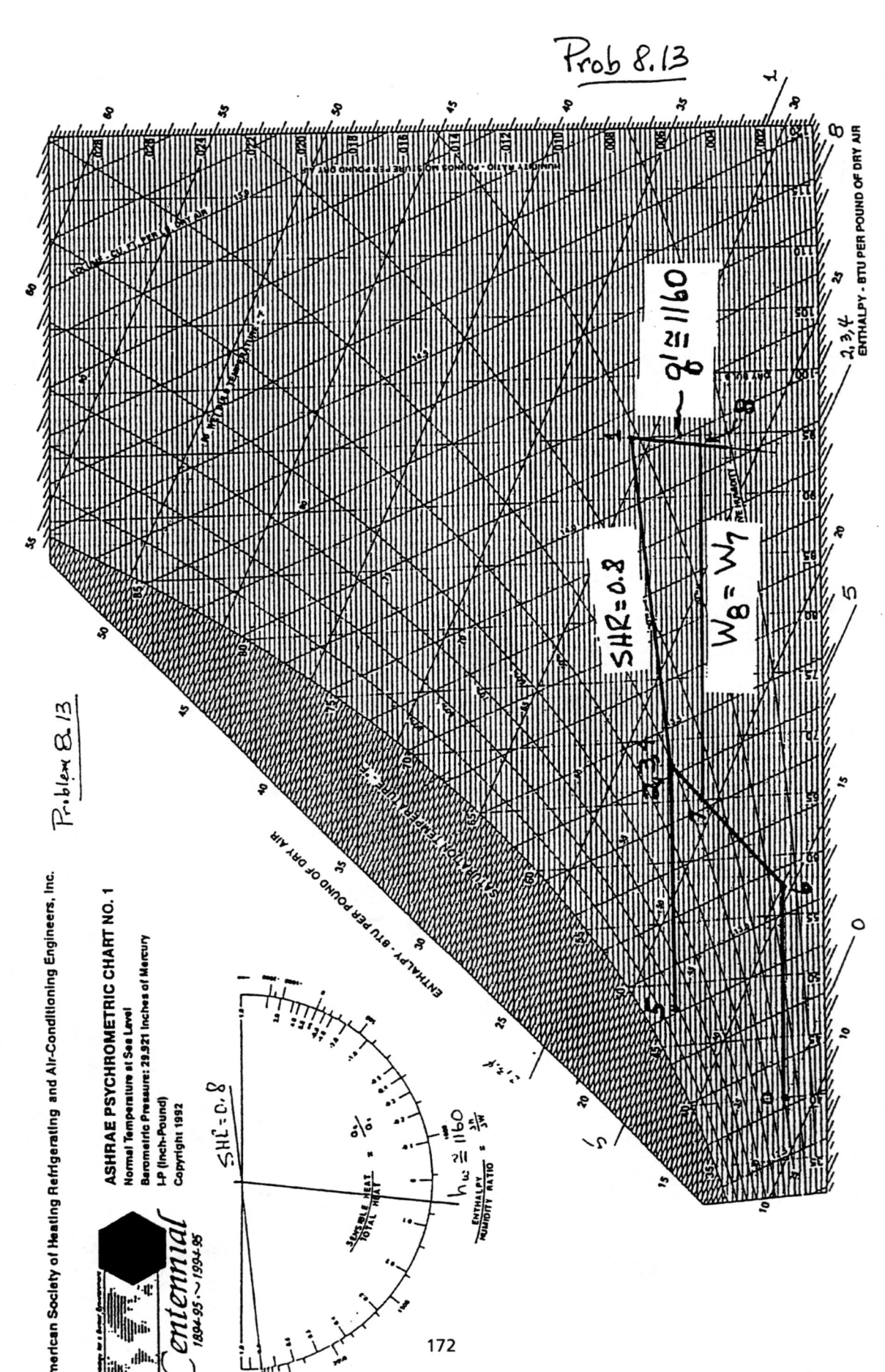

Problem 8.14

a.) By Eqn 8.28

$$\dot{Q}_S = \dot{m}_a \bar{C}_p (t_2 - t_1) \quad \text{and} \quad \bar{C}_p \approx 0.245 \frac{Btu}{lbma \cdot °F}$$

$$\therefore \dot{m}_a = \frac{-16,000 \text{ Btu/hr}}{\left(0.245 \frac{Btu}{lbma \cdot °F}\right)(70-120 °F)} = 1306 \frac{lbma}{hr}$$

Note: State 1 is supply air & State 2 is room air.

b.) By Eqn 8.29

$$\dot{Q}_L = \dot{m}_a \bar{h}_g (W_2 - W_1) \quad \text{& use } \bar{h}_g = 1100 \text{ Btu/lbmw}$$

By Eqn 8.30

$$\dot{Q}_L = \dot{Q}_S \left[\frac{1}{SHR} - 1\right] = -16,000 \left[\frac{1}{0.8} - 1\right] = -4000 \frac{Btu}{hr}$$

The furnace process is sensible heating.
Therefore, W_1 = W-value entering the furnace
i.e. W @ t = 50°F, ϕ = 0.4

By Eqn 7.18 and 7.27

$$W = 0.622 \frac{\phi P_{ws}}{P - \phi P_{ws}}$$

From Table A.1E @ 50°F, P_{ws} = 0.17805 psia

$$\therefore W_1 = 0.622 \frac{0.4(0.17805)}{13.5 - (0.4)(0.17805)} = 0.0034 \frac{lbmw}{lbma}$$

By Eqn 8.29

$$W_2 = W_1 + \frac{\dot{Q}_L}{\dot{m}_a \bar{h}_g} = 0.0034 + \frac{-4000}{(1306)(1100)}$$

$$\underline{W_2 = 0.0006 \text{ lbmw/lbma}}$$

Prob 8.15

To find state 1 draw the SHR = 0.6 line through state 2, 3 and calculate t_1 or h_1

$$\sum \dot{Q}_s' = (SHR)\left[\sum \dot{Q}_s' + \sum \dot{Q}_L'\right] = 0.6(98,000) = 58,800 \frac{Btu}{hr}$$

$$\sum \dot{Q}_s' = \dot{m}_{a,1} \bar{C}_p (t_1 - t_2)$$

$$t_1 = \frac{58,800 \text{ Btu/hr}}{(15,000 \frac{lbma}{hr})(0.245 \frac{Btu}{lbma \cdot °F})} + 70 = \underline{\underline{86 °F}}$$

Locate state 5 along the mixing line between 2 & 4

$$t_5 \cong \frac{\dot{m}_{a,4}}{\dot{m}_{a,5}} t_4 + \frac{\dot{m}_{a,2}}{\dot{m}_{a,5}} t_2 = 0.4(45) + 0.6(70) = 60°F$$

Humidifier has $h_w = 1164 \frac{Btu}{lbm_w}$ (Table A.1E) Sat steam @ 250°F

Locate 6 at intersection of const. W from "5" &
$g' = h_w$ through "1"

From the chart, $t_6 \cong 84 °F$, $W_5 = W_6 = 0.0050 \frac{lbm_w}{lbma}$

$W_1 \cong 0.0102 \text{ lbm}_w/\text{lbma}$

$$\therefore {}_5\dot{Q}_6 = \dot{m}_{a,1} \bar{C}_p (t_6 - t_5) = (15,000 \tfrac{lbma}{hr})(0.245 \tfrac{Btu}{lbm_a \cdot °F}) \times (84-60)°F$$

$$\boxed{{}_5\dot{Q}_6 = 88,200 \text{ Btu/hr}}$$

$$\dot{m}_w = \dot{m}_{a,1}(W_1 - W_6) = 15,000(0.0102 - 0.0050)$$

$$\boxed{\dot{m}_w = 78 \text{ lbm}_w/hr}$$

Problem 8.15

$q'_l = 1164$

$SHR = 0.6$

$W_s = W_6$

Problem 8.16

Let state 1 be the entering condition and state 2 be the exiting condition. For an air washer $t^* =$ constant $\therefore t_2^* = t_1^* = 65°F$

a) By Eqn 7.29
$$W_2 = \frac{W_{s,2}^* h_{fg}^* - C_{pa}(t_2 - t_2^*)}{h_g - h_f^*}$$

By Eqn 7.18
$$W_{s,2}^* = 0.622 \frac{P_{ws}^*}{P - P_{ws}^*} = 0.622 \frac{0.30569}{14.3 - 0.30569}$$

$$W_{s,2}^* = 0.01359 \text{ lbmw/lbma}$$

$$W_2 = \frac{0.01359(1057.04) - 0.24(70-65)}{1091.85 - 32.63} = 0.01243 \text{ lb/lb}$$

Note: Properties taken from Table A.1E

By Eqn 7.18
$$W_2 = 0.622 \frac{P_w}{14.3 - P_w} = 0.01243$$
$$P_w = 0.28017 \text{ psia}$$

By Eqn 7.27
$$\phi_2 = P_{w2}/P_{ws,2} = \frac{0.28017}{0.36324} = 0.77 = 77\%$$

b)
$$\dot{m}_w = \dot{m}_a (W_2 - W_1)$$
$$W_{s,1}^* = 0.622 \, P_{ws}^*/(P - P_{ws}^*) = 0.01359$$

$$W_1 = \frac{0.01359(1057.04) - 0.24(100-65)}{1104.83 - 32.63} = 0.00556$$

$$\dot{m}_w = 3000 \frac{\text{lbma}}{\text{hr}} (0.01243 - 0.00556) \frac{\text{lbmw}}{\text{lbma}}$$

$$\underline{\dot{m}_w = 20.6 \text{ lbmw/hr}}$$

Problem 8.17

Problem 8.17 Use the notation in Fig 8.18
$t_4 = 65°F$, $t_4^* = 65°F$ (Saturated)
$t_3 = 95°F$, $t_3^* = 65°F$ (t^* = const. for air washer)

a.) Read: $\phi_3 \cong 18\%$

b.) $SHR = 55{,}200/(55{,}200+16{,}800) = 0.77$

By Eqn 8.28
$t_2 = t_1 + \dfrac{\dot{Q}_s}{\dot{m}_a c_p} = 65 + \dfrac{55{,}200}{11{,}400(0.245)}$
$t_2 = 84.8°F$ Read $t_{d,2} = 68°F$

Problem 8.18   Use Chart C-8E

Let the supply air be state 1 and room air be state 2.

$$\dot{Q}_s = 82,000 \text{ Btu/hr}$$

$$\dot{Q}_L = \dot{m}_w h_w = 12 \frac{lbm_w}{hr}\left(1100 \frac{Btu}{lbm_w}\right) = 13,200 \frac{Btu}{hr}$$

$$\therefore SHR = \frac{82,000}{82,000 + 13,200} = 0.86$$

Locate State 2 at $t_2 = 75°F$, $\phi_2 = 0.5$

With the aid of the protractor draw the space condition line through state 2 with $SHR = 0.86$. State 1 is on the space condition line at $t_1 = 58°F$

a) Read: $t_{d,1} \cong 53.2°F$ and $t_1^* \cong 55.2°F$

b) We can find $\dot{m}_a$ by Eqns 8.25 or 8.28

Read: $h_2 = 28.1$ Btu/lbma and $h_1 = 23.3$ Btu/lbma

$$\therefore \text{By Eqn 8.25}$$

$$\dot{m}_a = \frac{\dot{Q}_s + \dot{Q}_L}{h_2 - h_1} = \frac{95,200}{(28.1 - 23.3)} = 19,800 \frac{lbma}{hr}$$

If we had used Eqn 8.28

$$\dot{m}_a = \frac{\dot{Q}_s}{c_p(t_2 - t_1)} = \frac{82,000}{0.245(75 - 58)} = 19,700 \frac{lbma}{hr}$$

By Eqn 8.10 and using the result for $\dot{m}_a$ from Eqn 8.25

$$\dot{V}_{std} = \frac{\dot{m}_a}{\rho_{std}} = \frac{(19,800 \text{ lbma/hr})\left(\frac{1}{60} \frac{hr}{min}\right)}{0.075 \text{ lbma/ft}^3}$$

$$\underline{\dot{V}_{std} = 4,400 \text{ ft}^3/min}$$

Problem 8.19

Use Chart C-8E

For the space condition line $SHR = \dfrac{180,000}{180,000+20,200} = 0.90$

- Locate state 2 on the chart, draw the space condition process line through state 2 and end at $t_1 = 100°F$
- Draw the mixing line from outdoor air (state 4) to state 2.

By Eqn: (8.9) & from $\dot{m}_{a,4} = 0.5\,\dot{m}_{a,1}$ & $\dot{m}_{a,5} = \dot{m}_{a,1}$

$$t_5 = 0.5(35) + 0.5(70) = 52.5°F$$

- Draw the const. W process lines through states 5 and 1.
- The process from 6→7 is constant t^* but we don't the value of t^*. By the air washer efficiency given on page 214

$$\eta_w = \dfrac{W_7 - W_6}{W_5^* - W_6} = 0.5$$

From the chart. $W_7 = 0.0064$ lbmw/lbma
$W_6 = 0.0050$ lbmw/lbma

$\therefore W_5^* = \dfrac{0.0064 - 0.0050}{0.5} + 0.0050 = 0.0078$ lbmw/lbma

Locating $W_5^* = 0.0078$ on the chart, Read $t^* \cong 50.2°F$ for the process from 6→7.

a) By Eqn (8.28): $\dot{m}_{a,1} = \dfrac{180,000\ \text{Btu/hr}}{(0.245\ \frac{Btu}{lbma\cdot°F})(100-70\,°F)} = 24,500$ lbma/hr.

By Eqn (8.10) $\dot{V}_{std} = \left(\dfrac{24,500}{0.075}\right)\dfrac{ft^3}{hr}\left(\dfrac{1\ hr}{60\ min}\right) = 5440\ \dfrac{ft^3}{min}$

b.) The spray water temp is $t_6^* = t_7^* = 50.2°F$

c) $\dot{m}_w = \dot{m}_{a,6}(W_7 - W_6) = 24,500(0.0064 - 0.0050)$
   $= 34.3$ lbmw/hr

d.) By Eqn. (7.45) & Reading $t_6 = 62.5°F$
$\therefore \dot{Q}_6 = (24,500)(0.245)(62.5 - 52.5) = 60,000$ Btu/hr.

Read: $t_7 = 56.5°F$
$\dot{Q}_8 = (24,500)(0.245)(100 - 56.5) = 261,000$ Btu/hr

Note: par d.) could also be solved by looking h-values on the chart & using Eqn (8.11).

Problem 8.19

Problem 8.20

$$SHR = \frac{260}{260+29} = 0.90$$

Locate 5 on the mixing line

$$t_5 \cong \frac{\dot{m}_{a,4}}{\dot{m}_{a,5}} t_4 + \frac{\dot{m}_{a,2}}{\dot{m}_{a,5}} t_2 = 0.45(3) + 0.55(20) = 12.4°C$$

Locate 1 on the SHR = 0.9 line through state 2 & $t_1 = 35°C$ given

Locate 7 at $W_7 = W_1$ & $\phi_7 = 80\%$ (given)

Locate 6 at intersection of const t^* from 7 and const W from 5

$$\dot{m}_{a,1} = \frac{|\Sigma \dot{Q}_s + \Sigma \dot{Q}_L|}{h_1 - h_2} = \frac{(260+29)\,kW}{(52-34.9)\,kJ/kg_a} = 16.9 \; \frac{kg_a}{s}$$

or $\dot{m}_{a,1} = \frac{|\Sigma \dot{Q}_s|}{\bar{c}_p(t_1-t_2)} = \frac{260\,kW}{(1.02 \frac{kJ}{kg_a \cdot °C})(35-20)°C} = 17.0 \; \frac{kg_a}{s}$

$$_5\dot{Q}_6 = \dot{m}_{a,1} \bar{c}_p (t_6 - t_5) = (17 \tfrac{kg_a}{s})(1.02 \tfrac{kJ}{kg_a \cdot °C})(17.4 - 12.4)°C$$

$$\boxed{_5\dot{Q}_6 = 87 \; kW}$$

$$\dot{m}_w = \dot{m}_{a,1}(W_7 - W_6) = (17 \tfrac{kg_a}{s})(0.0065 - 0.0040)\tfrac{kg_w}{kg_a}$$

$$\boxed{\dot{m}_w = 0.043 \; kg_w/s} \quad (153 \; kg_w/hr)$$

$$_7\dot{Q}_8 = \dot{m}_{a,1} \bar{c}_p (t_8 - t_7) = (17 \tfrac{kg_a}{s})(1.02 \tfrac{kJ}{kg_a \cdot °C})(35-11)°C$$

$$\boxed{_7\dot{Q}_8 = 416 \; kW}$$

Prob 8.20

Problem 8.21 Use Chart C-8E

$\dot{Q}_S = 84,000$ Btu/hr. $\dot{Q}_L = \dot{m}_w h_w = (20 \frac{lbm_w}{hr})(1120 \frac{Btu}{lbm_w})$

$\dot{Q}_L = 22,400$ Btu/hr

$\therefore SHR = \frac{84,000}{84,000 + 22,400} = 0.79$

Using the given data, locate States 2, 3 and 4 on the chart.

Note: Use the notation of Fig 8.15

With the aid of the protractor, draw the space condition line with a SHR = 0.79 through state 2, 3. State 1, 6 is on the condition line and with given temperature $t_6 = t_1 = 60°F$.

Read: $h_6 = 25.0$ Btu/lbma and $W_6 = 0.0098$ lbmw/lbma

By Eqn 8.28

$\dot{m}_{a,1} = \dot{Q}_s / C_p(t_2-t_1) = \frac{84,000}{0.245(76-60)} = 21,430 \frac{lbma}{hr}$

By Eqn 8.10

$\dot{m}_{a,4} = \rho_{std} \dot{V}_{std} = (0.075 \frac{lbma}{ft^3})(60 \frac{min}{hr})(825 \frac{ft^3}{min}) = 3710 \frac{lbma}{hr}$

$\therefore \dot{m}_{a,2} = 21,430 - 3710 = 17,720$ lbma/hr

By Eqn 8.9

$t_5 = \frac{17,720}{21,430}(76) + \frac{3710}{21,430}(95) = 79.3°F = t_5$

$\therefore$ Locate State 5 on the mixing line connecting States 2 and 4 and at $t_5 = 79.3°F$

Read: $t_5^* \cong 67.5°F$, $h_5 = 32.0 \frac{Btu}{lbma}$ and $W_5 = 0.0117 \frac{lbm_w}{lbma}$

b.) By Eqns. 8.14 and 8.15

$\dot{Q}_R = \dot{m}_{a,5}[(h_5 - h_6) - (W_5 - W_6)h_{f,6}]$

$= 21,430[(32.0 - 25.0) - (0.0117 - 0.0098)(27.63)]$

$\dot{Q}_R = 149,000$ Btu/hr $= (149,000 \text{ Btu/hr})/(12,000 \text{ Btu/hr-ton})$

$\dot{Q}_R = 12.4$ tons

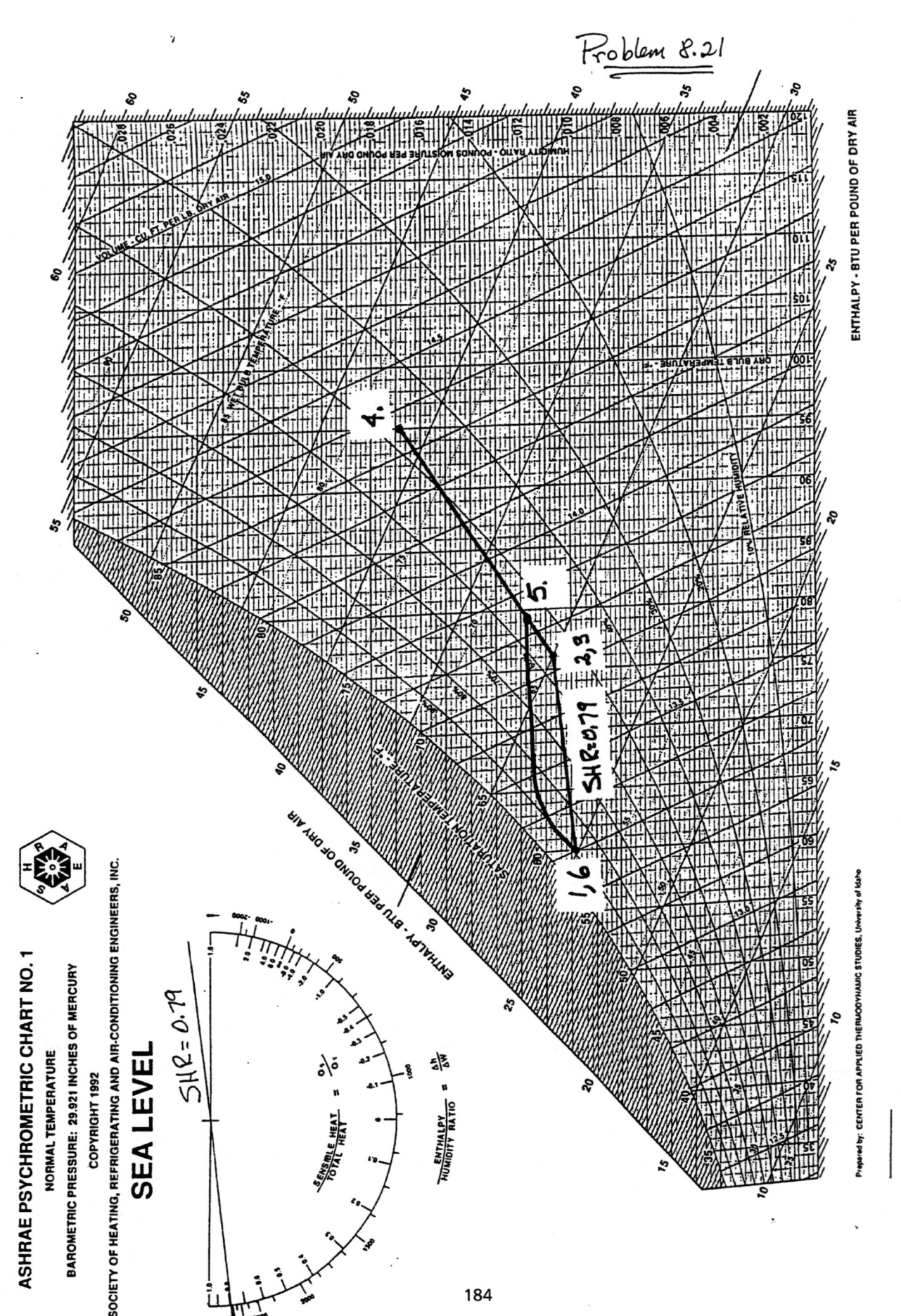

Prob. 8.22

$$SHR = \frac{73,500}{73,500 + 16,500} = 0.82$$

Eqn 8.10

$\dot{V}_1 = 7,200 \text{ ft}^3/\text{min}$ of std. air.; $\dot{m}_{a,1} = (0.075 \frac{lbma}{ft^3})\frac{(7200 \times 60)}{ft^3/hr}$

$\dot{V}_4 = 1,800 \text{ ft}^3/\text{min}$ of std air, $\dot{m}_{a,4} = 0.075(1800)(60)$

$\dot{m}_{a,1} = 32,400 \text{ lbma}/\text{hr}$; $\dot{m}_{a,4} = 8,100 \text{ lbma}/\text{hr}$.

$\phi_6 = 90\%$

States 6, & 1, 7 fall on the line through 2, 3 that has the SHR = 0.82 ∴ Draw SHR = 0.82 through State 2 & locate state 6 @ $\phi_6 = 0.9$

Locate state 1 on that line at t_1

$$t_1 = \frac{-\Sigma \dot{Q}_s}{\dot{m}_{a,1} C_p} + t_2 = \frac{-73,500}{(32,400)(0.245)} + 80$$

$$t_1 = 70.7°F$$

State 5 is a mix of 2a and 4 ∴ need $\dot{m}_{a,2a}$
From the mix of 2b & 6 to get 7 we can find the fractions of 2b & 6 (note $\dot{m}_{a,6} = \dot{m}_{a,5}$)
From mixing eqn:

$$\frac{\dot{m}_{a,2b}}{\dot{m}_{a,7}} = \frac{t_7 - t_6}{t_{2b} - t_6} = \frac{70.7 - 60.5}{80 - 60.5} \quad \left(\begin{array}{c} t_6 \cong 60.5 \\ \text{from chart} \end{array}\right)$$

∴ $\frac{\dot{m}_{a,2b}}{\dot{m}_{a,7}} = 0.52$ ∴ $\frac{\dot{m}_{a,6}}{\dot{m}_{a,7}} = 0.48$

or $\dot{m}_{a,2b} = 0.52 (32,400) = 16,850 \text{ lbma}/\text{hr}$

$\dot{m}_{a,6} = 15,550 \text{ lbma}/\text{hr}$

$\dot{m}_{a,2a} = \dot{m}_{a,1} - \dot{m}_{a,3} - \dot{m}_{a,2b} = 32,400 - 8,100 - 16,850$

$\dot{m}_{a,2a} = 7,450 \text{ lbma}/\text{hr}$.

__Locate state 5__ = mix of 2a & 4.

$\dot{m}_{a,5} t_5 = \dot{m}_{a,2a} t_{2a} + \dot{m}_{a,4} t_4$ (Note $\dot{m}_{a,5} = \dot{m}_{a,6}$)

$$t_5 = \frac{7450}{15550}(80) + \frac{8,100}{15550}(95) = \underline{87.8°F}$$

a.) From the chart $t_1 = 70.7°F$, $t_1^* \cong 63°F$

b.) $\dot{V}_{2b} = \frac{\dot{m}_{a,2b}}{\rho_{std}} = \frac{16,850 \frac{lbma}{hr}}{(0.075 \frac{lbma}{ft^2})(60 \frac{min}{hr})} = \underline{3,740 \text{ ft}^3/\text{min of std. air}}$

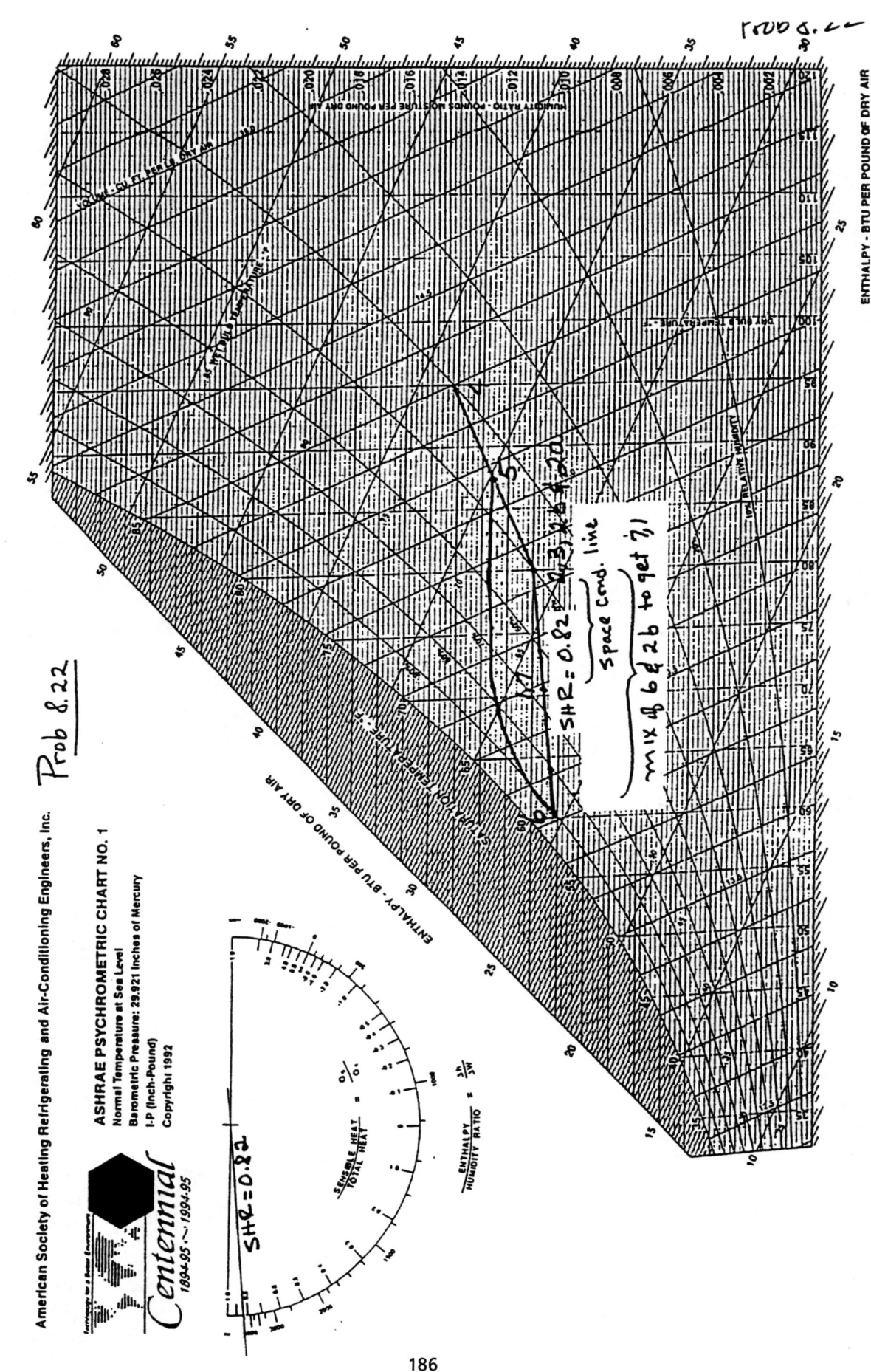

Problem 8.23

$$SHR = \frac{13}{13 + 8.5} = 0.60$$

Locate State 2 by observing that States 6, 1, 7, & 2 are on the same line & this line has a SHR = 0.60.
∴ Draw the SHR = 0.6 through the given state 6 & locate 2 on that line at the given $t_2 = 27°C$

(c.) Read $\phi_2 \cong 52\%$

Calculate either h_1 or t_1 to find state 1 on the line connecting 6 & 2b (use either Eqn 8.25 or 8.28)
In problem 8.22 we used the approx. eqn 8.28. In this problem let's use Eqn 8.25 so you get a look at both approaches.

$$\sum \dot{Q}_s + \sum \dot{Q}_L = \dot{m}_{a,1}(h_2 - h_1)$$

$$h_2 \cong 56.8 \tfrac{kJ}{kga} \text{ from the chart}$$

$$h_1 = 56.8 - \frac{13 + 8.5 \, kW}{1.1 \, kga/s} = 37.3 \, kJ/kga$$

(b.) Read $t_1 \cong 15.7°C$, $\phi_1 \cong 76\%$

From states 6, 1, & 2b determine the various mass flow rates:

$$\frac{\dot{m}_{a,6}}{\dot{m}_{a,1}} = \frac{h_{2b} - h_1}{h_{2b} - h_6} = \frac{56.8 - 37.3}{56.8 - 22.8} = 0.57$$

$$\therefore \dot{m}_{a,6} = 0.57(1.1 \, kga/s) = 0.63 \, kga/s$$

$$\therefore \dot{m}_{a,2b} = 1.1 - 0.63 = 0.47 \, kga/s \; ; \; \dot{m}_{a,2a} = 0.63 - 0.28 = 0.35 \, \tfrac{kga}{s}$$

Locate state 5 on the mixing line of 2a & 4

$$h_5 = \frac{\dot{m}_{a,2a}}{\dot{m}_{a,5}} h_{2a} + \frac{\dot{m}_{a,4}}{\dot{m}_{a,5}} h_4$$

$$h_5 = \frac{0.35}{0.63}(56.8) + \frac{0.28}{0.63}(69.7) = 62.5 \, \tfrac{kJ}{kga}$$

$$\dot{Q}_R = \dot{m}_{a,5}(h_5 - h_6) = (0.63 \, \tfrac{kga}{s})(62.5 - 22.8) \, \tfrac{kJ}{kga}$$

$$\boxed{\dot{Q}_R = 25 \, kW}$$

Prob 8.23

Problem 8.24

$\sum \dot{Q}_L = \dot{m}_w \bar{h}_w = (330 \, \frac{lbm_w}{hr})(1100 \, \frac{Btu}{lbm_w})$
$= 363,000 \, Btu/hr$

$SHR = \dfrac{377,000}{377,000 + 363,000} = 0.51$

Find State 1 at the intersection of $SHR = 0.51$ through State 2 and constant W from State 6.

(a) Read: $t_1 \cong 60.8°F$, $t_1^* \cong 55°F$

By Eqn 8.29: $\dot{m}_{a,1} = \sum \dot{Q}_L / \bar{h}_g (W_2 - W_1)$

Use $\bar{h}_g = 1100 \, Btu/lbm_w$ (see page 216)
Read W_2 & W_1 on chart.

$\dot{m}_{a,1} = (363,000 \, Btu/hr) / (1100 \, \frac{Btu}{lbm_w})(0.010 - 0.008)_r \, \frac{lbm_w/lbm_a}{}$

$\dot{m}_{a,1} \cong 110,000 \, lbm_a/hr$

$_6\dot{Q}_7 = \dot{m}_{a,1} \bar{c}_p (t_1 - t_6) =$
$= (110,000 \, \frac{lbm_a}{hr})(0.245 \, \frac{Btu}{lbm_a \cdot °F})(60.8 - 54)°F$

$_6\dot{Q}_7 = 183,000 \, Btu/hr$

SHR ≅ 0.51

$SHR = 0.51$

$W_7 = W_6$

Problem 8.25

a.) The system would be similar schematically to Fig 8.15, except with no cooling coil (State 5 = State 1)

Since $\dot{m}_w = 0$, $W_1 = W_2 = W_4 = W_5 = W$

From Chart C-7E $W = 0.00047$ lbmw/lbma

By Eqn 8.25

$$\dot{m}_{a,1} = \frac{60,000}{h_2 - h_1} = \frac{60,000}{(0.24 + 0.444(0.00047))(72-55)}$$

$$\dot{m}_{a,1} = \frac{60,000}{0.240(72-55)} = 14,706 \text{ lbma/hr}$$

By Eqn 8.10 $\dot{V} = \frac{14,706}{0.075(60)} = 3,270 \text{ ft}^3/\text{min}$ of Std. air.

b.) For the mixing of recirculated and outdoor air we have from Eqns 8.2 and 8.4

$$\frac{\dot{m}_{a,4}}{\dot{m}_{a,1}} = \frac{h_2 - h_5}{h_2 - h_4}$$

and since c_p is constant this is equal to (with no approximations).

$$\frac{\dot{m}_{a,4}}{\dot{m}_{a,1}} = \frac{t_2 - t_5}{t_2 - t_4} = \frac{72-55}{72+10} = 0.207$$

∴ 20.7% of the supply air is outdoor air.

Problem 8.26

$SHR = 180,000 / (180,000 + 19,700) = 0.90$ & State 2 given

Locate State 1 on $SHR = 0.90$ line through state 2 & $t_1 = 100°F$

Locate State 6 at $W_6 = W_1$ & $\phi_6 = 100\%$ (given)

Locate State 3 at $W_3 = W_2$ & $t_3 = 45°F$ (given)

Locate State 5:

$W_5 = W_4$; From Chart C-7E at $t_4 = 20°F$ & $\phi_4 = 100\%$

$W_4 = W_5 = 0.00215$ lbmw/lbma

(Note: Since pressure is Standard Atmospheric Pressure and since state 4 is saturated we could read $W_4 = W_5$ at 20°F from Table A4E as an alternative to using Chart C-7E)

Assume the heat recovery process is 100% efficient

$\therefore \dot{m}_{a,3} \bar{C}_p (t_2 - t_3) = \dot{m}_{a,4} \bar{C}_p (t_5 - t_4)$ & $\dot{m}_{a,3} = \dot{m}_{a,4}$

$\therefore t_5 = t_4 + (t_2 - t_3) = 20 + (70 - 45) = \underline{45°F}$

a) Read: $t_1^* \cong 66.7°F$, $t_5^* \cong 35.5°F$, $t_2^* \cong 55°F$ (this was given)

$t_6 \cong 45.5°F$

b) From Eqn 8.23

$h_w = q'$ for the process from 5-to-6

$\therefore$ From the chart $h_w \cong 1120$ Btu/lbmw

Prob 8.26

8.27

Zone A: $SHR = \dfrac{48,000}{48,000 + 32,000} = 0.60$

By Eqn 8.28

$$\dot{m}_{a,1} = \dfrac{48,000 \text{ Btu/hr}}{(0.245 \frac{\text{Btu}}{\text{lbma} \cdot °F})(95-78)°F} = 11,500 \text{ lbma/hr}$$

Zone B: $SHR = \dfrac{24,000}{24,000 + 6000} = 0.80$

By Eqn 8.28

$$\dot{m}_{a,4} = \dfrac{24,000 \text{ Btu/hr}}{(0.245 \frac{\text{Btu}}{\text{lbma} \cdot °F})(95-68)°F} = 3,630 \text{ lbma/hr}$$

15% exhaust

∴ $\dot{m}_{a,2} = 0.85(\dot{m}_{a,1}) = 0.85(11,500) = 9,780$ lbma/hr

$\dot{m}_{a,5} = 0.85(\dot{m}_{a,4}) = 0.85(3,630) = 3,090$ lbma/hr

$\dot{m}_{a,3} = 11,500 - 9,780 = 1,720$ lbma/hr

$\dot{m}_{a,6} = 3,630 - 3,090 = 540$ lbma/hr

By Eqn 8.9:

$$t_7 = \left(\dfrac{9,780}{9780+3090}\right)(78) + \left(\dfrac{3090}{9780+3090}\right)(68) = 75.6 \text{ °F}$$

$$t_9 = (0.15)(40) + (0.85)(75.6) = 70.3 \text{ °F}$$

Locate points 7 & 9 on the corresponding mixing lines.
Locate state 11 by drawing a const W line from state 12 to $\phi_{11} = 95\%$
Locate state 10 as the intersection of a const. W process from 9 and a const. t^* process from 11.

b) By Eqn 8.11 $\quad \dot{q}Q_{10} = (11,500 + 3630) \frac{\text{lbma}}{\text{hr}} (26.2 - 25.1) \frac{\text{Btu}}{\text{lbma}}$
$= 16,600$ Btu/hr

or By Eqn 8.28 $\dot{q}Q_{10} = (11,500 + 3630) \frac{\text{lbma}}{\text{hr}} (0.245 \frac{\text{Btu}}{\text{lbma} \cdot °F})(74.7 - 70.3 °F)$
$= 16,300$ Btu/hr.

c.) $\dot{m}_w = \dot{m}_{a,12}(W_{11} - W_{12}) = (15,130 \frac{\text{lbma}}{\text{hr}})(0.0106 - 0.0076) \frac{\text{lbmw}}{\text{lbma}} = 45 \frac{\text{lbmw}}{\text{hr}}$

Prob 8.27

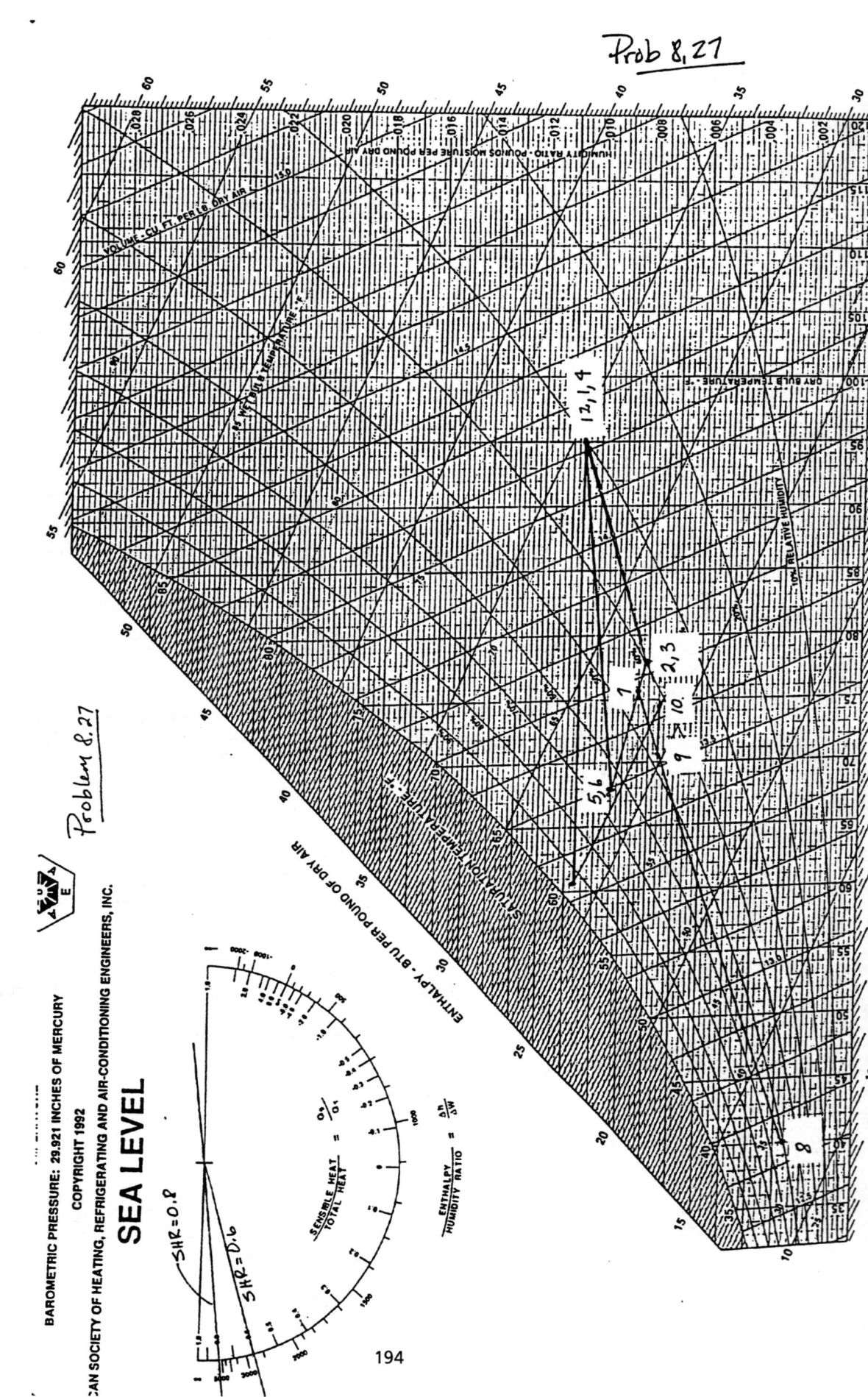

Problem 8.28   Use Chart C-8E

a.) Connect states 4 & 5

Use Eqns 8.3, 8.4 or 8.9 to locate state 6 on the mixing line.

By Eqn 8.9: $\dot{m}_{a,6} t_6 = \dot{m}_{a,4} t_4 + \dot{m}_{a,5} t_5$

$\dot{m}_{a,5} = \dot{m}_{a,2} = 6000 \text{ lb}_{ma}/\text{hr}$

$\dot{m}_{a,4} = 4000 \text{ lb}_{ma}/\text{hr}$

∴ $t_6 = 0.4(72) + 0.6(35) = 49.8°F$

Read $\phi_6 \cong 40\%$

b.) State 7 is located along the constant W-line starting at state 6 and ending at 80°F.
Use Eqn 8.11 or 8.12 to find $_6\dot{Q}_7$

By Eqn 8.12: $_6\dot{Q}_7 = \dot{m}_{a,6}(C_{pa} + C_{pw} W_6)(t_7 - t_6)$

$_6\dot{Q}_7 = 10,000 [0.24 + (0.444)(0.003)](80 - 49.8) = 72,900 \frac{\text{Btu}}{\text{hr}}$

c.) Process line from state 7 to state 8 is constant t^* and state 8 is saturated (Given condition)
From the chart: $t_7^* = t_8^* = 54°F$ and $t_8 = 54°F$.

d.) $\dot{m}_w = \dot{m}_{a,6}(W_8 - W_7) = 10,000 \frac{\text{lb}_{ma}}{\text{hr}} (0.0088 - 0.003) \frac{\text{lb}_{mw}}{\text{lb}_{ma}}$

$\dot{m}_w = 58 \text{ lb}_{mw}/\text{hr}$

e.) State 1 is on a constant W-line from state 8.

By Eqn 8.28: $(\dot{Q}_s)_{Zone\,A} = \dot{m}_{a,1} \bar{C}_p (t_1 - t_2)$

$32,000 = 6000(0.245)(t_1 - 68)$

$t_1 = 89.8°F$, Read $\phi_1 \cong 30\%$

f.) Locate state 3 using SHR = 0.7 from state 4 and at the intersection of constant W from state 8.
Read $t_3 \cong 94.5°F$, By Eqn 8.12

$_8\dot{Q}_3 = 4,000 [0.24 + (0.444)(0.0088)](94.5 - 54) = 39,500 \frac{\text{Btu}}{\text{hr}}$

Problem 8.28

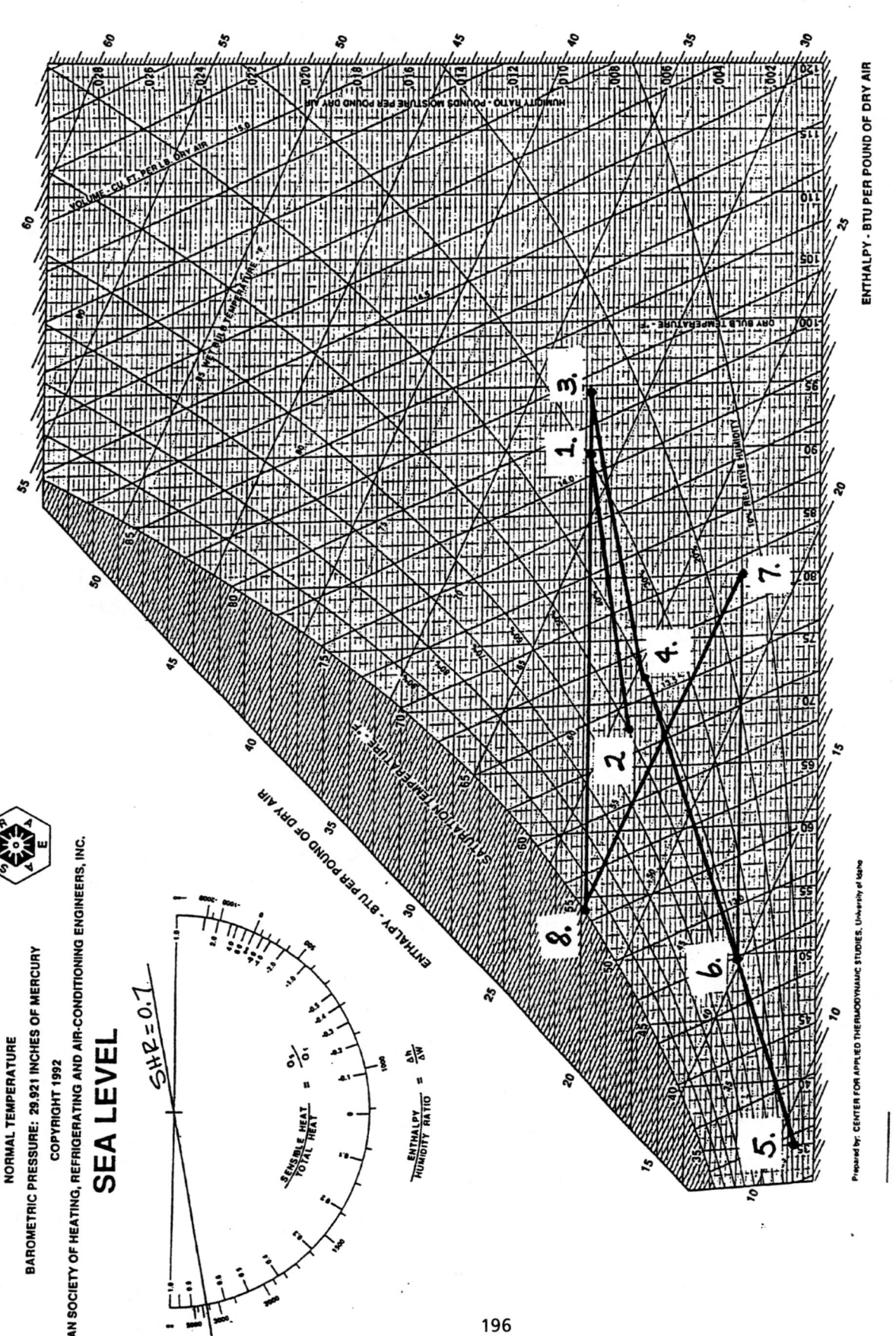

Problem 8.29 Use Chart C-8E

From the data given:

$$(SHR)_A = \frac{20,000}{20,000 + 16,400} = 0.55$$

$$(SHR)_B = 1.0$$

Use the notation in Fig 8.22
- Locate States 2A, 2B, 6 and 8 on the chart.
- Draw the mixing line connecting states 6 and 8
- Locate State 1A by drawing the space condition line with the aid of the protractor, $(SHR)_A = 0.55$, through state 2A. Where the space condition line intersects the mixing line of states 6 & 8 establishes state 1A.

Read: $t_{1A} = 48.5°F$, $h_{1A} = 17.6$ Btu/lbma, $\phi_{1A} \cong 0.78$

b.)
- Similarly locate state 1B at the intersection of the state condition line for Zone B and the mixing line of states 6 & 8

Read: $t_{1B} \cong 83°F$, $h_{1B} = 27.2$ Btu/lbma, $\phi_{1B} \cong 0.27$

a.) By Eqn 8.25

$$\dot{m}_a = \frac{\dot{Q}_S + \dot{Q}_L}{\Delta h}$$

For Zone A Read: $h_{2A} = 27.0$ Btu/lbma

and $\dot{m}_{a,1A} = 36,400/(27.0-17.6) = \underline{3,900}$ lbma/hr

For Zone B Read: $h_{2B} = 24.5$ Btu/lbma

and $\dot{m}_{a,1B} = 25,000/(27.2-24.5) = \underline{9,300}$ lbma/hr.

c.) By Eqns 8.2 and 8.9 along the mixing line

$$\dot{m}_{a,8A} = \dot{m}_{a,1A}\left(\frac{t_{1A}-t_6}{t_8-t_6}\right) = 3900(48.5-40)/(95-40)$$

$$\dot{m}_{a,8A} = \underline{600 \text{ lbma/hr}}$$

Problem 8.29

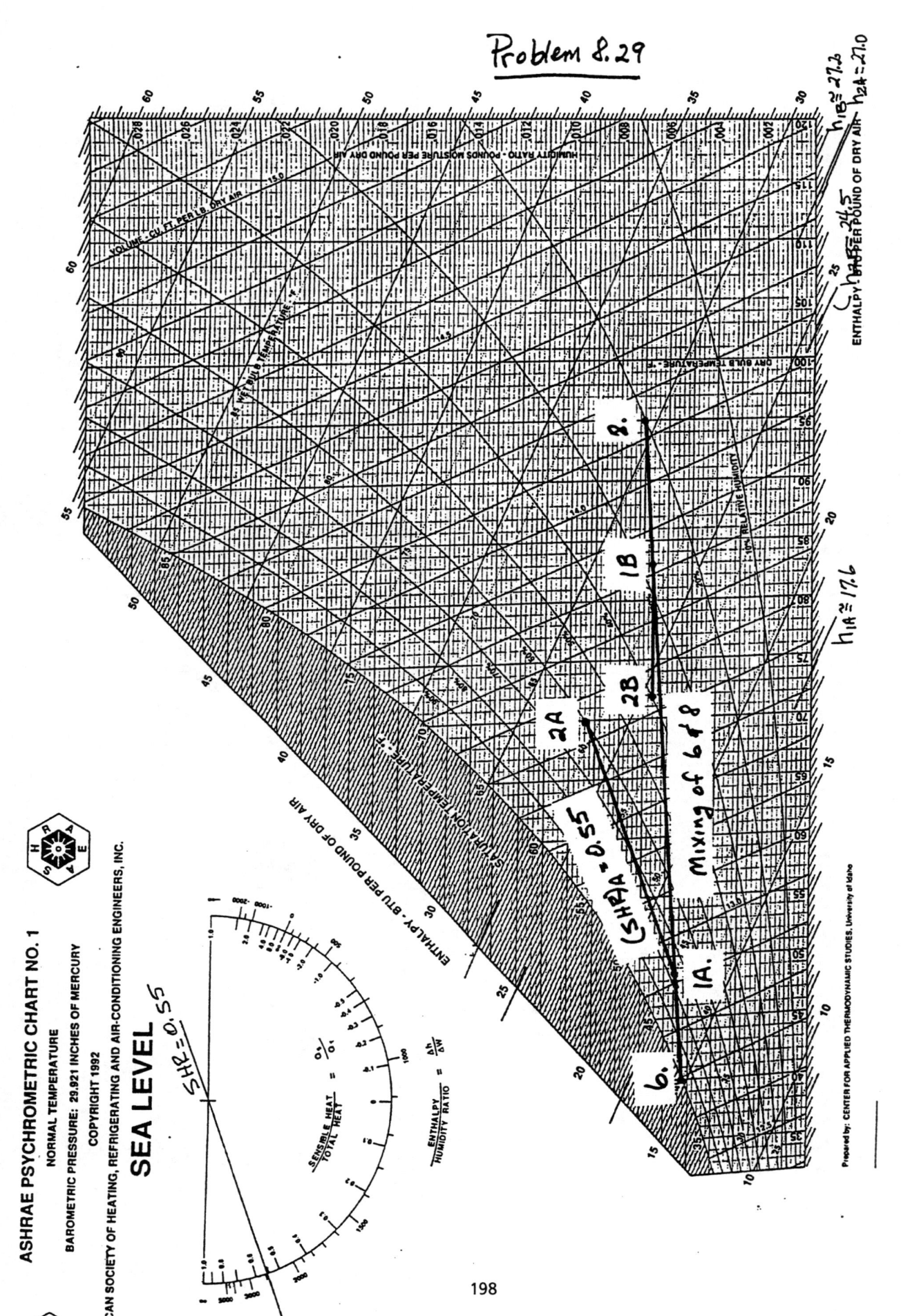

Problem 8.30 Use Chart C-8E and the notation of Fig 8.22. Note that conditions for Zones A & B in the figure are identical. Therefore drop the A & B notation. i.e. $1A = 1B = 1$, $8A = 8B = 8$, etc.

- Locate States 2, 0 and 6 (Given conditions)
- Draw the mixing line connecting 2 and 0
- From the problem statement, $\dot{m}_{a,0}/\dot{m}_{a,4} = 0.25$ and $\dot{m}_{a,2B}/\dot{m}_{a,4} = 0.75$ ∴ By Eqn 8.9
$$t_4 = 0.25(95) + 0.75(75) = 80°F$$
- Locate State 4 on the 2,0 mixing line at $t_4 = 80°F$
- Locate State 8 by a sensible process from State 4 to the temp. $t_8 = 105°F$
- Draw the mixing line connecting States 6 and 8
- With the aid of the protractor, draw the space condition line with SHR = 0.6 from State 2 to where it intersects the mixing line of 6 and 8 to establish State 1.
- For the balance of the problem properties are available either as given or from the chart.

a.) By Eqn 8.25 for 5 zones $\dot{m}_{a,1} = 5(209,000)/(h_2 - h_1)$
$\dot{m}_{a,1} = 5(209,000)/(28.2 - 24.8) = 294,000$ lbma/hr.
By Eqns 8.2 and 8.9 along the State 6 - State 8 mixing line
$$\dot{m}_{a,8} = \dot{m}_{a,1}\left(\frac{t_1 - t_6}{t_8 - t_6}\right) = 294,000(66.5 - 50)/(105 - 50)$$

$$\underline{\dot{m}_{a,8} = 88,200 \text{ lbma/hr}}$$

b) By Eqn 8.11
$$_sQ_8 = \dot{m}_{a,8}(h_8 - h_5) = 88,200(36.8 - 30.8)$$
$$\underline{_sQ_8 = 529,000 \text{ Btu/hr.}}$$

c.) $\dot{m}_{a,6} = \dot{m}_{a,1} - \dot{m}_{a,8} = 294,000 - 88,200$
$$\underline{\dot{m}_{a,6} = 205,800 \text{ lbma/hr.}}$$

d.) By Eqn 8.14 and 8.15 and neglecting the condensate term.
$$\dot{Q}_R = \dot{m}_{a,6}(h_5 - h_6) = 205,800(30.8 - 19.4) = 2.35 \times 10^6$$
$$\underline{\dot{Q}_R = 2.35 \times 10^6 \text{ Btu/hr.}}$$

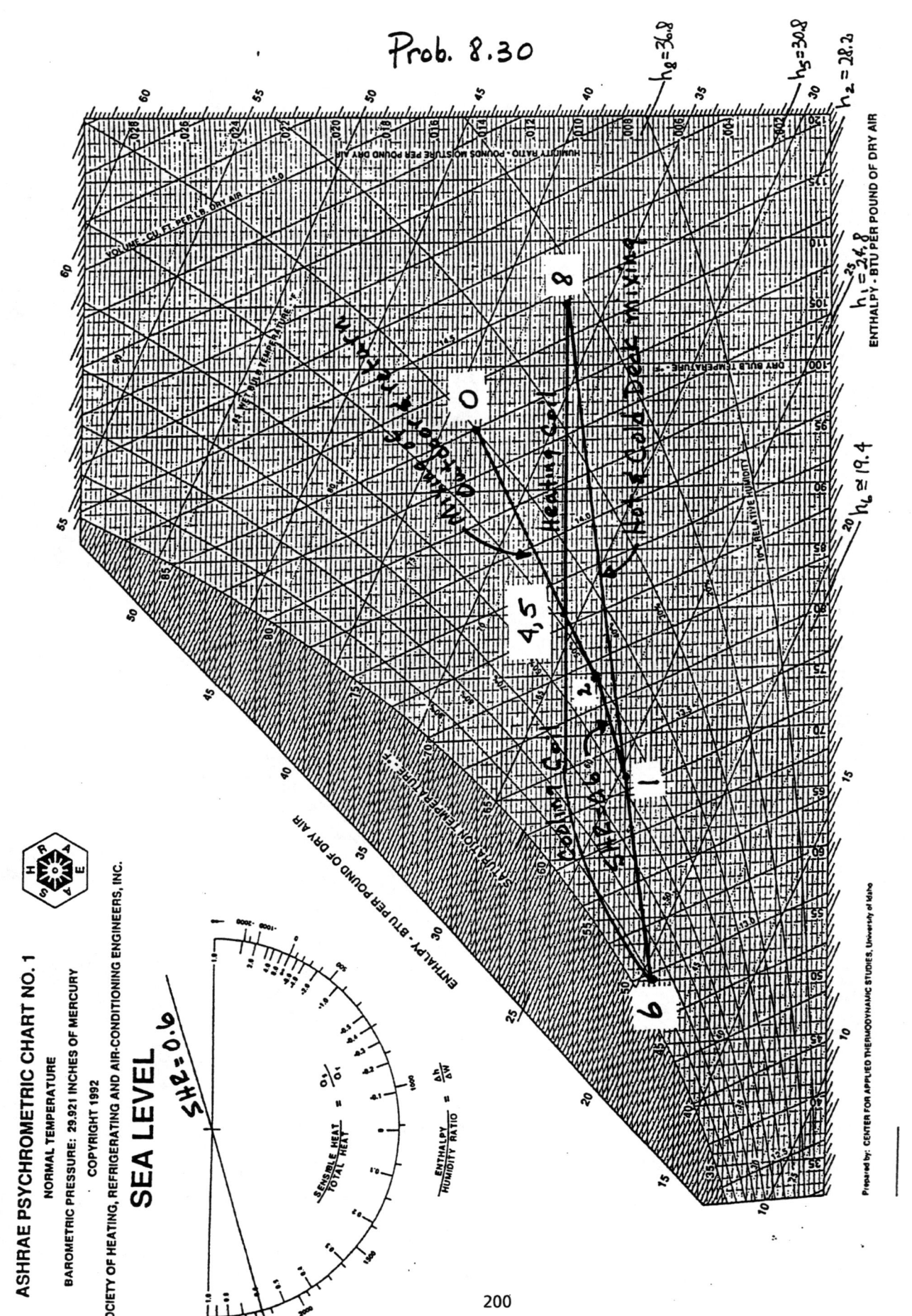

9.1

We have $t_{db} = 90°F$, $t_{wb} = 70°F$

By Table A.1E, $h_{fg,wb} = 1091.85 - 37.65 = 1054.20$ Btu/lb$_m$

$P_{w,s,wb} = 0.36324$ psia

By Eq. (7.18)

$$W_{s,wb} = \frac{(0.622)(0.36324)}{(13.00 - 0.36324)} = 0.01788 \frac{lb_{mw}}{lb_{ma}}$$

By Table 9.1, $d/D = 0.850$ (sat. at 70°F)
By Eq. (9.7), $Le = (0.850)^{2/3} = 0.897$
By Eq. (9.14) and Table 7.3

$$K = \frac{0.240 + (0.444)(0.01788)}{\frac{1054.20}{0.897} + \frac{(0.444)(90-70)}{2}} = 2.102 \times 10^{-4} \frac{lb_w}{lb_{ma} \cdot °F}$$

By Eq. (9.9)
$W = 0.01788 - 2.102 \times 10^{-4}(90-70) = 0.01368 \frac{lb_{mw}}{lb_{ma}}$

By Eq. (7.18)

$$P_w = \frac{1.608(P)W}{1 + 1.608W} = \frac{1.608(13.0)0.01368}{1 + 1.608(0.01368)} = 0.2798 \text{ psia}$$

By Table A.1E, $P_{w,s} = 0.69890$ psia (at 90°F)

By Eq. (7.27), $\phi = \dfrac{0.2798}{0.69890} = 0.400$ or 40.0%

9.2

We have $t_{db} = 25°C$, $t_{wb} = 18°C$

By Table A.1SI, $h_{fg,wb} = 2533.96 - 74.51 = 2459.45 \frac{kJ}{kg}$

$P_{w,s,wb} = 0.002064$ MPa

By Eq. (7.18)

$$W_{s,wb} = \frac{(0.622)(0.002064)}{(0.105 - 0.002064)} = 0.01247 \frac{kg_w}{kg_a}$$

By Table 9.1, $\alpha/D = 0.851$ (sat at 18°C)
By Eq. (9.7), $Le = (0.851)^{2/3} = 0.898$
By Eq. (9.14) and Table 7.3

$$K = \frac{1.00 + (1.86)(0.01247)}{\frac{2459.45}{0.898} + \frac{(1.86)(25-18)}{2}} = 3.727 \times 10^{-4} \frac{kg_w}{kg_a \cdot °C}$$

By Eq. (9.9)

$$W = 0.01247 - 3.727 \times 10^{-4}(25-18) = 0.00986 \frac{kg_w}{kg_a}$$

By Eq. (7.18)

$$P_w = \frac{1.608(P)W}{1 + 1.608 W} = \frac{1.608(0.105)0.00986}{1 + 1.608(0.00986)} = 0.00164 \text{ MPa}$$

By Table A.1SI, $P_{w,s} = 0.003169$ MPa (at 25°C)

By Eq. (7.27), $\phi = \frac{0.00164}{0.003169} = 0.517$ or 51.7%

9.3

$\bar{t}_r = t_{db} = 120°F$ so we can use Eq. (9.13)

By Table A.1E, $h_{fg,wb} = 1094.03 - 42.67 = 1051.36 \ \frac{Btu}{lb_m}$

$P_{w,s,wb} = 0.87806$ in Hg

By Table 9.1, $\alpha/D = 0.849$ (sat. at 75°F)

By Eq. (9.7), $L_e = (0.849)^{2/3} = 0.897$

Fig. 9.3 applies, we find $h_{r,t}/h_c = 0.07$

By Eq. (7.18)
$$W_{s,wb} = \frac{0.622(0.87806)}{(27.56 - 0.87806)} = 0.02047 \ \frac{lb_{mw}}{lb_{ma}}$$

By Eq. (9.13) and Table 7.3

$$K = \frac{0.24 + 0.444(0.02047)}{\frac{1051.36}{0.897(1+0.07)} + \frac{0.444(120-75)}{2}} = 2.253 \times 10^{-4} \ \frac{lb_{mw}}{lb_{ma} \cdot °F}$$

By Eq. (9.9)

$W = 0.02047 - 2.253 \times 10^{-4}(120 - 75)$

$W = 0.01033 \ \frac{lb_{mw}}{lb_{ma}}$

9.4

By Fig. 9.3, $(h_{r,t}/h_c)_{db} = 0.109$ (120°F surroundings, 90°F dry bulb temp.)

By Eq. (9.8)
$$t = 90 - (0.109)(120-90) = \underline{86.73°F}$$

Use Eq. (9.17) for K

By Table A.4E, $W_{s,wb} = 0.011087$ lbmw/lbma

By Table A.1E, $h_{fg,wb} = 1087.48 - 27.63 = 1059.85$ Btu/lbm

By Fig. 9.3, $(h_{r,t}/h_c)_{wb} = 0.092$

By Table 9.1, $\alpha/D = 0.852$
By Eq. (9.7), $Le = (0.852)^{2/3} = 0.899$

By Eq. (9.17)

$$K = \frac{0.240 + 0.444(0.011087)}{0.899 \left[1 + \frac{0.092(579.67^4 - 519.67^4)}{(546.40^4 - 519.67^4)} \right]} + \frac{0.444(86.73-60)}{2}$$

$$K = 2.53 \times 10^{-4} \text{ lbmw/lbma °F}$$

By Eq. (9.9), $W = 0.011087 - 2.53 \times 10^{-4}(86.73 - 60)$
$$W = 0.00432 \text{ lbmw/lbma}$$

By Chart C-8E, $\underline{t^* = 58.8°F}$

9.5

Figure 9.3 was generated assuming a dry bulb sensor emissivity of 0.9. For $\varepsilon = 0.05$ we have

$$\left(\frac{h_{r,t}}{h_c}\right)_{db} = \frac{0.05}{0.9}\left(\frac{h_{r,t}}{h_c}\right)_{db, \text{Fig. 9.3}}$$

Assuming $t_{db} \cong 87°F$, By Fig. 9.3
$h_{r,t}/h_c = 0.109$

Thus

$$\left(\frac{h_{r,t}}{h_c}\right)_{db} = \frac{0.05(0.109)}{0.9} = 0.00606$$

By Eq. (9.8) (see Prob. 9.4 for value of t)

$$t_{db} = \frac{t + (h_{r,t}/h_c)(t_s - t)}{1 + (h_{r,t}/h_c)}$$

$$t_{db} = \frac{86.73 + (0.00606)(120 - 86.73)}{1 + 0.00606}$$

$$t_{db} = 86.41°F$$

9.6

Since we have $\bar{t}_r = t_{db}$, $t = t_{db} = 250°F$

Eq. (9.13) applies for solution of K

By Table A.1E, $h_{fg,wb} = 1104.83 - 67.87 = 1036.96$ Btu/lb$_m$

By Table A.4E, $W_{s,wb} = 0.043219$ lb$_{mw}$/lb$_{ma}$

By Table 9.1, $\lambda/D = 0.843$ (sat. at 100°F)

By Eq. (9.7), $Le = (0.843)^{2/3} = 0.892$

By Eq. (9.15),

$$h_r = \frac{(0.1714 \times 10^{-8})(0.9)(709.67^4 - 559.67^4)}{(250 - 100)}$$

$h_r = 1.599$ Btu/hr ft^2 °F

By Table A.5E, at an air film temperature of $\frac{250-100}{2} = 175°F$, $k = 0.01747$ Btu/hr ft °F

$\mu = 0.05043$ lb$_m$/ft hr, $\rho = 0.06251$ lb$_m$/ft^3
$Pr = 0.696$

Then $Re = \rho d V / \mu$

$$= \frac{(0.06251 \text{ lb}_m/\text{ft}^3)\, 0.3 \text{in}\, (900 \text{ ft/min})\, 60 \text{min/hr}}{(0.05043 \text{ lb}_m/\text{ft hr})\, 12 \text{in/ft}}$$

$Re = 1673$

Thus Eq. (9.16) applies

9.6 con't

$$h_c = \frac{0.01747 \text{ Btu/hr ft°F}}{0.3 \text{ in } (\text{ft}/12\text{in})} \left[0.4(1673)^{0.5} + 0.06(1673)^{2/3} \right] 0.696^{0.4}$$

$$h_c = 15.0 \text{ Btu/hr ft}^2\text{°F}$$

Thus, $\dfrac{h_r}{h_c} = \dfrac{1.599}{15.0} = 0.107$

By Eq. (9.13)

$$K = \frac{0.240 + 0.444(0.043219)}{\dfrac{1036.96}{0.892(1+0.107)} + \dfrac{0.444(250-100)}{2}} = 2.391 \times 10^{-4} \dfrac{\text{lbm}_w}{\text{lbm}_a}$$

By Eq. (9.9)

$$W = 0.043219 - 2.391 \times 10^{-4}(250-100)$$
$$W = 0.00735 \text{ lbm}_w/\text{lbm}_a$$

By Chart C-9E at $t = 250°F$, $W = 0.00735 \dfrac{\text{lbm}_w}{\text{lbm}_a}$

$t^* = 99.3 \text{ °F}$

9.7
a) $d = 0.1$ in.

We must satisfy the relation

$$Le\left(1 + \frac{h_{r,t}}{h_c}\right) = 1.0$$

By Table 9.1, $d/D = 0.850$
By Eq. (9.7)
$$Le = (0.850)^{2/3} = 0.897$$

Thus $h_{r,t}/h_c = 1/Le - 1 = 1/0.897 - 1 = 0.115$

By Fig. 9.4, the required air velocity is approximately 100 ft/min

b) $d = 0.5$ in.

We do not have a figure to read $h_{r,t}/h_c$ directly so we will modify h_c and use Fig. 9.3

By Eq. (9.16), we can approximate h_c

$$h_c \approx \frac{k}{d}\left[A\left(\frac{dV\rho}{\mu}\right)^{0.6}\right] Pr^{0.4}$$

or $h_c \propto B d^{-0.4}$ where B is a constant

Thus $\dfrac{h_{r,t}}{h_c} = \left(\dfrac{h_{r,t}}{h_c}\right)^* \left(\dfrac{0.5}{0.3}\right)^{0.4} = \left(\dfrac{h_{r,t}}{h_c}\right)^* 1.227$

9.7 Cont'd

where * indicates the quantity is evaluated at $d = 0.3$ in.

Thus we need

$$\left(\frac{h_{r,t}}{h_c}\right)^* = \frac{1}{1.227}\left(\frac{1}{Le} - 1\right) = 0.0936$$

From Fig. 9.3, we need a velocity of approximately 600 ft/min.

9.8

We have for the throttling process, $h_1 = h_2$, $W_1 = W_2$, and for an ideal gas, $t_1 = t_2$, $P_{w,s,1} = P_{w,s,2}$

By Eq. (7.18)

$$\frac{P_{w,s,1}}{P_1 - P_{w,s,1}} = \frac{P_{w,2}}{P_2 - P_{w,2}}$$

or

$$\frac{1}{\frac{P_1}{P_{w,s,1}} - 1} = \frac{1}{\frac{P_2}{P_{w,2}} - 1}$$

Thus $\dfrac{P_{w,2}}{P_{w,s,1}} = \dfrac{P_2}{P_1} = \dfrac{P_L}{P_H}$

From the ideal gas relation above,

$$\frac{P_{w,2}}{P_{w,s,1}} = \frac{P_{w,2}}{P_{w,s,2}} = \phi_2 \quad \text{by Eq. (7.27)}$$

Thus $\phi_2 = P_L / P_H$

9.9

By Eq. (7.27), $\phi_2 = P_{w,s}/P_{w,s,2}$

In a sensible heating process, $W_1 = W_2$ and $P_{w,s,1} = P_{w,2}$

From Table A.1SI, $P_{w,s,1} = 1.705 \text{ kPa}$ (15°C)
$P_{w,s,2} = 3.169 \text{ kPa}$ (25°C)

Thus $\phi_2 = \dfrac{1.705 \text{ kPa}}{3.169 \text{ kPa}} = 0.538$ or 53.8%

10.1 Air Washer

Inlet (1): 30°C, 50% RH air
Outlet (2): air

- Face velocity = 1.5 m/s
- Face area = 2 m²
- NTU = 0.8
- Standard Sea Level Pressure

a) washer efficiency

By Eqn. 10.10, $\eta_w = 1 - e^{(-NTU)}$

$\eta_w = 1 - e^{(-0.8)} = 0.551$

b) leaving air dry bulb temperature and humidity ratio

By Eqn. 10.9, $\eta_w = \dfrac{W_2 - W_1}{W_s^* - W_1}$

From chart C-8SI, $W_1 = 13.4 \, g_w/kg_a$, $W_s^* = 16.7 \, g_w/kg_a$

solving for W_2, $W_2 = \eta_w (W_s^* - W_1) + W_1$

$= 0.551(16.7 - 13.4) + 13.4 = 15.2 \, g_w/kg_a$

By Chart C-8SI, $t_2 = 25.8$ °C

c) mass of water evaporated

water mass balance (see Eq. 7.11)

$\dot{m}_w = \dot{m}_a (W_2 - W_1) = \dfrac{\vec{V}_1 A_1 (W_2 - W_1)}{v_1}$

By chart C-8SI, $v_1 = 0.877 \, m^3/kg_a$

$\dot{m}_w = \dfrac{1.5 \, m/s \, (2 \, m^2)(15.2 - 13.4) \, g_w/kg_a}{0.877 \, m^3/kg_a} \left(\dfrac{kg}{10^3 g}\right) = 0.0062 \, kg_w/s$

10.2 Air Washer

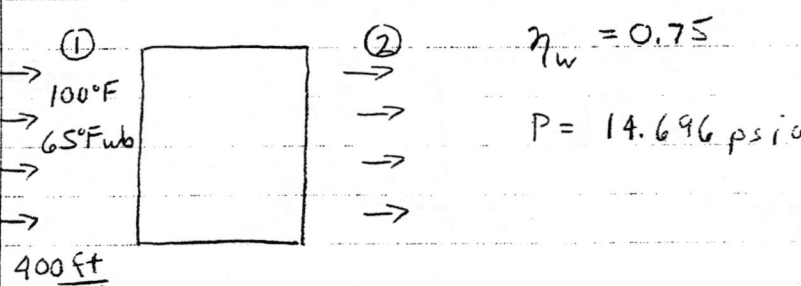

$\eta_w = 0.75$

$P = 14.696 \text{ psia}$

Determine t_2 and t_2^*

By Eq. 10.11, $t_2 = t_1 - \eta_w(t_1 - t^*)$

$t_2 = 100°F - 0.75(100-65)°F = 73.7°F$

$t_2^* = t_1^* = 65°F$

10.3

Heating Coil | Air washer

① → 30°F (sat) → | → 500 fpm ② | ③ 70°F db, $\mu = 0.6$

$h_D A_v = 400 \, \text{lb}_{mw}/\text{hr·f} \cdot (\text{lb}_{mw}/\text{lb}_{ma})$

10,000 cfm

a) t_2 and t_2^*

$W_3 = \mu_3 W_s (70°F)$

By Table A.4E, $W_s(70°F) = 0.015832 \, \text{lb}_{mw}/\text{lb}_{ma}$

$W_1 = W_s(30°F) = 0.0034552 \, \text{lb}_{mw}/\text{lb}_{ma}$

$W_3 = 0.6 (0.015832) = 0.00950 \, \text{lb}_{mw}/\text{lb}_{ma}$

By Chart C-8E, $t_3^* = t_2^* = 61.2°F$

$W_1 = W_2 = 0.0034552 \, \text{lb}_{mw}/\text{lb}_{ma}$

Read t_2 at known t_2^* and W_2, $t_2 = 96.3°F$

b) Capacity of heating coil, Btu/hr

$$\dot{Q}_{1-2} = \frac{\dot{V}_1}{v_1}(h_2 - h_1)$$

By Chart C-8E, $h_2 = 27.0 \, \text{Btu}/\text{lb}_{ma}$

By Table A.4E, $v_1 = 12.406 \, \text{ft}^3/\text{lb}_{ma}$, $h_1 = 10.917 \, \frac{\text{Btu}}{\text{lb}_{ma}}$

$$\dot{Q}_{1-2} = \frac{10,000 \, \text{ft}^3/\text{min}}{12.406 \, \text{ft}^3/\text{lb}_{ma}} \left(\frac{60 \, \text{min}}{\text{hr}}\right)(27.0 - 10.917) \frac{\text{Btu}}{\text{lb}_{ma}}$$

$\dot{Q}_{1-2} = 778,000 \, \text{Btu/hr}$

10.3 cont'd

c) Quantity of makeup water in spray washer

$$\dot{m}_w = \dot{m}_a (W_3 - W_2)$$

$$\dot{m}_a = \frac{\dot{V}_1}{v_1} = \frac{10{,}000 \text{ ft}^3/\text{min}}{12.406 \text{ ft}^3/\text{lb}_{ma}} = 806.1 \text{ lb}_{ma}/\text{min}$$

$$\dot{m}_w = 806.1 \frac{\text{lb}_{ma}}{\text{min}} (0.00950 - 0.0034552) \frac{\text{lb}_{mw}}{\text{lb}_{ma}} \left(\frac{\text{ft}^3}{62.4 \text{ lb}_m}\right)\left(\frac{264.2 \text{ gal}}{35.32 \text{ ft}^3}\right) = 0.579 \text{ gpm}$$

d) Required contact volume, ft^3

By Eqns. 10.10 and 10.11 (can use 10.9 also)

$$\exp(-NTU) = 1 - \frac{(t_2 - t_3)}{(t_2 - t_2^*)} = 1 - \frac{(96.3 - 70)}{(96.3 - 61.2)} = 0.251$$

$$NTU = \ln(1/0.251) = 1.38$$

$$NTU = h_D A_v V / \dot{m}_a$$

$$V = NTU \, \dot{m}_a / h_D A_v$$

$$= \frac{1.38 (806.1 \text{ lb}_{ma}/\text{min})}{400 \frac{\text{lb}_{mw}/\text{hr ft}^3}{(\text{lb}_{mw}/\text{lb}_{ma})}} \left(\frac{60 \text{ min}}{\text{hr}}\right)$$

$$V = 167 \text{ ft}^3$$

Problem 10.3

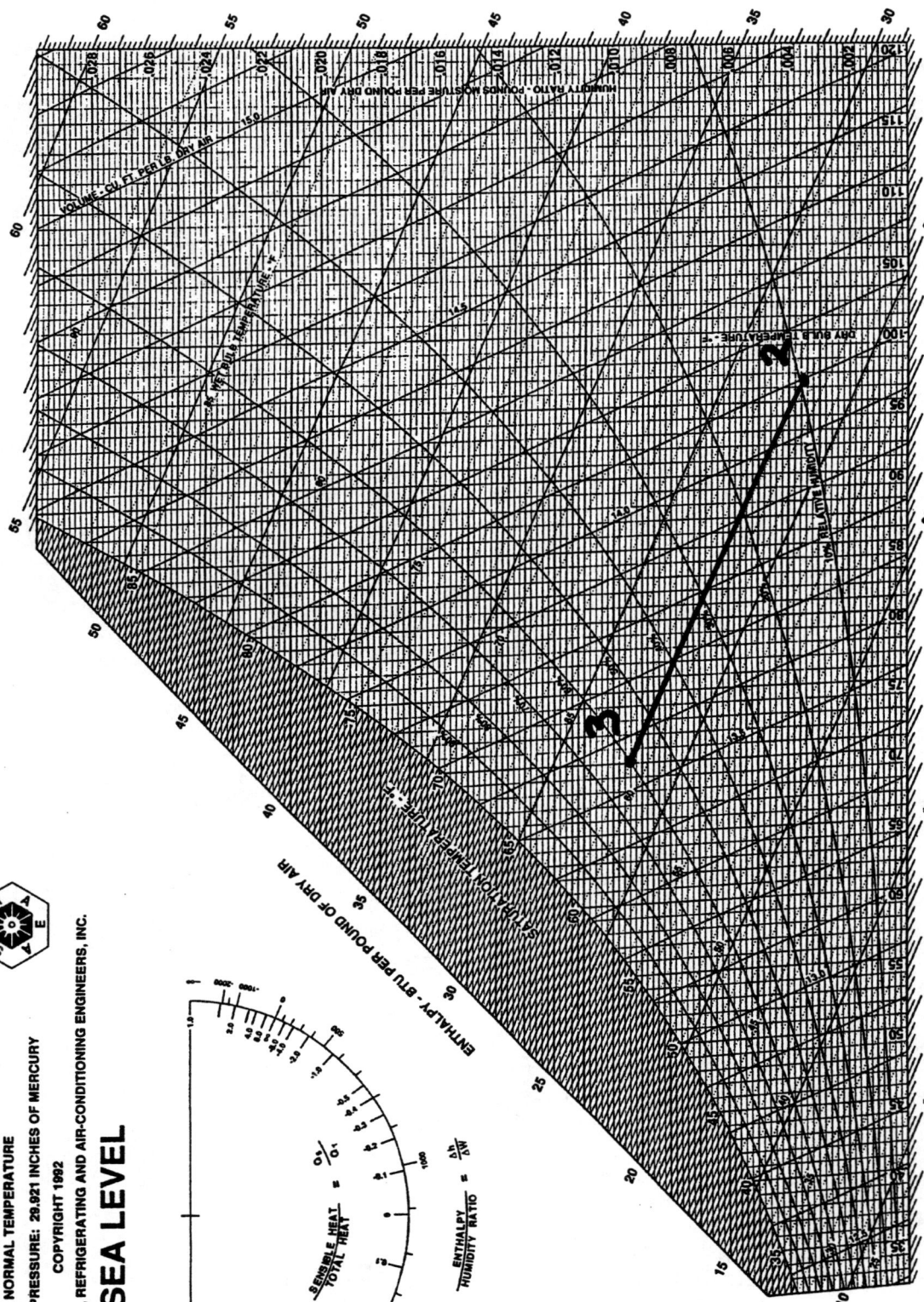

10.4

water ⓦ1 ↓ 100 L/s, 30°C

air ↑ ②

↓ water 25°C ⓦ2

air ↑ ①
25°C db
20°C wb

$\dot{m}_a = \dot{m}_w$

$h_D A_V = \dfrac{0.5\ kg_w/s\ m^3}{kg_w/kg_a}$

a) leaving air enthalpy (assumed to be twice as close to saturation as the incoming air, $1-\phi_2 = \tfrac{1}{2}(1-\phi_1)$, $\phi_2 = 82\%$)
By Eq. 10.20 we solve for h_2 and W_2 by trial (a more rigorous approach would be to plot the complete moist air process curve on a psychrometric chart)

$$h_2 = h_1 + \dfrac{\dot{m}_w c_w}{\dot{m}_a}(t_{w,1} - t_{w,2}) + (W_2 - W_1)h_{f,w,2}$$

By Chart C-8SI, $h_1 = 57.5\ kJ/kg_a$, $W_1 = 12.7\ g_w/kg_a$

By Table A.1 SI, $h_{f,w,2} = h_f(25°C) = 103.93\ kJ/kg_w$

By Table A.6 SI, $c_w = 4.183\ kJ/kg_w\ °C$

$$h_2 = 57.5\ \dfrac{kJ}{kg_a} + 1.0\ \dfrac{kg_w}{kg_a} \cdot \dfrac{4.183\ kJ}{kg_w\ °C}(30-25)°C + (W_2 - 12.7)\dfrac{g_w}{kg_a} \cdot \dfrac{103.93\ kJ/kg}{kg_w}\left(\dfrac{1}{1000}\right)$$

$h_2 = 57.5 + 20.92 + (W_2 - 12.7)\,0.104$

$h_2 = 78.42 + (W_2 - 12.7)\,0.104$

By Trial: $h_2 = 79.2\ kJ/kg_a$, $W_2 = 19.9\ g_w/kg_{ma}$

10.4 Con't

b) $\dot{m}_w' = \dfrac{\dot{m}_w c_{p,w}}{b_s}$

$b_s = \dfrac{100.006 - 76.592 \text{ kJ/kg}_w}{(30-25)\,°C} = \dfrac{4.68 \text{ kJ}}{\text{kg}_w\,°C}$

Then $\dot{m}_w' = \dfrac{100 \text{ L/s} (10^{-3} \text{ m}^3/\text{L})}{0.001005 \text{ m}^3/\text{kg}_w} \dfrac{4.183 \text{ kJ/kg}_w\,°C}{4.68 \text{ kJ/kg}_w\,°C} = 88.94 \dfrac{\text{kg}}{s}$

$\dot{m}_a = \dot{m}_w = \dfrac{100 \text{ L/s} (10^{-3} \text{ m}^3/\text{L})}{0.001005 \text{ m}^3/\text{kg}_w} = 99.5 \dfrac{\text{kg}_a}{s}$

$\dot{m}_w' < \dot{m}_a$ so $\dot{m}_{min} = \dot{m}_w'$ and $C = \dot{m}_w'/\dot{m}_a$

$C = 88.94/99.5 = 0.894$

$\varepsilon = \dfrac{t_{w,1} - t_{w,2}}{t_{w,1} - t_{wb,1}} = \dfrac{30-25}{30-20} = 0.5$

By Eq. (10.43)

$NTU = -\ln\left(\dfrac{1-0.5}{1-0.5(0.894)}\right)/(1-0.894) = 0.950$

By Eq. (10.40)

$V = \dfrac{0.950\,(88.94 \text{ kg}_w/s)}{0.5 \text{ kg}_w/s\,\text{m}^3 (\text{kg}_w/\text{kg}_a)} = 169 \text{ m}^3$

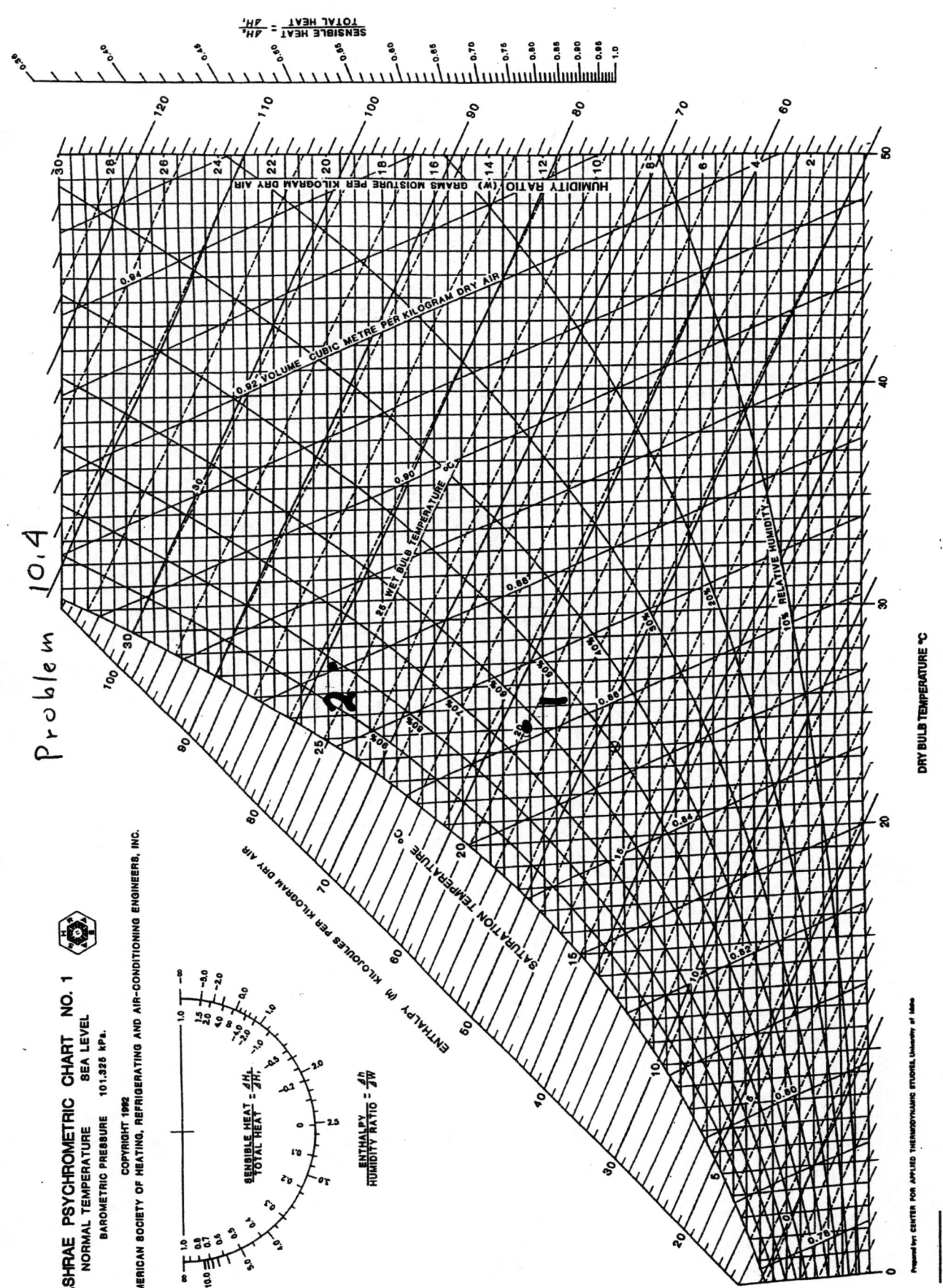

10.5

a) We need to generate the moist air process curve similar to that obtained in Ex. 10.3

By Table A.1SI at 30°C, $v_w = 0.001005$ m³/kg

$$\dot{m}_w = 100 \, L/s \left(\frac{10^{-3} m^3}{L}\right)\left(\frac{kg}{0.001005 \, m^3}\right) = 99.5 \, kg/s$$

Since $\dot{m}_w = \dot{m}_a$, $\dot{m}_a = 99.5$ kg/s

Locate inlet air state 1 on Chart C-8SI and saturated moist air state (s,w,2) at 30°C. (see also Table A.4SI)

By Table A.1SI, $h_{s,w} = 103.93$ kJ/kg, $h_{g,w} = 2546.73$ kJ.
By Eq. (9.7) and Table 9.1 at 25°C, degree of saturation $= 1.0$, we find $Le = 0.896$. With the chart protractor or using numerical values, we obtain

$$\frac{h_{s,w,2} - h_1}{W_{s,w,2} - W_1} = \frac{76.59 - 57.5}{20.21 - 12.6} = 2.51 \, kJ/g_w$$

By Eq. (10.18)

$$\frac{dh}{dw} = 0.896 \, (2.51) \frac{kJ}{g_w} + 2.54673 \frac{kJ}{g_w} - (2.501) \frac{kJ}{g_w} (0.896)$$

$$\frac{dh}{dw} = 2.55 \, kJ/g_w = \frac{\Delta h}{\Delta W}$$

Follow a line in this direction on Chart C-8SI and arbitrarily stop at $h_a = 62.0$ kJ/kg$_a$,

$$t_{w,a} = 25°C + \frac{kg_a/kg_w}{4.183 \, \frac{kJ}{kg_w \, °C}} \left[(62.0 - 57.5)\frac{kJ}{kg_a} - (14.3 - 12.6)\frac{g_w}{kg_a} \, \frac{103.93 \, kJ}{kg_w}\right]$$

220

10.5 con't

$t_{w,a} = 26.0°C$

Repeating the steps above to find state b:

$$\frac{80.798 - 62.0}{21.448 - 14.3} = 2.63 \text{ kJ/g}_w$$

$$\frac{dh}{dw} = 0.896(2.63) + 2.54855 - 2.501(0.896) = 2.66$$

let $h_b = 66.0$ kJ/kg$_a$, $t_b = 25.2°C$

$$t_{w,b} = 26°C + \frac{1}{4.183}\left[(66.0 - 62.0) - (15.9 - 14.3)\frac{108.15}{1000}\right] = 26.9°C$$

Repeating for state c:

$$\frac{84.928 - 66.0}{22.70 - 15.9} = 2.78 \text{ kJ/g}_w$$

$$\frac{dh}{dw} = 0.896(2.78) + 2.55018 - 2.501(0.896) = 2.80$$

let $h_c = 71.0$ kJ/kg$_a$, $t_c = 25.8°C$

$$t_{w,c} = 26.9°C + \frac{1}{4.183}\left[(71.0 - 66.0) - (17.7 - 15.9)\frac{111.94}{1000}\right] = 28.0°C$$

Repeating for state d:

$$\frac{89.976 - 71.0}{24.226 - 17.7} = 2.91$$

$$\frac{dh}{dw} = 0.896(2.91) + 2.55218 - 2.501(0.896) = 2.92$$

let $h_d = 76.0$ kJ/kg$_a$, $t_d = 26.4°C$

$$t_{w,d} = 28.0 + \frac{1}{4.183}\left[(76 - 71) - (19.3 - 17.7)\frac{116.39}{1000}\right] = 29.2°C$$

10.5 Con't

Repeating for state e:

$$\frac{95.994 - 76.0}{26.09 - 19.3} = 2.95$$

$$\frac{dh}{dw} = 0.896(2.95) + 2.55435 - 2.501(0.896) = 2.95$$

let $h_e = 81.0 \, kJ/kg_a$, $t_e = 27.2\,°C$

$$t_{w,e} = 29.2\,°C + \frac{1}{4.183}\left[(80.0 - 76.0) - (20.7 - 19.3)\frac{121.55}{1000}\right] = 30.1\,°C$$

This is nearly equal to the inlet water temperature (30°C) so let state ⓔ = state ②

Therefore,

$$h_2 = 80.0 \, kJ/kg_a$$

b) Following the procedure in Ex. 10.4

Using state c as an average value

$$\left.\frac{h_{f,w}}{dh/dw}\right|_m = \frac{116.39 \, kJ/kg_a}{2.92 \, kJ/g_w} = 39.86 \, g_w/kg_a$$

By Eq. (10.28)

$$h_2 = 57.5 \, \frac{kJ}{kg_a} + \frac{4.183 \, kJ/kg_w\,°C}{1 - 0.03986}\left(\frac{kg_w}{kg_a}\right)(5\,°C) = 79.3 \, \frac{kJ}{kg_a}$$

10.5 Con't

$$h_m = 57.5 + \frac{4.183\,(2.5)}{1 - 0.03986} = 68.4 \text{ kJ/kg}_a$$

At $t_{w,1} = 30°C$, $h_{s,w,1} = 100.0$ kJ/kg$_a$
at $t_{w,m} = 27.5°C$, $h_{s,w,m} = 87.7$ kJ/kg$_a$
at $t_{w,2} = 25°C$, $h_{s,w,2} = 76.6$ kJ/kg$_a$

Thus:
$$y_1 = 100.0 - 79.3 = 20.7 \text{ kJ/kg}_a$$
$$y_2 = 76.6 - 57.5 = 19.1 \text{ kJ/kg}_a$$
$$y_m = 87.7 - 68.4 = 19.3 \text{ kJ/kg}_a$$

$y_m/y_1 = 0.93$, $y_m/y_2 = 1.01$

By Fig. 10.7, $f = 1.01$

By Eqn. (10.27)

$$0.896 + \frac{2436 \text{ kJ/kg}_w - 2501 \text{ kJ/kg}_w (0.896)}{2.91 \text{ kJ/g}_w \,(1000)} = 0.963$$

By Eqs. (10.27) and (10.29)

$$V = \frac{99.5 \text{ kg}_w/s\,(4.183 \text{ kJ/kg}_w°C)\,(5°C)/(1.01)(19.3 \text{ kJ/kg}_a)}{0.5 \text{ kg}_w/s\,m^3\,(kg_w/kg_a)\;\;0.963}$$

$$V = 222 \text{ m}^3$$

10.5 con't

Air state	t °C	h kJ/kg a	W gw/kga	tw °C	$h_{f,w}$ kJ/kgw	Le	$\frac{h_{s,w}-h}{W_{s,w}-W}$ kJ/gw	$h_{g,w}$ kJ/kgw	dh/dW kJ/gw
1	25	57.5	12.6	25	103.93	0.896	2.51	2546.73	2.55
a	25.1	62.0	14.3	26.0	108.15	0.896	2.63	2548.55	2.66
b	25.2	66.0	15.9	26.9	111.94	0.896	2.78	2550.18	2.80
c	25.8	71.0	17.7	28.0	116.39	0.896	2.91	2552.18	2.92
d	26.4	76.0	19.3	29.2	121.55	0.896	2.95	2554.35	2.95
2	27.0	80.0	20.71	30.1					

Problem 10.5

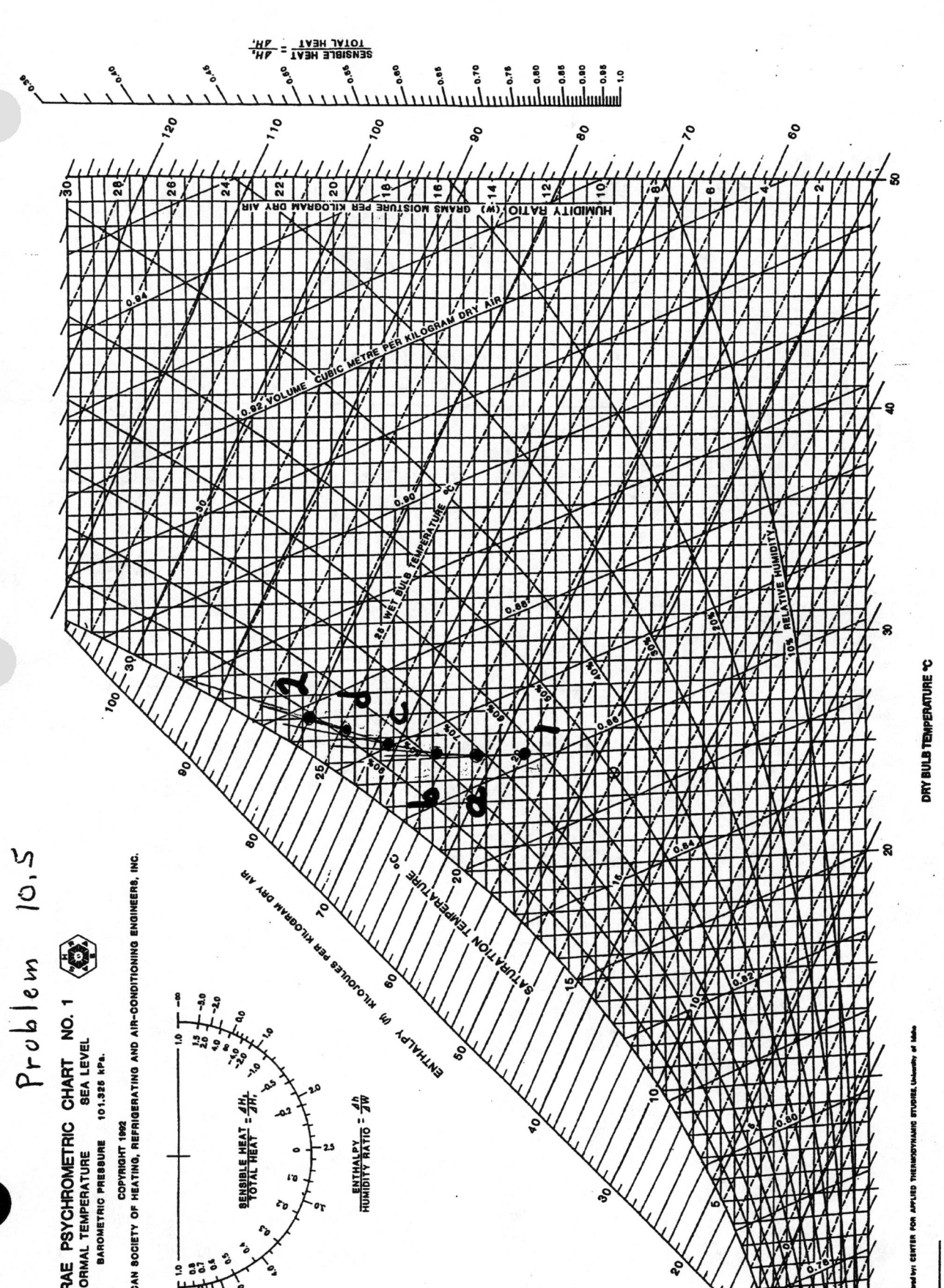

Chart C-8SI

10.6

a) The solution is identical to that for part a) in Problem 10.5

$h_2 = 80.0 \text{ kJ/kg}_a$, $W_2 = 20.7 \text{ g}_w/\text{kg}_a$, $t_2 = 27.0°C$

b) Follow the procedure in part c) of Example 10.3

Air state	t_w °C	W g_w/kg_a	$W_{s,w}$ g_w/kg_a	$W_{s,w} - W$ g_w/kg_a	$f(w)$ g_w/kg_a
1	25	12.6	20.21	7.61	0.131
a	26.0	14.3	21.45	7.15	0.140
b	26.9	15.9	22.70	6.80	0.147
c	28.0	17.7	24.23	6.53	0.153
d	29.2	19.3	26.09	6.79	0.147
2	30	20.7	27.33	6.63	0.151

$$F(w) = \frac{(14.3-12.6)(0.140+0.131)}{2} + \frac{(15.9-14.3)(0.147+0.141)}{2}$$

$$+ \frac{(17.7-15.9)(0.153+0.147)}{2} + \frac{(19.3-17.7)(0.147+0.153)}{2}$$

$$+ \frac{(20.7-19.3)(0.151+0.147)}{2} = 1.179$$

By Eq. (10.21)

$$V = \frac{99.5 \text{ kg}_a/s \, (1.179)}{0.5 \text{ kg}_w/s \, m^3 (\text{kg}_w/\text{kg}_a)} = 235 \text{ m}^3$$

This value is 39% larger than the volume obtained using the approximate NTU analysis in Problem 10.4.

10.7

Assume the conditions between the two sections are at state d as determined in Ex. 10.3.

Section I (top)

$\dot{m}_w + \dot{m}_a$ remain unchanged

$$b_s = \frac{h_{s,w,1} - h_{s,w,d}}{t_{w,1} - t_{w,d}} = \frac{71.761 - 59.20}{100 - 92.2} = 1.61$$

$$\dot{m}_w' = \frac{\dot{m}_a \, 1.0}{1.61} = 0.621 \, \dot{m}_a$$

$$C = 0.621$$

$$\varepsilon = \frac{t_{w,1} - t_{w,d}}{t_{w,1} - t_{wb,d}} = \frac{100 - 92.2}{100 - 82.2} = 0.438$$

$$NTU = -\ln\left(\frac{1 - 0.438}{1 - 0.438(0.621)}\right) / (1 - 0.621) = 0.68$$

$$V_I = \frac{0.68 (0.621)(746,000)}{120} = 2625 \, ft^3$$

Section II (bottom)

$$b_s = \frac{h_{s,w,d} - h_{s,w,2}}{t_{w,d} - t_{w,2}} = \frac{59.20 - 49.445}{92.2 - 85} = 1.35$$

$$\dot{m}_w' = \frac{\dot{m}_a \, 1.0}{1.35} = 0.741 \, \dot{m}_a$$

$$C = 0.741$$

10.7 Con't

$$\varepsilon = \frac{t_{w,d} - t_{w,2}}{t_{w,d} - t_{wb,1}} = \frac{92.2 - 85}{92.2 - 75} = 0.419$$

$$NTU = -\ln\left(\frac{1 - 0.419}{1 - 0.419(0.741)}\right) / (1 - 0.741) = 0.661$$

$$V_{II} = \frac{0.661\,(0.741)(746{,}000)}{120} = 3045 \text{ ft}^3$$

$$V_{Total} = V_I + V_{II} = 5670 \text{ ft}^3$$

This value is 21% less than the value obtained from the detailed analysis in Ex. 10.3 but 12% more than the single step NTU method results given in Ex. 10.5.

10.8

By chart C-8E
$h_A = 39.38$ Btu/lbma, $W_A = 0.0150$ lbmw/lbma
$h_B = 50.67$ Btu/lbma, $W_B = 0.02525$ lbmw/lbma
$v_B = 15.28$ ft³/lbma

By Table A.1E
$h_1 = 62.82$ Btu/lbmw, $v_1 = 0.01612$ ft³/lbma
$h_2 = 49.63$ Btu/lbmw,
$h_3 = 47.70$ Btu/lbmw, $v_3 = 0.01608$ ft³/lbma
$h_4 = 49.63$ Btu/lbmw

Performing a mass and energy balance on the Tower as shown in Fig. 10.13

$\dot{m}_{a_A} = \dot{m}_{a_B}$, $\dot{m}_4 = 0.2\,\dot{m}_3$, $\dot{m}_1 = \dot{m}_2$

$\dot{m}_1 + \dot{m}_{a_A} W_A + \dot{m}_3 = 0.01\,\dot{m}_1 + \dot{m}_{a_B} W_B + 0.2\,\dot{m}_3 + \dot{m}_2$

combining these
$\dot{m}_a (W_B - W_A) = 0.8\,\dot{m}_3 - 0.01\,\dot{m}_1$

$\dot{m}_a = (0.8\,\dot{m}_3 - 0.01\,\dot{m}_1)/(W_B - W_A)$ (1)

$\dot{m}_1 h_1 + \dot{m}_3 h_3 + \dot{m}_{a_A} h_A = \dot{m}_{a_B} h_B + \dot{m}_2 h_2 + \dot{m}_4 h_4 + 0.01\,\dot{m}_1$

combining with mass balance relations above

$0.99\,\dot{m}_1 h_1 + \dot{m}_3 h_3 + \dot{m}_a h_A = \dot{m}_a h_B + \dot{m}_1 h_2 + 0.2\,\dot{m}_3 h_2$ (2)

By Eqns. (1) and (2)

10.8 con't

$$0.99 \dot{m}_1 h_1 + \dot{m}_3 h_3 = \left(\frac{0.8 \dot{m}_3 - 0.01 \dot{m}_1}{W_B - W_A}\right)(h_B - h_A)$$

$$\dot{m}_3 = \dot{m}_1 \frac{\left[\frac{0.99 h_1 + 0.01(h_B - h_A)}{(W_B - W_A)} - h_2\right] + \dot{m}_1 h_2 + 0.2 \dot{m}_3 h_2}{\left[\frac{0.8(h_B - h_A)}{(W_B - W_A)} + 0.2 h_2 - h_3\right]}$$

$$\dot{m}_1 = \dot{m}_2 = \frac{2000 \text{ gal/min}}{0.01612 \text{ ft}^3/\text{lbm}_w}\left(\frac{35.32 \text{ ft}^3}{264.2 \text{ gal}}\right) = 16{,}590 \frac{\text{lbm}_w}{\text{min}}$$

$$\dot{m}_3 = 16{,}590 \left[\frac{\frac{0.99(62.82) + 0.01(50.67 - 39.38)}{(0.02525 - 0.0150)} - 49.63}{\frac{0.8(50.67 - 39.38)}{(0.02525 - 0.0150)} + 0.2(49.63) - 47.70}\right]$$

$$\dot{m}_3 = 464 \text{ lbm}_w/\text{min}$$

By Eq. (1), $\dot{m}_a = [0.8(464) - 0.01(16{,}590)]/(0.02525 - 0.0150)$

$$\dot{m}_a = 20{,}030 \text{ lbm}_a/\text{min}$$

$$\dot{V}_B = \dot{m}_a v_B = 20{,}030 \frac{\text{lbm}_a}{\text{min}}\left(\frac{15.28 \text{ ft}^3}{\text{lbm}_a}\right) = 306{,}060 \frac{\text{ft}^3}{\text{min}}$$

b) $\dot{V}_3 = \dot{m}_3 v_3 = 464 \frac{\text{lbm}_w}{\text{min}}\left(\frac{0.01608 \text{ ft}^3}{\text{lbm}_w}\right)\left(\frac{264.2 \text{ gal}}{35.32 \text{ ft}^3}\right)$

$$\dot{V}_3 = 55.8 \text{ gal/min}$$

10.9

a) By Table 10.1, use air state (e) as the mean value ($\approx 95°F$)

$$\left.\frac{h_{s,w}}{dh/dW}\right|_m \approx \frac{62.1 \text{ Btu/lbm}_w}{1163 \text{ Btu/lbm}_w} = 0.0534$$

By Eq. (10.28)

$$h_2 = 38.39 \text{ Btu/lbm}_a + \frac{10°F}{0.947 °F/\text{Btu/lbm}_a} = 48.95 \frac{\text{Btu}}{\text{lbm}_a}$$

$$h_m = 38.39 + 5/0.947 = 43.67 \text{ Btu/lbm}_a$$

By Table A.4E

$h_{s,w,1} = 71.761 \text{ Btu/lbm}_a$, $h_{s,w,m} = 63.343 \text{ Btu/lbm}_a$

$h_{s,w,2} = 55.951 \text{ Btu/lbm}_a$

Thus $y_1 = 71.761 - 48.95 = 22.81$
$y_2 = 55.951 - 38.39 = 17.56$
$y_m = 63.343 - 43.67 = 19.67$

$y_m/y_1 = 0.862$, $y_m/y_2 = 1.120$
By Fig. 10.7, $f = 1.02$
By Table 10.1 using air state (e)

$$\left[Le + \frac{h_{fg,w} - 1061\, Le}{(h_{s,w} - h)/(W_{s,w} - W)}\right]_m = \left[0.895 + \frac{1039.9 - 1061(0.895)}{\frac{(63.343 - 48.00)}{(0.036757 - 0.0233)}}\right]$$

$= 0.974$

By Eqns. (10.27) and (10.29)

10.9 Con't

$$\dot{V} = \frac{(746{,}000)(1)(10)}{120\,(0.974)(1.02)(19.67)} = 3181 \text{ ft}^3$$

b) By Table 10.1, use air state (c) as the mean value ($\approx 90°F$)

$$\left.\frac{h_{f,w}}{dh/dw}\right|_m \approx \frac{58.3}{1118} = 0.0521$$

$h_2 = 38.39 + 20/0.947 = 59.51$
$h_m = 38.39 + 10/0.947 = 48.95$
$h_{s,w,1} = 71.761,\quad h_{s,w,m} = 55.951$
$h_{s,w,2} = 43.701$
$Y_1 = 71.761 - 59.51 = 12.25$
$Y_2 = 43.701 - 38.39 = 5.31$
$Y_m = 55.951 - 48.95 = 7.00$
$Y_m/Y_1 = 0.572,\ Y_m/Y_2 = 1.318,\ f = 1.01$

$$\frac{0.895 + 1043 - 1061(0.895)}{55.951 - 44.00} = \frac{0.985}{0.031203 - 0.01975}$$

$$\dot{V} = \frac{(746{,}000)(1)(20)}{120\,(0.985)(1.01)(7.00)} = 17{,}850 \text{ ft}^3$$

10.10a) leaving water Temperature = 27°C
$h_{f,w,2} = 112.36$ kJ/kg$_a$

Procedure follows that of Prob. 10.5
$$\frac{h_{s,w,2} - h_1}{W_{s,w,2} - W_1} = \frac{85.387 - 57.5}{22.84 - 12.6} = 2.72 \text{ kJ/g}_w$$

$$\frac{dh}{dw} = 0.896(2.72) + 2.55036 - 2.501(0.896) = 2.75$$

let $h_a = 62.0$ kJ/kg$_a$, $t_a = 25.4$°C

$$t_{w,a} = 27.0 + \frac{1}{4.183}\left[(62.0 - 57.5) - (14.3 - 12.6)\frac{112.36}{1000}\right] = 28.0°C$$

Repeat for state b:
$$\frac{89.976 - 62.0}{24.226 - 14.3} = 2.82$$

$$\frac{dh}{dw} = 0.896(2.82) + 2.55218 - 2.501(0.896) = 2.85$$

let $h_b = 66.0$ kJ/kg$_a$, $t_b = 25.9$°C

$$t_{w,b} = 28.0 + \frac{1}{4.183}\left[(66.0 - 62.0) - (15.7 - 14.3)\frac{116.39}{1000}\right] = 28.9°C$$

Repeat for state c:
$$\frac{94.49 - 66.0}{25.62 - 15.8} = 2.90$$

$$\frac{dh}{dw} = 0.896(2.90) + 2.55318 - 2.501(0.896) = 2.91$$

let $h_c = 70.5$ kJ/kg$_a$, $t_c = 26.5$°C

$$t_{w,c} = 28.9 + \frac{1}{4.183}\left[(70.5 - 66.0) - (17.2 - 15.7)\frac{120.26}{1000}\right] = 29.9°C \approx 30°C$$

Thus $t_2 \approx 26.5$°C, $h_2 \approx 70.5$ kJ/kg$_a$

10.10 a

Air State	t °C	h kJ/kg_a	W g_w/kg_a	t_w °C	h_{s,w} kJ/kg_w	Le	$\frac{h_{s,w}-h}{W_{s,w}-W}$ kJ/g_w	h_{g,w} kJ/kg_w	dh/dW kJ/g_w
1	25.0	57.5	12.6	27.0	112.36	0.896	2.72	2550.36	2.75
a	25.4	62.0	14.3	28.0	116.39	0.896	2.82	2552.18	2.89
b	25.9	66.0	15.7	28.9	120.26	0.896	2.90	2553.18	2.91
2	26.5	70.5	17.2	30		0.896			

10.10 a) Cont'd

Following the procedure used in Example 10.4 and in Problem 10.5, with state b as an average value, $t_b \approx t_{ave} = 28.9°C$

$$\left[\frac{h_{s,w}}{dh/dw}\right]_m = \frac{120.26}{2.91 \times 10^3} = 0.0413$$

$$h_m = 57.5 + \frac{4.183(1.5)}{1 - 0.0413} = 64.0 \text{ kJ/kg}_a$$

At $t_{w,1} = 30°C$, $h_{s,w,1} = 100.0$ kJ/kg$_a$
at $t_{w,m} = 28.5°C$, $h_{s,w,m} = 92.5$ kJ/kg$_a$
at $t_{w,2} = 27°C$, $h_{s,w,2} = 85.4$ kJ/kg$_a$

Thus: $y_1 = 100.0 - 70.5 = 29.5$ kJ/kg$_a$
$y_2 = 85.4 - 57.5 = 27.9$ kJ/kg$_a$
$y_m = 92.5 - 64.0 = 29.5$ kJ/kg$_a$

$y_m/y_1 = 1.0$, $y_m/y_2 = 1.06$

By Fig. 10.7, $f = 0.99$

By Eqn. (10.27)

$$0.896 + \frac{2435 - 2501(0.896)}{2.90(1000)} = 0.963$$

$$V = \frac{99.5(4.183)(3)/(0.99)(29.5)}{0.5(0.963)} = 89 \text{ m}^3$$

Problem 10.10 a)

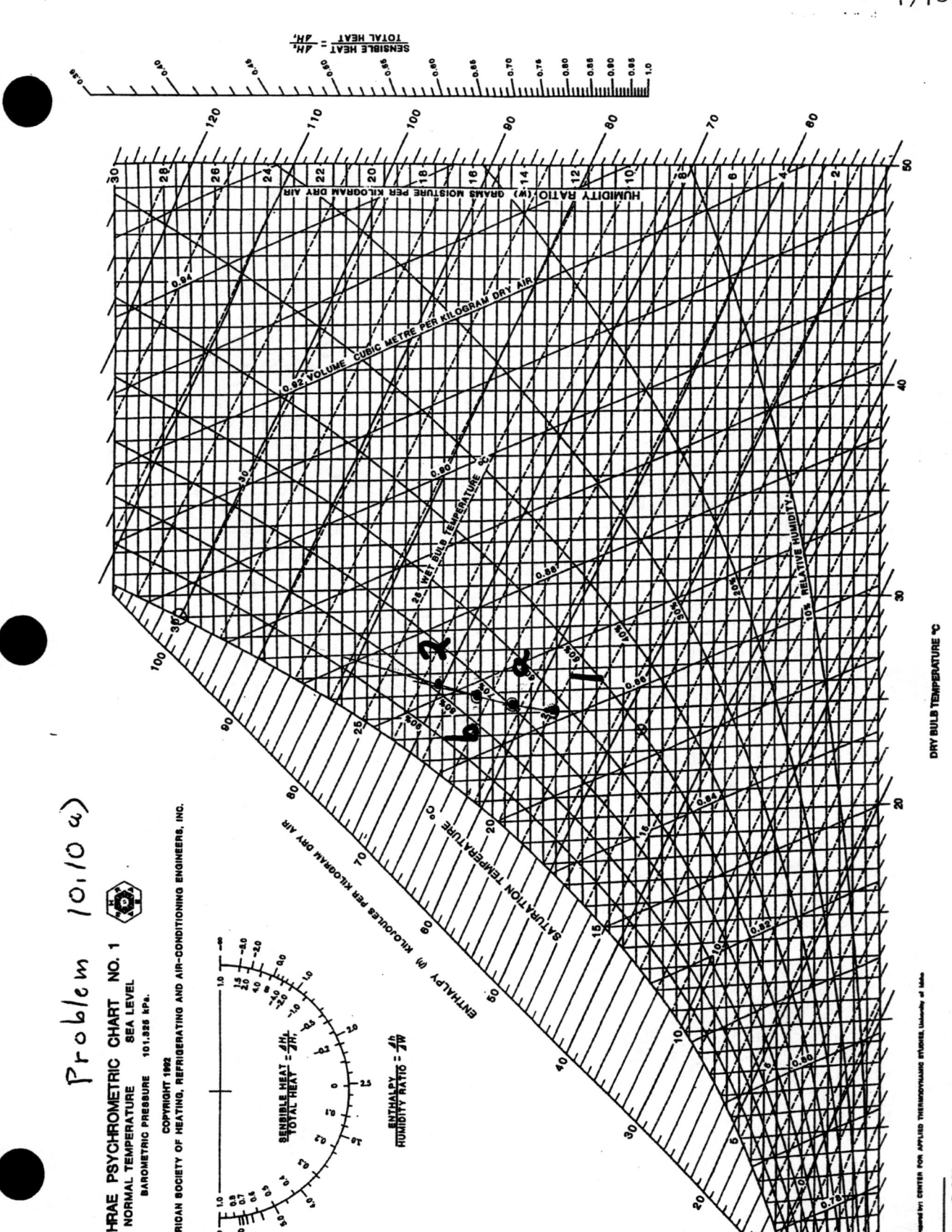

10.10b) leaving water temperature = 23°C

Procedure follows that of Prob. 10.5

$$\frac{h_{s,w,2} - h_1}{w_{s,w,2} - w_1} = \frac{68.52 - 57.5}{17.85 - 12.6} = 2.10$$

$$\frac{dh}{dw} = 0.896(2.10) + 2.54309 - 2.501(0.896) = 2.18$$

let $h_a = 60.0$ kJ/kg$_a$, $t_a = 24.7$°C

$$t_{w,a} = 23.0 + \frac{1}{4.183}\left[(60 - 57.5) - (13.8 - 12.6)\frac{95.51}{1000}\right] = 23.6 \text{ °C}$$

Repeat for state b:

$$\frac{70.84 - 60.0}{18.52 - 13.8} = 2.30$$

$$\frac{dh}{dw} = 0.896(2.30) + 2.54418 - 2.2409 = 2.36$$

let $h_b = 63.0$ kJ/kg$_a$, $t_b = 24.4$°C

$$t_{w,b} = 23.6 + \frac{1}{4.183}\left[(63.0 - 60) - (15.1 - 13.8)\frac{98.04}{1000}\right] = 24.3 \text{ °C}$$

Repeat for state c:

$$\frac{73.65 - 63.0}{19.34 - 15.1} = 2.51$$

$$\frac{dh}{dw} = 0.896(2.51) + 2.54546 - 2.2409 = 2.56$$

let $h_c = 66.0$ kJ/kg$_a$, $t_c = 24.4$°C

$$t_{w,c} = 24.3°C + \frac{1}{4.183}\left[(66.0 - 63.0) - (16.3 - 15.1)\frac{100.98}{1000}\right] = 25.0 \text{ °C}$$

10.10 b) cont'd

Repeat for state d:

$$\frac{76.59 - 66.0}{20.21 - 16.3} = 2.71$$

$$\frac{dh}{dw} = 0.896(2.71) + 2.54673 - 2.2409 = 2.74$$

let $h_d = 70.0$ kJ/kg$_a$, $t_d = 24.8°C$

$$t_{w,d} = 25.0 + \frac{1}{4.183}\left[(70.0-66.0) - (17.8-16.3)\frac{103.93}{1000}\right] = 25.9°C$$

Repeat for state e:

$$\frac{80.38 - 70.0}{21.32 - 17.8} = 2.95$$

$$\frac{dh}{dw} = 0.896(2.95) + 2.54837 - 2.2409 = 2.95$$

let $h_e = 76.0$ kJ/kg$_a$, $t_e = 25.7°C$

$$t_{w,e} = 25.9 + \frac{1}{4.183}\left[(76.0-70.0) - (19.7-17.8)\frac{107.73}{1000}\right] = 27.3°C$$

Repeat for state f:

$$\frac{86.76 - 76.0}{23.25 - 19.7} = 3.03$$

$$\frac{dh}{dw} = 0.896(3.03) + 2.55091 - 2.2409 = 3.02$$

let $h_f = 82.0$ kJ/kg, $t_f = 26.7°C$

$$t_{w,f} = 27.3 + \frac{1}{4.183}\left[(82.0-76.0) - (21.7-19.7)\frac{113.57}{1000}\right] = 28.7°C$$

10.10b) cont'd

Repeat for state g:

$$\frac{93.49 - 82.0}{25.31 - 21.7} = 3.18$$

$$\frac{dh}{dw} = 0.896(3.18) + 2.55091 - 2.2409 = 3.16$$

let $h_g = 88.0 \text{ kJ/kg}_a$, $t_g = 27.8°C$

$$t_{w,g} = 28.7 + \frac{1}{4.183}\left[(88.0 - 82.0) - (23.6 - 21.7)\frac{119.40}{1000}\right] = 30.1°C$$

As this is essentially the incoming water temperature, state ⑨ = state ②

Thus $t_2 \approx 27.8°C$, $h_2 \approx 88.0 \text{ kJ/kg}_a$

To determine the tower volume, let state ⓓ represent the average state in the tower;
$t_m \approx 24.8°C$

$$\left[\frac{h_{f,w}}{dh/dw}\right]_m = \frac{107.73}{2.95 \times 10^3} = 0.0365$$

$$h_m = 57.5 + \frac{4.183(3.5)}{1 - 0.0365} = 72.70 \text{ kJ/kg}_a$$

At $t_{w,1} = 30°C$, $h_{s,w,1} = 100.0 \text{ kJ/kg}_a$
$t_{w,m} = 26.5°C$, $h_{s,w,m} = 83.1 \text{ kJ/kg}_a$
$t_{w,2} = 23°C$, $h_{s,w,2} = 68.5 \text{ kJ/kg}_a$

10.10 b) Cont'd

$y_1 = 100.0 - 88.0 = 12.0$
$y_2 = 68.5 - 57.5 = 11.0$
$y_m = 83.1 - 72.7 = 10.4$

$y_m/y_1 = 0.867, \quad y_m/y_2 = 0.945$

By Fig. 10.7, $f = 1.03$

$$0.896 + \frac{2438 - 2501(0.896)}{2.95(1000)} = 0.963$$

$$V = \frac{99.5(4.183)(7)/(1.03)(10.4)}{0.5(0.963)} = 565 \text{ m}^3$$

10.10b

Air state	t °C	h kJ/kg$_a$	W g$_w$/kg$_a$	t_w °C	$h_{c,w}$ kJ/kg$_a$	Le	$\dfrac{h_{s,w}-h}{W_{s,w}-W}$ kJ/g$_w$	$h_{g,w}$ kJ/kg$_w$	dh/dW kJ/g$_w$
1	25	57.5	12.6	23	95.51	0.896	2.10	2543.09	2.18
a	24.7	60.0	13.8	23.6	98.04	0.896	2.30	2544.18	2.36
b	24.4	63.0	15.1	24.3	100.98	0.896	2.51	2545.46	2.56
c	24.4	66.0	16.3	25.0	103.93	0.896	2.71	2546.73	2.74
d	24.8	70.0	17.8	25.9	107.73	0.896	2.95	2548.37	2.95
e	25.7	76.0	19.7	27.3	113.57	0.896	3.03	2550.91	3.02
f	26.7	82.0	21.7	28.7	119.40	0.896	3.18	2553.45	3.16
2	27.8	88.0	23.6	30		0.896			

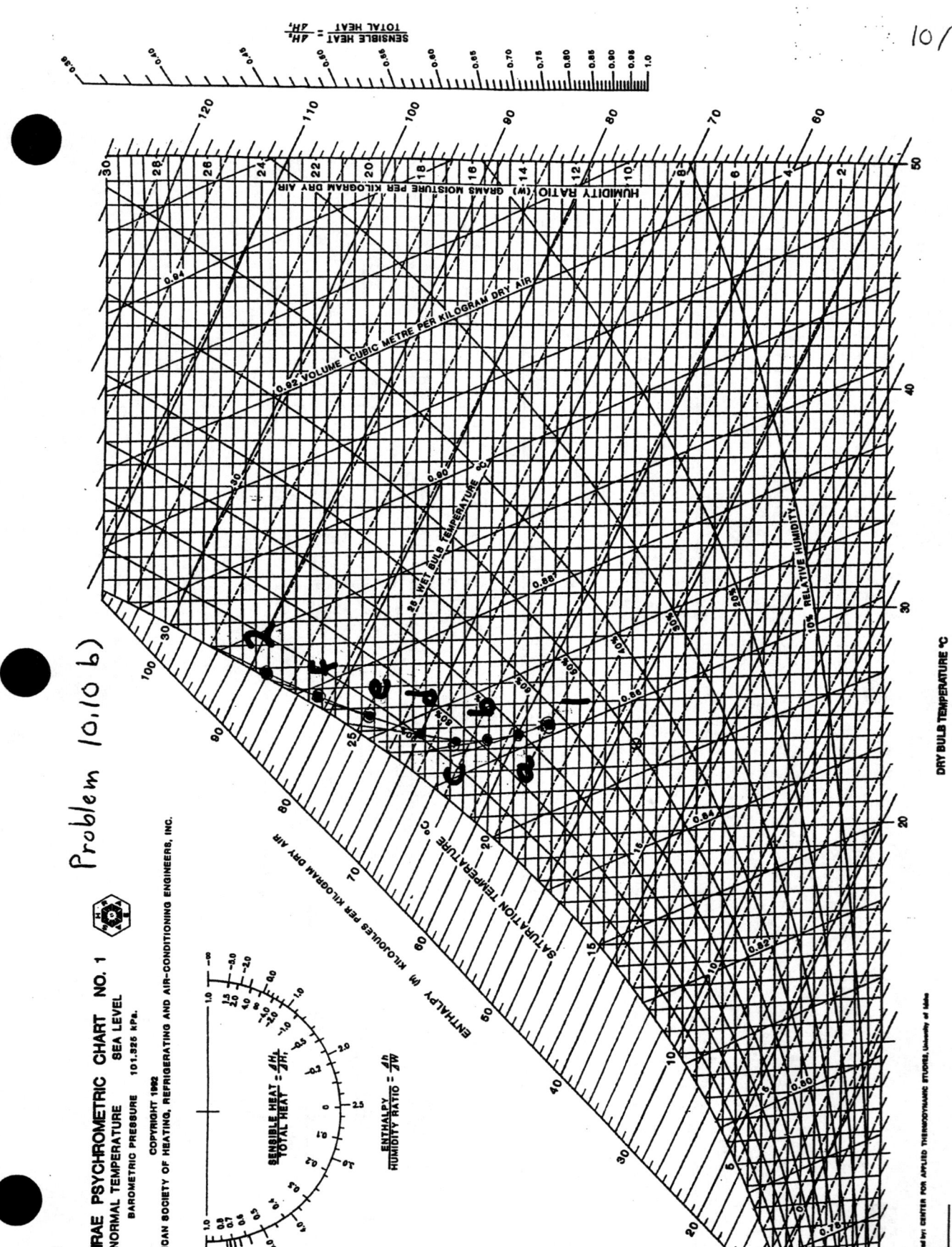

10.12

By Table A.1E at $45°F$, $v_{f,w} = 0.01602 \text{ ft}^3/\text{lbm}_w$

$$\dot{m}_w = \frac{(50 \text{ gal/min})}{(0.01602 \text{ ft}^3/\text{lbm}_w)} \left(\frac{35.32 \text{ ft}^3}{264.2 \text{ gal}}\right) = 417.2 \text{ lbm}_w/\text{min}$$

From chart C-8E, $v_{a,1} = 14.0 \text{ ft}^3/\text{lbm}_a$

$$\dot{m}_a = \frac{5000 \text{ ft}^3/\text{min}}{14.0 \text{ ft}^3/\text{lbm}_a} = 357 \text{ lbm}_a/\text{min}$$

$$\frac{\dot{m}_a}{\dot{m}_w c_w} = \frac{357 \text{ lbm}_a/\text{min}}{417.2 \text{ lbm}_w/\text{min} \, (1.0 \text{ Btu/lbm}_w°F)} = 0.856 \frac{\text{lbm}_a °F}{\text{Btu}}$$

The process line may be drawn on Chart C-8E using procedures similar to those shown in Example 10.3. Calculations are tabulated on the next page.

Thus $t_2 = 55.7°F$ and $t_2^* = 53.3°F$

Air State	t	h	W	t_w	$h_{f,w}$	Le	$\Delta h/\Delta W$	$h_{g,t}$	dh/dW
	°F	Btu/lbm$_a$	lbm$_w$/lbm$_a$	°F	Btu/lbm$_w$		Btu/lbm$_w$	Btu/lbm$_w$	Btu/lbm$_w$
1	85.0	33.95	0.01228	55.00	22.65	0.917	3550	1098	3375
a	79.8	32.00	0.01173	53.34	21.00	0.917	3250	1096	3103
b	74.6	30.00	0.01108	51.64	19.30	0.917	3020	1094	2891
c	69.4	28.00	0.01035	49.94	17.62	0.917	2850	1092	2729
d	64.5	26.00	0.00960	48.24	15.94	0.916	2710	1089	2597
e	59.9	24.00	0.00885	46.54	14.25	0.916	2600	1087	2494
2	55.7	22.19	0.00812	45.00	12.73				

10.13

We have $W_m = \dfrac{(W_1 + W_2)}{2} = \dfrac{0.01228 + 0.00812}{2} = 0.01020 \ \dfrac{lb_{mw}}{lb_{ma}}$

Estimate $t_{w,m} \approx 50°F$ at value of W_m above.

Thus $W_{s,w,m} = 0.007661 \ \dfrac{lb_{mw}}{lb_{ma}}$ from Table A.4E

By Eqns. (10.22 - 10.24)
$Y_1 = 0.006334 - 0.00812 = -0.00179 \ lb_{mw}/lb_{ma}$
$Y_2 = 0.009223 - 0.01228 = -0.00306$ "
$Y_m = 0.007661 - 0.01020 = -0.00254$ "

Thus $Y_m/Y_1 = 1.42$, $Y_m/Y_2 = 0.83$

By Fig. 10.7, $f = 0.961$

By Eq. (10.25)

$F(w) = \dfrac{(0.00812 - 0.01228)}{0.961(-0.00254)} = 1.704$

By Eq. (10.21)

$V = \dfrac{357 \ lb_{ma}/min \ (1.704)(60 \ min/hr)}{600 \ lb_{mw}/hr \ ft^3 (lb_{mw}/lb_{ma})} = 60.8 \ ft^3$

10.19

Procedure follows that given in Ex. 10.3

By Table A.1SI, $v_{w,1} = 0.001 \text{ m}^3/\text{kg}_w$

$$\dot{m}_w = \frac{3 \times 10^{-3} \text{ m}^3/\text{s}}{0.001 \text{ m}^3/\text{kg}_w} = 3.0 \text{ kg}_w/\text{s}$$

From chart C-8SI, $v_{a,1} = 0.854 \text{ m}^3/\text{kg}_a$

$$\dot{m}_a = \frac{2.5 \text{ m}^3/\text{s}}{0.854 \text{ m}^3/\text{kg}_a} = 2.93 \text{ kg}_a/\text{s}$$

By Table A.1SI, $h_{f,w} = 61.95 \text{ kJ/kg}_w$, $h_{g,w} = 2528.47 \text{ kJ/kg}_w$
By Eq. (9.7) and Table 9.1 we find $L_e = 0.896$
From Table A.4SI at 15°C, $h_{s,w,2} = 42.167 \text{ kJ/kg}_a$,
$W_{s,w,2} = 0.010713 \text{ kg}_w/\text{kg}_a$

$$\frac{h_{s,w,2} - h_1}{W_{s,w,2} - W_1} = \frac{42.167 - 57.4}{10.713 - 10.6} = -135 \text{ kJ/g}_w$$

By Eq. (10.18)
$$\frac{dh}{dw} = 0.896(-135) + 2.52847 - 2.501(0.896) = -121 \text{ kJ/g}_w$$

Follow a line in this direction from state ① and arbitrarily stop at $t_a = 26°C$ on chart C-8SI
By Eq. (10.19)
$$t_{w,a} = 15°C + \frac{2.93/3.0}{4.186}\left[(53.4 - 57.4) - (10.6 - 10.6) \, 61.95 \times 10^{-3}\right]$$

$t_{w,a} = 14.07°C$

10.14 Cont'd

Repeating above steps to find state b
From Table A.4SI at 14.07°C
$$h_{s,w,a} = 39.76 \text{ kJ/kg}_a, \quad W_{s,w,a} = 0.010110 \text{ kg}_w/\text{kg}_a$$

$$\frac{39.76 - 53.4}{10.110 - 10.6} = 27.84 \text{ kJ/g}_w$$

$$\frac{dh}{dw} = 0.896(27.84) + 2.52677 - (2.501)0.896 = 25.23 \text{ kJ/}$$

$$t_{w,b} = 14.07 + \frac{2.93/3}{4.186}\left[(48.1 - 53.4) - (10.15 - 10.6)58.07 \times 10^{-3}\right]$$

$$t_{w,b} = 12.84 \text{ °C}$$

Repeat to find state c
From Table A.4SI at 12.84°C
$$h_{s,w,b} = 36.36 \text{ kJ/kg}_a, \quad W_{s,w,b} = 0.009289 \text{ kg}_w/\text{kg}_a$$

$$\frac{36.36 - 48.1}{9.289 - 10.15} = 13.64 \text{ kJ/g}_w$$

$$\frac{dh}{dw} = 0.897(13.64) + 2.52451 - (2.501)0.897 = 12.51 \text{ kJ/g}_w$$

$$t_{w,c} = 12.84 + \frac{2.93/3}{4.186}\left[(42.9 - 48.1) - (9.7 - 10.15)52.94 \times 10^{-3}\right]$$

$$t_{w,c} = 11.63 \text{ °C}$$

Repeat to find state d
From Table A.4SI at 11.63°C
$$h_{s,w,c} = 33.29 \text{ kJ/kg}_a, \quad W_{s,w,c} = 0.008562 \text{ kg}_w/\text{kg}_a$$

$$\frac{33.29 - 42.9}{8.562 - 9.7} = 8.44 \text{ kJ/g}_w$$

10.14 Cont'd.

$$\frac{dh}{dw} = 0.899(8.44) + 2.52229 - (2.501)0.899 = 7.87 \text{ kJ/g}_w$$

$$t_{w,d} = 11.63 + \frac{2.93/3}{4.186}\left[(36.0 - 42.9) - (8.8 - 9.7)\,47.90 \times 10^{-3}\right]$$

$$t_{w,d} = 10.03 \,°C$$

As this nearly equals the incoming water temperature (10°C) let state ⓓ = state ②

From chart C-8SI

$t_2 = 13.8\,°C$

$t_{wb,2} = 12.8\,°C$

10.14 Cont'd

Air state	t °C	h kJ/kg$_a$	W g$_w$/kg$_a$	t_w °C	$h_{f,w}$ kJ/kg	Le	$\dfrac{h_{s,w}-h}{W_{s,w}^*-W}$ kJ/g$_w$	$h_{g,w}$ kJ/kg$_w$	dh/dW kJ/g$_w$
1	30	57.9	10.6	15	61.95	0.896	−135	2528.47	−121
a	26	53.4	10.6	14.07	58.07	0.896	27.84	2526.77	25.23
b	22	48.1	10.15	12.84	52.99	0.897	13.64	2524.51	12.51
c	18	42.9	9.7	11.63	47.90	0.899	8.44	2522.29	7.87
2	13.8	36.0	8.8	10.03					

248

Problem 10.14

ASHRAE PSYCHROMETRIC CHART NO. 1
NORMAL TEMPERATURE   SEA LEVEL
BAROMETRIC PRESSURE 101.325 kPa
COPYRIGHT 1992
AMERICAN SOCIETY OF HEATING, REFRIGERATING AND AIR-CONDITIONING ENGINEERS, INC.

Chart C-8SI

10.15

We will follow the procedure used in part c) of Example 10.3

Air state	t_w °C	W g_w/kg_a	$W_{s,w}$ g_w/kg_a	$W_{s,w} - W$ g_w/kg_a	$f(w)$ g_w/kg_a
1	15	10.6	10.71	0.11	9.09
a	14.07	10.6	10.06	-0.54	-1.85
b	12.84	10.15	9.29	-0.86	-1.16
c	11.63	9.7	8.56	-1.14	-0.88
2	10	8.8	7.66	-1.14	-0.88

$$F(w) = 0 + \frac{(10.15 - 10.6)(-1.16 - 1.85)}{2} + \frac{(9.7 - 10.15)(-0.88 - 1.16)}{2}$$

$$+ (8.8 - 9.7)(-0.88) = 1.93$$

By Eq. (10.21)

$$\dot{V} = \frac{2.93 \, kg_a/s \, (1.93)}{2.5 \, kg_w/s \, m^3 (kg_w/kg_a)} = 2.26 \, m^3$$

With an average air velocity of 1 m/s, the cross sectional area is

$$A = \dot{V}/\vec{v} = (2.5 \, m^3/s)/(1 \, m/s) = 2.5 \, m^2$$

The height becomes

$$H = V/A = 2.26 \, m^3 / 2.5 \, m^2 = 0.90 \, m$$

11.1

```
         50°F  ① saturated
      14.696 psia ↓ air (2000 cfm)
   water
30gpm ───→ ┌─────────┐ ──→ water
   180°F   │   hx    │    mixed Ⓑ
    Ⓐ     └─────────┘
           120°F ↓ air
              ② unmixed
```

a) leaving water temperature,
  from an energy balance on the hx

$$\dot{m}_{air}(h_2 - h_1) = \dot{m}_{water}(h_A - h_B) = \dot{m}_{water}\, c_{p\,water}(t_A - t_B)$$

solve for t_B

$$t_B = \frac{\dot{m}_{air}\, c_p\,(t_1 - t_2)}{\dot{m}_{water}\, c_{p\,water}} + t_A$$

$\dot{m}_{air} = \dot{V}_1 / v_1$, $v_1 = 13.0\ \text{ft}^3/\text{lbm}_a$ (chart C-8E)

$$= \frac{2000\ \text{ft}^3/\text{min}}{13.0\ \text{ft}^3/\text{lbm}_a} = 153.8\ \text{lbm}_a/\text{min}$$

$\dot{m}_{water} = \dot{V}_A / v_A$, $v_A \cong v_f(180°F) = 0.01651\ \text{ft}^3/\text{lbm}_w$
(Table A.1E)

$$= \frac{30\ \text{gal}}{\text{min}}\left(\frac{35.32\ \text{ft}^3}{264.2\ \text{gal}}\right) / \frac{0.01651\ \text{ft}^3}{\text{lbm}_w} = 242.9\ \frac{\text{lbm}_w}{\text{min}}$$

By Eq. (7-23), $c_p = c_{pa} + W\, c_{pw}$

From Fig. C-8E, $W_1 = W_2 = 0.0076\ \text{lbm}_w/\text{lbm}_a$

From Table 7.3, $c_{pa} = 0.240\ \text{Btu}/\text{lbm}_a\,°F$

$c_{pw} = 0.444\ \text{Btu}/\text{lbm}_w\,°F$

11.1 cont'd

$$c_p = 0.240 + 0.0076(0.444) = 0.2434 \text{ Btu/lbm}_a \, °F$$

use $c_{p_{water}} = 1 \text{ Btu/lbm}_w \, °F$

$$t_B = \frac{153.8 \text{ lbm}_a/\text{min} \,(0.2434 \text{ Btu/lbm}_a °F)(50-120)°F}{242.9 \text{ lbm}_w/\text{min} \,(1 \text{ Btu/lbm}_w °F)} + 180°$$

$$t_B = 169.2 \, °F$$

b) Value for F

By Fig. 11.5

$$R = \frac{t_A - t_B}{t_2 - t_1} = \frac{180 - 169.2}{120 - 50} = 0.154$$

$$P = \frac{t_2 - t_1}{t_A - t_1} = \frac{120 - 50}{180 - 50} = 0.538$$

$$F \cong 0.99$$

c) By Eq. (11.4)

$$\Delta t_{m,cf} = \frac{(t_2 - t_1)(R-1)}{\ln\left(\frac{1-P}{1-RP}\right)}$$

$$= \frac{(120-50)°F \,(0.154-1)}{\ln\left(\frac{1-0.538}{1-0.154(0.538)}\right)} = 86.36 \, °F$$

By Eq. (11.5) $\Delta t_m = 0.99(86.36 °F) = 85.5 °F$

d) Effectiveness, ε

Determine which fluid has the larger capacity rate

11.1 cont'd

$$\dot{m}_a c_p = 153.8 \frac{lb_{ma}}{min} \left(0.2434 \frac{Btu}{lb_{ma} \, °F}\right) = 37.43 \frac{Btu}{min \, °F}$$

$$\dot{m}_{water} \, c_{p_{water}} = 242.9 \frac{lb_{mw}}{min} \left(1 \, Btu/lb_{mw} °F\right) = 242.9 \frac{Btu}{min \, °F}$$

$C_{air} < C_{water}$ so $C_{min} = C_{air} = C_{cold}$

By Eqn. 11.19 $\varepsilon = P = 0.538$

e) $C_r = C_{min}/C_{max} = C_{air}/C_{water}$

$\quad = 37.43/242.9 = 0.154$

f) By Table 11.2

$$\varepsilon = \frac{1}{C_r} \cdot \left[1 - \exp\left(-C_r \left[1 - \exp(-NTU)\right]\right)\right]$$

Solve for NTU

$$1 - \varepsilon C_r = \exp\left(-C_r \left[1 - \exp(-NTU)\right]\right)$$

$$\frac{-1}{C_r} \ln(1 - \varepsilon C_r) = 1 - \exp(-NTU)$$

$$NTU = -\ln\left[1 + \frac{1}{C_r} \ln(1 - \varepsilon C_r)\right]$$

$$= -\ln\left[1 + \frac{1}{0.154} \ln(1 - 0.538(0.154))\right]$$

$$NTU = 0.825$$

11.1 cont'd

g) By Eq. 11.20 $NTU = U_o A_o / C_{air}$

$U_o A_o = NTU (C_{air})$

$= 0.825 \left(37.43 \dfrac{Btu}{min\,°F} \right) \left(\dfrac{60\,min}{hr} \right) = 1853 \dfrac{Btu}{hr\,°F}$

11.2

moist air ①
2 m³/s ↓ 21°C db, 13°C wb

water Ⓐ → [HX] → Ⓑ
70°C   60°C
        40°C ↓ ②

a) Mass flow rate of water
From an energy balance on the HX

$$\dot{m}_w c_{pw}(t_A - t_B) = \dot{m}_a c_{pa}(t_2 - t_1)$$

$$\dot{m}_w = \frac{\dot{m}_a c_{pa}}{c_{pw}} \frac{(t_2 - t_1)}{(t_A - t_B)}$$

$\dot{m}_a = \dot{V}_1 / v_1$, $v_1 = 0.841 \ m^3/kg_a$ (Chart C-8 SI)

$\dot{m}_a = (2 \ m^3/s)/(0.841 \ m^3/kg_a) = 2.38 \ kg_a/s$

By Eq. (7.23), Table 7.3 and Chart C-8 SI,

$$c_{pa} = \frac{1.00 \ kJ}{kg_a \cdot °C} + \left(0.006 \frac{kg_w}{kg_a}\right)\left(1.86 \frac{kJ}{kg_w \cdot °C}\right) = \frac{1.011 \ kJ}{kg_a \cdot °C}$$

By Table A.6 SI, $c_{pw} = 4.185 \ kJ/kg_w \cdot °C$

$$\dot{m}_w = \frac{2.38 \ kg_a/s \ (1.011 \ kJ/kg_a \cdot °C)}{4.185 \ kJ/kg_w \cdot °C} \frac{(40-21)°C}{(70-60)°C}$$

$\dot{m}_w = 1.092 \ kg_w/s$

11.2 Cont'd

b) $c_r = C_{min}/C_{max}$

$C_a = \dot{m}_a c_{pa} = 2.38\, kg_a/s\, (1.011\, kJ/kg_a\,°C) = 2.41\, kJ/s\,°C$

$C_w = \dot{m}_w c_{pw} = 1.092\, kg_w/s\, (4.185\, kJ/kg_w\,°C) = 4.57\, kJ/s\,°C$

$c_r = 2.41/4.57 = 0.527$

$C_{min} = C_a$

Assume HX is crossflow with C_{min} unmixed and C_{max} mixed

By Eq. (11.19)

$$\varepsilon = \frac{40-21}{70-21} = 0.388$$

By Table 11.2, cross-flow, C_{min} unmixed, C_{max} mixed, rearranging and solving for NTU

$$NTU = -\ln\left[1 + \frac{\ln(1-c_r\varepsilon)}{c_r}\right]$$

$$= -\ln\left[1 + \frac{\ln(1-0.527(0.388))}{0.527}\right]$$

$NTU = 0.569$

11.2 cont'd

c) $U_o A_o = NTU (C_a)$

$= 0.569 (2.41 \, kJ/s\,°C) = 1.37 \, kW/°C$

d) By Fig. 11.5

$R = \dfrac{70-60}{40-21} = 0.526$

$P = \dfrac{40-21}{70-21} = 0.388 = \varepsilon$

$F = 0.985$

e) $\Delta t_m = \Delta t_{m,cf} F = \dfrac{(t_B - t_1) - (t_A - t_2)}{\ln\left(\dfrac{t_B - t_1}{t_A - t_2}\right)} F$

$= \dfrac{(60-21) - (70-40)}{\ln\left(\dfrac{60-21}{70-40}\right)} \, 0.985 = 33.8\,°C$

f) $\dot{Q} = \dot{m}_a c_{pa} (t_2 - t_1)$ (or $\dot{m}_w c_{pw} (t_A - t_B)$)

$= \dfrac{2.38 \, kg_a}{s} \left(\dfrac{1.011 \, kJ}{kg_a \cdot °C}\right)(40-21)\,°C$

$\dot{Q} = 45.7 \, kW$

11.3

Diagram: saturated moist air entering at ① 50°F, 14.696 psia, 2000 scfm. Saturated steam 10 psig enters at (A). Saturated water 10 psig exits at (B). Exit ② at 120°F.

a) leaving water temperature,

$$t_B = t_{sat}(10\,psig) = t_{sat}(24.696\,psia)$$

By Table A.1E, $t_B = 239.38°F$

b) $F = 1.0$ as one of the fluids is isothermal

c) By Eq. (11.18)

$$\Delta t_m = \frac{(120-50)}{\ln\left(\frac{239.38-50}{239.38-120}\right)} = 151.7\,°F$$

d) By Eq. (11.19)

$$\varepsilon = \frac{120-50}{239.38-50} = 0.370$$

e) $C_r = 0$ as $C_{max} = \infty$ (condensing steam)

f) Rearranging Eq. (11.25)

$$NTU = -\ln(1-\varepsilon) = -\ln(1-0.370) = 0.462$$

11.3 cont'd

g) By Eq. (11.20)

$$U_o A_o = NTU (C_c)$$

$$C_c = \dot{m}_a c_{pa}$$

$$\dot{m}_a = \dot{V}_1 / v_1, \quad v_1 = 13.0 \text{ ft}^3/\text{lb}_m \quad (\text{chart C-8E})$$

$$\dot{m}_a = \frac{2000 \text{ ft}^3/\text{min}}{13.0 \text{ ft}^3/\text{lb}_{ma}} = 153.8 \text{ lb}_{ma}/\text{min}$$

By Eq. (7.23), $c_{pa} = c_{p,a} + W c_{p,w}$

From Chart C-8E, $W_1 = W_2 = 0.0076 \text{ lb}_{mw}/\text{lb}_{ma}$

From Table 7.3, $c_{p,a} = 0.240 \text{ Btu}/\text{lb}_{ma} \, °F$

$$c_{p,w} = 0.444 \text{ Btu}/\text{lb}_{mw} \, °F$$

$$c_{pa} = 0.240 + 0.0076(0.444) = 0.2434 \text{ Btu}/\text{lb}_{ma} \, °F$$

$$C_c = 153.8 \, \frac{\text{lb}_{ma}}{\text{min}} \left(0.2434 \, \frac{\text{Btu}}{\text{lb}_{ma} \, °F}\right) = 37.43 \, \frac{\text{Btu}}{\text{min} \, °F}$$

$$U_o A_o = 0.462 \left(37.43 \, \frac{\text{Btu}}{\text{min} \, °F}\right)\left(\frac{60 \text{ min}}{\text{hr}}\right) = 1038 \, \frac{\text{Btu}}{\text{hr} \, °F}$$

11.4

moist air ① 21°C db, 13°C wb
2 m³/s ↓ 101.325 kPa

R-22 sat vapor 1.942 MPa → [HX] → R-22 mixed sat liq. 1.942 MPa
(A) ↓ (B)
moist air ② 40°C unmixed

a) Mass flow rate of R-22
from an energy balance on the HX

$$\dot{m}_{R-22}(h_A - h_B) = \dot{m}_a c_p (t_2 - t_1)$$

$$\dot{m}_{R-22} = \frac{\dot{m}_a c_p (t_2 - t_1)}{h_A - h_B}$$

$\dot{m}_a = \dot{V}_1 / v_1$, $v_1 = 0.841$ m³/kg$_a$ (chart C-8SI)

$= 2 \frac{m^3}{s} / 0.841$ m³/kg$_a$ = 2.38 kg$_a$/s

By Eq. (7.23), Table 7.3 and Chart C-8SI,

$$c_{pa} = 1.00 \frac{kJ}{kg_a \cdot °C} + \left(0.006 \frac{kg_w}{kg_a}\right)\left(1.86 \frac{kJ}{kg_w \cdot °C}\right) = 1.011 \frac{kJ}{kg_a \cdot °C}$$

By Table A.3SI, $h_A = 262.43$ kJ/kg, $h_B = 107.85$ kJ/kg
(50°C sat)

$$\dot{m}_{R-22} = \frac{2.38 \text{ kg}_a/s (1.011 \text{ kJ/kg}_a°C)(40-21)°C}{(262.43 - 107.85) \text{ kJ/kg}}$$

$\dot{m}_{R-22} = 0.296$ kg/s

11.4 Cont'd

b) As the refrigerant remains isothermal, $c_r = 0$

By Eq. 11.25, $NTU = -\ln(1-\varepsilon)$

$$\varepsilon = \frac{t_2 - t_1}{t_A - t_1} = \frac{40-21}{50-21} = 0.655$$

$NTU = -\ln(1 - 0.655) = 1.064$

c) $NTU = U_o A_o / C_{min}$ or $U_o A_o = NTU(C_{min})$

$C_{min} = C_{air} = \dot{m}_a c_p = 2.38 \frac{kg_a}{s} (1.011 \, kJ/kg_a \cdot °C) = 2.41 \frac{kW}{°C}$

$U_o A_o = 1.064 \left(2.41 \frac{kW}{°C}\right) = 2.56 \frac{kW}{°C}$

d) As the refrigerant is isothermal, $F \equiv 1$

e) $\Delta t_m = \Delta t_{m,cf} = \dfrac{t_2 - t_1}{\ln\left(\dfrac{t_B - t_1}{t_A - t_2}\right)} = \dfrac{(40-21)°C}{\ln\left(\dfrac{50-21}{50-40}\right)} = 17.8 °C$

f) $\dot{Q} = \dot{m}_a c_p (t_2 - t_1)$ (or $\dot{m}_{R-22}(h_A - h_B)$)

$= C_{air}(t_2 - t_1)$

$= 2.41 \frac{kW}{°C}(40-21)°C = 45.8 \, kW$

11.5

a) copper fins

By Fig. 11.11, with $\frac{r_2}{r_1} = \frac{1+0.5}{0.5} = 3.0$

$L\sqrt{h_{c,o}/ky} = 0.70$

Rearranging to solve for y

$$y = \frac{h_{c,o}}{k}\left(\frac{L}{0.7}\right)^2$$

By Table 11.3, $k = 223$ Btu/hr ft °F

$y = \frac{10 \text{ Btu/hr ft}^2{}°F}{223 \text{ Btu/hr ft °F}}\left(\frac{1.0 \text{ in}}{0.7}\right)^2\left(\frac{ft}{12 \text{ in}}\right) = 0.0076 \text{ in}$

or fin thickness, $2y$, is 0.0152 in.

b) By Table 11.3, $k = 120$ Btu/hr ft °F

$y = \frac{10}{223}\left(\frac{1.0}{0.7}\right)^2 \frac{1}{12} = 0.0142 \text{ in}$

or $2y = 0.0284$ in

11.6 Find row spacing for rectangular plate fin so fin efficiency is 0.7, $r_1 = 4$ mm, tube spacing (c) = 2 cm, y = 0.05 mm and $h_{c,o} = 50$ W/m²°C

a) copper fins (k = 223 Btu/hr ft°F = 386 W/m°C)
(see Example 11.3)

By Fig. 11.11, find value of r_2 so $\phi = 0.7$

1st Trial: let $r_2/r_1 = 4.0$, $L\sqrt{h_{c,o}/ky} = 0.75$

$L = r_2 - r_1 = 0.75 / \sqrt{\dfrac{50 W}{m^2 °C} / \dfrac{386 W}{m°C} \cdot 0.05 mm \left(\dfrac{m}{10^3 mm}\right)}$

$r_2 - r_1 = 0.75/50.9 = 0.0147 m = 14.7 mm$
$r_2 = 14.7 + 4 = 18.7 mm$
check: $r_2/r_1 = 18.7/4 = 4.68$

2nd Trial: let $r_2/r_1 = 4.5$, $L\sqrt{h_{c,o}/ky} = 0.7$

$L = r_2 - r_1 = 0.7/50.9 = 0.0138 m = 13.8 mm$
$r_2 = 13.8 + 4 = 17.8 mm$
check: $r_2/r_1 = 17.8/4 = 4.5$

By Eq. 11.32, $a = \dfrac{r_2^2 \pi}{c}$

$a = \dfrac{(17.8 mm)^2 \pi}{2 cm} \left(\dfrac{cm^2}{100 mm^2}\right) = 5.0 cm$

11.6 Cont'd

b) Aluminum fins ($k = 120$ Btu/hr ft°F = 208 W/m°C)
(see Ex. 11.3)

1st Trial: $r_2/r_1 = 4.0$, $L\sqrt{h_{c,o}/ky} = 0.75$

$$L = r_2 - r_1 = 0.75 / \sqrt{\frac{50 W}{m^2 °C} / \frac{208 W}{m °C} \cdot 0.05 mm \left(\frac{m}{10^3 mm}\right)}$$

$r_2 - r_1 = 0.75 / 69.3 = 0.0108\ m = 10.8\ mm$

$r_2 = 10.8 + 4 = 14.8\ mm$
check: $r_2/r_1 = 14.8/4 = 3.7$

2nd Trial: $r_2/r_1 = 3.5$, $L\sqrt{h_{c,o}/ky} = 0.8$

$r_2 - r_1 = 0.8 / 69.3 = 0.0115\ m = 11.5\ mm$

$r_2 = 11.5 + 4 = 15.4\ mm$
check: $r_2/r_1 = 15.4/4 = 3.85$

let $r_2/r_1 = 3.75$, $r_2 = 3.75 \times 4\ mm = 15\ mm$

By Eq. 11.32

$$a = \frac{(15 mm)^2 \pi}{2 cm}\left(\frac{cm^2}{100 mm^2}\right) = 3.5\ cm$$

11.7 Find U_o for finned tube HX, circular plate fins, Aluminum, $D_2 = 2$ in, $D_1 = 0.530$ in, $D_i = 0.480$ in, copper tube, 84 fins/ft, $h_o = 9$ Btu/hr ft²°F, $h_i = 1000$ Btu/hr ft²°F, $y = 0.005$ in

Need terms included in Eq. 11.39 (see Ex. 11.3)

Fin efficiency: By Fig. 11.11
$$r_2/r_1 = D_2/D_1 = 2\text{in}/0.530\text{in} = 3.77$$

$$L\sqrt{h_{c,o}/ky} = \frac{(2-0.530)\text{in}}{2}\sqrt{\frac{9\text{Btu}}{\text{hr ft}^2°F} \bigg/ \frac{120\text{Btu}}{\text{hr ft}°F} \cdot 0.005\text{in}} \left(\frac{12\text{in}}{\text{ft}}\right)$$

$$= 0.822, \quad \phi = 0.68$$

Take a 1 ft long section to determine areas

$$A_{P,i} = \pi D_i \ell = \pi \, 0.480\text{in} \,(1\text{ft})\left(\frac{\text{ft}}{12\text{in}}\right) = 0.126 \text{ ft}^2/\text{ft}$$

$$A_{P,m} = \frac{\pi(D_1+D_i)}{2}\ell = \frac{\pi(0.530+0.480)}{2}\cdot\frac{1}{12} = 0.132 \text{ ft}^2/\text{ft}$$

$$A_{P,o} = \pi D_1 (\ell - 84(2y)) = \pi \, 0.530\text{in}\,(12\text{in} - 84(2\times 0.005\text{in}))\left(\frac{\text{ft}^2}{144\text{in}^2}\right)$$
$$= 0.129 \text{ ft}^2/\text{ft}$$

$$A_F = \frac{\pi(D_2^2 - D_1^2)}{4}\,2\,(84) = \frac{\pi(2^2 - 0.530^2)\text{in}^2}{4}\,2\,(84)\left(\frac{\text{ft}^2}{144\text{in}^2}\right)$$

$$= 3.408 \text{ ft}^2/\text{ft}$$

$$A_o = A_{P,o} + A_F = 0.129 + 3.408 = 3.537 \text{ ft}^2/\text{ft}$$

11.7 Cont'd

Substituting into Eq. 11.39

$$U_o = \left[\frac{3.537}{0.126(1000)} + \frac{3.537(0.530-0.480)}{0.132(2)(12)(223)} + \frac{1-0.68}{9\left(\frac{0.129}{3.408}+0.68\right)} + \frac{1}{9} \right]^{-1}$$

$$U_o = \left[0.028 + 2.5 \times 10^{-4} + 0.050 + 0.111 \right]^{-1}$$

$$U_o = 0.189^{-1} = 5.28 \frac{Btu}{hr\ ft^2\ °F}$$

11.8

By Eq. (11.39) following the procedure used in Example 11.3

Fin Efficiency: By Fig. 11.13 and Eq. (11.32)

$$r_2 = \sqrt{\frac{ac}{\pi}}$$

$$= \sqrt{\frac{2.0\,cm \cdot 2.5\,cm}{\pi}} = 1.26\,cm, \quad \frac{r_2}{r_1} = \frac{1.26}{0.5} = 2.52$$

$$L\sqrt{h_{c,o}/ky} = 0.76\,cm \sqrt{\frac{50\,W/m^2\,°C}{\left(\frac{W/m\,°C}{0.5778\,Btu/hr\,ft°F}\right)\left(\frac{m}{1000\,mm}\right)\left(\frac{m}{100\,cm}\right)}} = 0.53$$

By Fig. 11.11, $\phi = 0.88$

Take a 1-m long section to determine areas

$$A_{P,i} = \pi D_i \ell = \pi (1.0 - 0.1)\,cm\,(1\,m)(1\,m/100\,cm) = 0.0283\,m^2$$

$$A_{P,m} = \pi \frac{(D_i + D_o)}{2} \ell = \pi \frac{(0.9 + 1)}{2}\,cm\,(1\,m)(1\,m/100\,cm) = 0.0298\,m^2$$

$$A_{P,o} = \pi D_o (\ell - 400(2y)) = \pi (1\,cm)(1\,m - 400(2)0.1\,mm\,(1\,m/1000\,mm))(1\,m/100\,cm) = 0.0289\,m^2$$

11.8 con't

$$A_F = \left(ac - \frac{\pi D_o^2}{4}\right)(2)(400)$$

$$= \left(2.0\,cm\,(2.5\,cm) - \frac{\pi(1.0\,cm)^2}{4}\right)(2)(400)\left(\frac{m^2}{10^4\,cm^2}\right)$$

$$= 0.3372\ m^2$$

$$A_o = A_{P,o} + A_F = 0.0289 + 0.3372 = 0.3661\ m^2$$

By Eq. (2.52)

$$h_i = \frac{k}{D_i}\ 0.023\left(\frac{\vec{V} D_i}{\nu}\right)^{0.8}\left(\frac{\mu c_p}{k}\right)^{0.3}$$

By Table A.6 SI at 60°C
$c_p = 4.183\ kJ/kg\,°C$
$\mu = 466.5 \times 10^{-6}\ kg/m\,s$
$k = 0.6543\ W/m\,°C$

By Table A.1 SI at 60°C
$\rho = 983.3\ kg/m^3$

$$\nu = \mu/\rho = \frac{466.5 \times 10^{-6}\ kg/m\,s}{983.3\ kg/m^3} = 4.744 \times 10^{-7}\ m^2/s$$

$$\frac{\vec{V} D_i}{\nu} = Re = \frac{1\,m/s\,(0.9\,cm)}{4.744 \times 10^{-7}\ m^2/s}\left(\frac{1\,m}{100\,cm}\right) = 18,971$$

$$\frac{\mu c_p}{k} = Pr = \frac{466.5 \times 10^{-6}\ kg/m\,s\,(4.183\ kJ/kg\,°C)}{0.6543 \times 10^{-3}\ kJ/s\,m\,°C} = 2.98$$

11.8 Cont.

$$h_i = \frac{0.6543 \text{ W/m}°\text{C}}{0.9 \text{ cm } (1\text{m}/100\text{cm})} 0.023 (18,971)^{0.8} (2.98)^{0.3}$$

$$h_i = 6138 \text{ W/m}^2°\text{C}$$

Substituting values into Eq. (11.39)

$$U_o = \left[\frac{0.3661}{0.0283(6138)} + \frac{0.3661(5 \times 10^{-4})}{0.0298(223/0.5778)} \right.$$

$$\left. + \frac{1 - 0.88}{50((0.0289/0.3372) + 0.88)} + \frac{1}{50} \right]^{-1}$$

$$U_o = \left[2.108 \times 10^{-3} + 1.592 \times 10^{-5} + 2.485 \times 10^{-3} + 2.00 \times 10^{-2} \right]^{-1}$$

$$U_o = 40.6 \text{ W/m}^2°\text{C}$$

11.9

All quantities remain as determined in Prob. 11.8 except for those recalculated here

a) $A_{P,o} = \pi(1cm)\left(1m - \dfrac{200(2)0.1mm}{1000mm/m}\right)\left(\dfrac{1m}{100cm}\right) = 0.0302\ m$

$A_F = 0.3372 \div 2 = 0.1686\ m^2$

$A_o = 0.0302 + 0.1686 = 0.1988\ m^2$

$U_o = \left[\dfrac{0.1988}{0.0283(6138)} + \dfrac{0.1988(5\times 10^{-4})}{0.0298(223/0.5778)} + \dfrac{1-0.88}{70((0.0302/0.1686)+0.88)} + \dfrac{1}{70}\right]^{-1}$

$U_o = \left[1.144\times 10^{-3} + 8.643\times 10^{-6} + 1.619\times 10^{-3} + 1.429\times 10^{-2}\right]^{-1}$

$U_o = 58.6\ W/m^2\ °C$

b) $\dfrac{\vec{V}D_i}{\nu} = R_e = 18,971 \times 2 = 37,942$

$h_i = \dfrac{0.6543(0.023)(37,942)^{0.8}(2.98)^{0.3}}{0.9/100} = 10,686\ W/m^2$

$U_o = \left[\dfrac{0.3661}{0.0283(10,686)} + 1.592\times 10^{-5} + 2.485\times 10^{-3} + 2.00\times 10^{-2}\right]^{-1}$

$U_o = 42.2\ W/m^2\ °C$

11.9 Cont'd.

c) $\sqrt{\frac{ac}{\pi}}/r_1 = \sqrt{\frac{3(2.5)}{\pi}}/0.5 = 3.09$

$L\sqrt{h_{c,o}/k_y} = \frac{1.05(0.53)}{0.76} = 0.73$

By Fig. 11.11, $\phi = 0.76$

$A_F = \left(3.0(2.5) - \frac{\pi}{4}\right)\frac{(2)(400)}{10^4} = 0.5372 \text{ m}^2$

$A_o = 0.0289 + 0.5372 = 0.5661 \text{ m}^2$

$U_o = \left[\frac{0.5661}{0.0283(6138)} + \frac{0.5661(5\times10^{-4})}{0.0298(223/0.5778)}\right.$

$\left. + \frac{1-0.76}{50((0.0289/0.5372)+0.76)} + 2.00\times10^{-2}\right]^{-1}$

$U_o = \left[3.259\times10^{-3} + 2.461\times10^{-5} + 5.898\times10^{-3} + 2.00\times10^{-2}\right]^{-1}$

$U_o = 34.3 \text{ W/m}^2\text{°C}$

11.10

We will evaluate the various areas based on one row of tubes and one ft^2 face area

$$A_F = (2)(96)\left[\frac{(2\,in)(1\,ft)}{(12\,in/ft)} - \frac{(8\,tubes)\,\pi\,(0.5\,in)^2}{(144\,in^2/ft^2)\,4}\right]$$

$$= 29.91\ ft^2/ft^2_{face}/row$$

$$A_{P,o} = \frac{(8\,tubes)\,\pi\,(0.5\,in)}{(12\,in/ft)}\left(1\,ft - \frac{(96\,fins)(0.01\,in/fin)}{12\,in/ft}\right)$$

$$A_{P,o} = 0.96\ ft^2/ft^2_{face}/row$$

$$A_{P,i} = \frac{(8\,tubes)\,\pi\,(0.44\,in)(1\,ft)}{12\,in/ft} = 0.92\ ft^2/ft^2_{face}/row$$

$$A_o = A_F + A_{P,o} = 29.91 + 0.96 = 30.87\ ft^2/ft^2_{face}/row$$

Thus $A_o/A_{P,i} = 33.55$, $A_{P,o}/A_F = 0.0321$

By Eq. (11.32)

$$r_2 = \sqrt{\frac{2\,in\,(1.5\,in)}{\pi}} = 0.977\,in$$

$$r_2/r_1 = 0.977/0.25 = 3.91$$

$$L\sqrt{\frac{h_o}{k\,y}} = \frac{(0.977-0.25)\,in}{12\,in/ft}\sqrt{\frac{(10\,Btu/hr\,ft^2\,°F)\,2(12\,in/ft)}{(120\,Btu/hr\,ft\,°F)(0.01\,in)}}$$

$$= 0.857$$

11.10 Cont'd

By Fig. 11.11, $\phi = 0.67$

By Eq. (11.40)

$$U_o = \left[\frac{33.55}{1200 \frac{Btu}{hr\,ft^2\,°F}} + \frac{33.55}{1/0.001\,hr\,ft^2\,°F/Btu} + \frac{1-0.67}{10 \frac{Btu}{hr\,ft^2\,°F}(0.0321+0.67)} + \frac{1}{10 \frac{Btu}{hr\,ft^2\,°F}} \right]$$

$U_o = 4.79\ Btu/hr\,ft^2\,°F$

By Table A.4E, $v_1 = 12.127\ ft^3/lb_{m_a}$
$W_1 = 0.0021531\ lb_{m_w}/lb_{m_a}$

Thus $\dot{m}_a = \frac{(500\,ft/min)(1\,ft^2)}{12.127\,ft^3/lb_{m_a}} \left(\frac{60\,min}{hr} \right) = 2474\ \frac{lb_{m_a}}{hr\,ft^2_{face}}$

By Eq. (7.23) and Table 7.3
$c_p = 0.240 + 0.444(0.0021531) = 0.241\ \frac{Btu}{lb_{m_a}\,°F}$

By Eq. (11.8) we have for one row

$$K_1 = \frac{(4.79\ Btu/hr\,ft^2\,°F)(30.87\ ft^2/ft^2_{face})}{2474\ \frac{lb_{m_a}}{hr\,ft^2_{face}} \left(0.241\ \frac{Btu}{lb_{m_a}\,°F} \right)}$$

$K_1 = 0.248$

By Table A.1E, $t_{steam} = t_{sat}(20\,psia) = 227.91°F$

11.10 Cont'd

By Eq. (11.7) we have

$$t_2 = 20°F + (227.91 - 20)°F \left(1 - \exp(-0.248)\right)$$

$$t_2 = 65.7 °F$$

For 2 rows, $K_1 = 2(0.248) = 0.496$

$$t_2 = 20 + 207.91\left(1 - \exp(-0.496)\right) = 101.3 °F$$

For 4 rows, $K_1 = 4(0.248) = 0.992$

$$t_2 = 20 + 207.91\left(1 - \exp(-0.992)\right) = 150.8 °F$$

11.11

```
            water
             Ⓐ ↓ 180°F
  moist air  ┌───┐
    ① →     │   │  ② →
  sat, 60°F │   │  110°F
  6000 cfm  └───┘
  600 fpm      ↓ water
              Ⓑ 170°F
```

a) Determine the required water volumetric flow rate and then calculate the coil dimensions based on number of tubes needed per row and the total face area of the coil.

Air-side energy balance (sensible heating)

$$\dot{Q} = \dot{m}_a c_p (t_2 - t_1)$$

$$\dot{m}_a = \dot{V}_1 / v_1$$

By Table A.4E, $v_1 = 13.329 \text{ ft}^3/\text{lbm}$

$W_1 = W_2 = 0.011087 \text{ lbm}_w/\text{lbm}_a$

$$\dot{m}_a = \frac{6000 \text{ ft}^3/\text{min}}{13.329 \text{ ft}^3/\text{lbm}_a} \left(\frac{60 \text{ min}}{\text{hr}}\right) = 27{,}000 \text{ lbm}_a/\text{hr}$$

By Eq. (7.23) and Table 7.3

$$c_p = 0.240 + 0.444(0.011087) = 0.245 \text{ Btu}/\text{lbm}_a \cdot °F$$

$$\dot{Q} = 27{,}000 \frac{\text{lbm}_a}{\text{hr}} \cdot 0.245 \frac{\text{Btu}}{\text{lbm}_a \cdot °F} (110-60)°F = 330{,}750 \frac{\text{Btu}}{\text{hr}}$$

Water-side energy balance (sensible cooling)

11.11 Cont'd

$$\dot{m}_w = \dot{Q}/(c_{p,w}\Delta t_w)$$

By Table A.6E at 175°F, $c_{p,w} = 1.0016$ Btu/lbm$_w$°F

$$\dot{m}_w = \frac{330{,}750 \text{ Btu/hr}}{(1.0016 \text{ Btu/lbm}_w\text{°F})(10\text{°F})} = 33{,}022 \text{ lbm}_w/\text{hr}$$

Water volumetric flowrate
$$\dot{V}_w = \dot{m}_w \upsilon_w$$

By Table A.1E at 175°F, $\upsilon_w = 0.01648$ ft^3/lbm$_w$

$$\dot{V}_w = 33{,}022 \frac{\text{lbm}_w}{\text{hr}}\left(0.01648 \frac{\text{ft}^3}{\text{lbm}_w}\right)\left(\frac{1 \text{ hr}}{3600 \text{ s}}\right) = 0.151 \frac{\text{ft}^3}{\text{s}}$$

Total cross sectional area needed
$$A_w = \dot{V}_w/\vec{V}_w = (0.151 \text{ ft}^3/\text{s})/(4 \text{ ft/s}) = 0.0378 \text{ ft}^2$$

Cross sectional flow area per tube
$$A_t = 0.0557 \frac{\text{ft}^2}{\text{row}}/(32 \text{ tubes/row}) = 0.00174 \text{ ft}^2/\text{tube}$$

Number of tubes needed per row (flow in parallel)
#Tubes = A_w/A_t = 0.0378 ft^2/(0.00174 ft^2/tube)
#Tubes = 21.7 → 22 tubes/row

Coil height: H = (#Tubes)(tube spacing)
= 22 (1.5 in) = 33 in.

Coil width: W = $A_{face}/H = (\dot{V}_a/\vec{V}_a)/H$

$$= \left(\frac{6000 \text{ ft}^3/\text{min}}{600 \text{ ft/min}}\right)/33 \text{ in}\left(\frac{1 \text{ ft}^2}{144 \text{ in}^2}\right) = 43.6 \text{ in.}$$

11.11 Cont'd

b) We will solve for A_o using Eq. (11.33) and then determine the depth of the coil

Log-mean Temperature difference approach is used as all four temperatures in and out are known

$$\Delta t_{m,cf} = \frac{(170-60)-(180-110)}{\ln\left(\frac{170-60}{180-110}\right)} = 88.5°F$$

$R = (180-170)/(110-60) = 0.20$
$P = (110-60)/(180-60) = 0.417$
By Fig. 11.7, $F \cong 1.0$ so $\Delta t_m = \Delta t_{m,cf}$

Use Eq. (11.40) To determine U_o
By Example 11.4, $A_o/A_{P,i} = 19.31$, $A_{P,o}/A_F = 0.105$

By Eq. (2.52)

$$h_i = \frac{k}{D_i} 0.023 (Re)^{0.8} (Pr)^{0.3}$$

By Table A.6E at 175°F
  $k = 0.3868$ Btu/hr ft °F
  $\mu = 0.866$ lbm/ft hr
  $Pr = 2.24$

By Table A.1E at 175°F
  $\rho = 60.68$ lbm/ft^3

$$Re = \frac{\rho \vec{V} D_i}{\mu} = \frac{(60.68 \text{ lbm/ft}^3)(4 \text{ ft/s})(0.565 \text{ in})}{(0.866 \text{ lbm/hr ft})(12 \text{ in/ft})} \left(\frac{3600 s}{hr}\right) = 47,500$$

11.11 Cont'd

$$h_i = \frac{0.3868 \text{ Btu/hr ft °F} \cdot 0.023(47,500)^{0.8}(2.24)^{0.3}}{(0.565 \text{ in})(1 \text{ ft}/12 \text{ in})}$$

$h_i = 1327$ Btu/hr ft² °F

We will use $1/h_{d_i} = 0.0005$ hr ft² °F/Btu as in Example 11.4.

We will estimate h_o from Fig. 11.15
By Table A.5E at the mean air temperature,

$t_m = (110 + 60)/2 = 85$°F
$\mu = 0.04496$ lbm/hr ft
$Pr = 0.707$

Mass flow of moist air
$\dot{m} = \dot{m}_a (1+W) = 27,000(1+0.011087) = 27,300$ lbm/hr

By Table 11.5, $A_c = 0.497$, $4r_h = 0.01268$ ft

$$G = \dot{m}/(A_c A_{\text{Face}}) = \frac{27,300 \text{ lbm/hr}}{0.497(10 \text{ ft}^2_{\text{face}})} = 5493 \frac{\text{lbm}}{\text{hr ft}^2_{\text{face}}}$$

$$Re = \frac{4r_h G}{\mu} = \frac{0.01268 \text{ ft}(5493 \text{ lbm/hr ft}^2)}{0.04496 \text{ lbm/hr ft}} = 1549$$

By Fig. 11.15, $\left(\dfrac{h_o}{G c_p}\right) Pr^{2/3} = 0.0074$

$h_o = 0.0074(5493 \text{ lbm/hr ft}^2)(0.245 \text{ Btu/lbm °F})(0.707)$

$h_o = 12.5$ Btu/hr ft² °F

11.11 Cont'd

By Example 11.4 we have $r_2 = 0.914$ in., $L = 0.576$ in., $r_2/r_1 = 2.70$

$$\frac{L}{\sqrt{\frac{h_o}{k\,y}}} = \frac{(0.576\text{ in})}{(12\text{ in}/\text{ft})}\sqrt{\frac{12.5\text{ Btu}/\text{hr ft}^2\text{°F}\,(2)}{120\text{ Btu}/\text{hr ft °F}\,(0.016\text{ in})}}\left(\frac{12\text{ in}}{\text{ft}}\right)$$

$$= 0.600$$

By Fig. 11.11, $\phi = 0.84$

By Eq. (11.40)

$$U_o = \left[\frac{19.31}{1327} + 19.31(0.0005) + \frac{1-0.84}{12.5\,(0.105+0.84)} + \frac{1}{12.5}\right]^{-1}$$

$$U_o = 8.49\text{ Btu}/\text{hr ft}^2\text{°F}$$

By Eq. (11.33)

$$A_o = \frac{(330{,}750\text{ Btu}/\text{hr})}{8.49\text{ Btu}/\text{hr ft}^2\text{°F}\,(88.5\text{°F})} = 440\text{ ft}^2$$

By Table 11.5, Surface II has an outside surface area of $22.86\text{ ft}^2/\text{ft}^2_{\text{face}}/\text{row}$.

$$\text{\# rows} = \frac{440\text{ ft}^2}{10\text{ ft}^2_{\text{face}}\,(22.86\text{ ft}^2/\text{ft}^2_{\text{face}}/\text{row})} = 1.92$$

or # rows = 2

11.12

$h_{c,o} = 5 \, W/m^2 \, °C$

1 m/s, 60°C, 1.8 cm, 2.0 cm

$\varepsilon = 0.2$

surroundings = 20°C

Heat loss from the water to the ambient

$$\dot{Q} = U_o A_o \Delta t$$

t_w 60°C — R_w (water) — R_p (Pipe) — [$R_{c,o}$ / $R_{r,o}$] (air) — t_{amb} 20°C

By Eq. (2.85)

$$U_o = \left[\frac{D_o}{D_i h_i} + \frac{D_o \ln(D_o/D_i)}{2 k_p} + \frac{1}{h_o} \right]^{-1}$$

where $1/h_o = R_o = [R_{c,o}^{-1} + R_{r,o}^{-1}]^{-1}$

$= [h_{c,o} + h_{r,o}]^{-1}$

By Eq. (2.78), $h_{r,o} \cong \varepsilon \, 4\sigma \, T_{ave}^3$
with an estimate for $T_{ave} \cong 310°K$

$h_{r,o} = 0.2 \, (4)(5.670 \times 10^{-8} \, W/m^2 K^4)(310°K)^3$

$= 1.351 \, W/m^2 K$

$1/h_o = [5 + 1.35]^{-1} = 0.1574 \, m^2 °C/W$

11.12 Cont'd

By Eq. (2.52)
$$h_i = \frac{k_i}{D_i} 0.023 \, Re^{0.8} Pr^{0.3}$$

By Table A.6SI at 60°C
- $k_i = 0.6543$ W/m·°C
- $\mu = 466.5 \times 10^{-6}$ kg m/s
- $c_p = 4.183$ kJ/kg·°C
- $Pr = 2.98$

By Table A.1SI, $\rho = 983$ kg/m³

$$Re = \frac{1 \text{ m/s} (1.8 \text{ cm})(983 \text{ kg/m}^3)}{466.5 \times 10^{-6} \text{ kg m/s} (100 \text{ cm/m})} = 37,930$$

$$h_i = \frac{0.6543 \text{ W/m·°C}}{1.8 \text{ cm} (1 \text{ m}/100 \text{ cm})} 0.023 (37,930)^{0.8} (2.98)^{0.3}$$

$$h_i = 5342 \text{ W/m}^2\text{·°C}$$

By Example 11.3, $k_{copper} = \frac{223 \text{ Btu}}{\text{hr ft·°F}} \left(\frac{1 \text{ W/m·°C}}{0.5778 \text{ Btu/hr ft·°F}} \right)$
$$= 386 \text{ W/m·°C}$$

Substituting into Eq. (2.85)

$$U_o = \left[\frac{2}{1.8(5342)} + \frac{2 \text{ cm} \ln(2/1.8)}{2(386 \text{ W/m·°C})} \left(\frac{m}{100 \text{ cm}} \right) + 0.1574 \frac{\text{m}^2\text{·°C}}{\text{W}} \right]^{-1}$$

$$U_o = 6.34 \text{ W/m}^2\text{·°C}$$

11.12 Cont'd

Note the only effective thermal resistance is the convection and radiation in the air

$$\dot{Q} = 6.34 \frac{W}{m^2 {}^\circ C} \; \pi \, 2.0\,cm \left(\frac{1\,m}{100\,cm}\right)(100\,m)(60-20)\,{}^\circ C$$

$$= 1.593\,kW$$

Write an energy balance on the water
By Eq. (2.10)

$$(t_2 - t_1) = \dot{Q}/(\dot{m}_w c_p)$$

$$\dot{m}_w = \rho \, \frac{\pi D_i^2}{4} \, \vec{V}$$

$$= 983\,kg/m^3 \; \frac{\pi(1.8\,cm)^2}{4} \; 1\,m/s \left(\frac{m^2}{10^4\,cm^2}\right)$$

$$= 0.250\,kg/s$$

$$(t_2 - t_1) = 1.593\,kW / (0.250\,kg/s)(4.183\,kJ/kg\,{}^\circ C)$$

$$= 1.52\,{}^\circ C$$

11.13 $t_i = 250°F$, $t_o = 80°F$

Pipe I.D. = 2.059 in
Pipe O.D. = 2.375 in
k_{insul} = 0.042 Btu/hr ft °F
Insulation O.D. = 2.375 + 2(2) = 6.375 in

As saturated steam is condensing in the pipe and the pipe thermal conductivity is much greater than that of the insulation, we can neglect both internal and pipe resistances, leaving only the conduction through the insulation and the external convective resistance (neglect external infrared radiation as the external surface emissivity is nearly zero).

$$\dot{q}' = \frac{\dot{Q}}{L} = \frac{U_o A_o}{L}(t_i - t_o) \approx \frac{(t_i - t_{so})}{R'_{insul}} = \frac{(t_{so} - t_o)}{R'_{conv}}$$

By Eq. 2.49

$$R'_{insul} = \left[\frac{2\pi k_{insul}}{\ln(r_o/r_i)}\right]^{-1} = \left[\frac{2\pi \cdot 0.042 \text{ Btu/hr ft °F}}{\ln\left(\frac{6.375/2}{2.375/2}\right)}\right]^{-1}$$

$R'_{insul} = 3.74$ hr ft °F/Btu

By Eq. 2.50

$$R'_{conv} = \frac{L}{h_o A_o} = \frac{L}{0.27 \, D_o^{-.25}(t_{so} - t_o)^{0.25} \pi D_o L}$$

$$R'_{conv} = \left[0.27 \pi D_o^{0.75}(t_{so} - t_o)^{0.25}\right]^{-1}$$

$$= \left[0.27 \pi \left(\frac{6.375}{12}\right)^{0.75}(t_{so} - 80)^{0.25}\right]^{-1}$$

$$= \left[0.528 \,(t_{so} - 80)^{0.25}\right]^{-1} \text{ hr ft °F/Btu}$$

11.13 Cont'd

substituting into eqn above for $\dot{q}'$

$$\frac{250 - t_{so}}{3.74} = 0.528 (t_{so} - 80)^{1.25}$$

solve for t_{so} by trial: $t_{so} \cong 110.2°F$

substituting into eqn for $\dot{q}'$

$$\dot{q}' = \frac{t_i - t_{so}}{R'_{insul}} = \frac{(250 - 110.2)°F}{3.74 \, hr \, ft \, °F/Btu}$$

$$\dot{q}' = 37.4 \, Btu/hr \, ft$$

11.14

a) $P_{atm} = 27.45 \text{ in Hg} \left(\dfrac{1.45 \times 10^{-4} \text{ psia}}{2.961 \times 10^{-4} \text{ in Hg}} \right) = 13.44 \text{ psia}$

$P_{steam} = 4.50 \text{ psig} + 13.44 \text{ psia} = 17.94 \text{ psia}$

By Table A.1E, $t_{steam} = 222.2 \, °F$

By procedures in Ch 7, $W_1 = 0.0074 \, lbm_w/lbm_a$, $v_1 = 15.0 \, ft^3/lbm_a$

The face velocity G is determined as

$G = \dfrac{(1+W_1) \dot{V}}{A_{face} \, v_1} = \dfrac{(1.0074 \, lbm_m/lbm_a)(\dot{V} \, ft^3/min)(60 \, min/hr)}{3 \, ft^2 (15 \, ft^3/lbm_a)}$

$G = 1.343 \, (\dot{V} \, ft^3/min) \, lbm/hr \, ft^2_{face}$ \hfill (1)

By Eq. (11.17)

$\exp(K_1) = \dfrac{t_h - t_{c,i}}{t_h - t_{c,o}} = \dfrac{222.2 - 80}{222.2 - t_2} = \dfrac{142.2}{222.2 - t_2}$ \hfill (2)

By Eq. (11.8), $U_o A_o = \dot{m}_a \, c_{p,a} \, K_1$

By Eq. (7.23) and Table 7.3

$c_{p,a} = 0.240 + 0.444(0.0074) = 0.243 \, Btu/lbm_a \, °F$

Also, $\dfrac{\dot{m}_a}{A_{face}} = \dfrac{(\dot{V} \, ft^3/min)(60 \, min/hr)}{A_{face} \, v_1} = \dfrac{(\dot{V} \, ft^3/min) 60 \, min/hr}{3 \, ft^2 \, \dfrac{15 \, ft^3}{lbm_a}}$

1/4

11.14 Cont'd

$$\frac{\dot{m}_a}{A_{face}} = \frac{4}{3}(\dot{V}\, ft^3/min) \quad lb_{ma}/hr\, ft^2_{face}$$

Thus

$$U_o A_o = \frac{4}{3}(\dot{V}\, ft^3/min)\frac{lb_{ma}}{hr\, ft^2_{face}}\left(\frac{0.243\, Btu}{lb_{ma}\,°F}\right) K_1$$

$$U_o A_o = 0.324\,(\dot{V}\, ft^3/min)\,\frac{Btu}{hr\, ft^2_{face}\,°F} \qquad (3$$

Using Eqns. (1-3) above we can generate the following table

Inlet CFM	t_2 F	K_1 e	K_1	G lb/(hr)(sq ft)	$U_o A_o$ Btu/(hr)(F)(sq ft)
900	171.0	2.770	1.02	1209	297
1200	159.0	2.250	0.81	1611	315
1500	151.0	1.994	0.69	2015	336
1800	145.0	1.840	0.61	2417	356
2100	140.0	1.728	0.55	2820	373
2400	136.0	1.648	0.50	3223	389

A plot of G vs $U_o A_o$ can be generated using the results in this Table

11.14 Cont'd

b) For these new operating conditions,
$P_{steam} = 24.7 \text{ psia}$, $t_{steam} = 239.4°F$

By Table A.4E, $W_1 = 0.0021531 \text{ lbm}_w/\text{lbm}_a$,
$v_1 = 12.127 \text{ ft}^3/\text{lbm}_a$

$$\frac{\dot{m}_a}{A_{face}} = \frac{\vec{V}_1}{v_1} = \frac{500 \text{ ft/min } (60 \text{ min/hr})}{12.127 \text{ ft}^3/\text{lbm}_a}$$

$$= 2474 \text{ lbm}_a / \text{hr ft}^2_{face}$$

$$c_{p,a} = 0.240 + 0.444(0.0021531) = 0.241 \frac{\text{Btu}}{\text{lbm}_a \text{ °F}}$$

$$G = \dot{m}_a(1+W_1) = \frac{2474 \text{ lbm}_a}{\text{hr ft}^2_{face}}(1 + 0.0021531)$$

$$= 2479 \text{ lbm/hr ft}^2_{face}$$

By linear interpolation of the table generated in part a), $U_o A_o = 358.6 \frac{\text{Btu}}{\text{hr ft}^2_{face} \text{ °F}}$

This is for a single-row coil, for a 2-row coil $U_o A_o = 2 \times 358.6$
$$= 717 \frac{\text{Btu}}{\text{hr ft}^2_{face} \text{ °F}}$$

11.14 Cont'd

$$K_1 = \dfrac{U_o A_o}{\dot{m}_a c_{p,a}/A_{face}} = \dfrac{717 \text{ Btu/hr ft}^2_{face}\text{°F}}{\dfrac{2474 \text{ lb}_{ma}}{\text{hr ft}^2_{face}} \cdot \dfrac{0.241 \text{ Btu}}{\text{lb}_{ma}\text{°F}}} = 1.203$$

$$t_2 = t_h - (t_h - t_{c,1})e^{-K_1}$$

$$= 239.4\text{°F} - (239.4 - 20\text{°F})e^{-1.203}$$

$$t_2 = 173.5\text{°F}$$

11.15

a) copper fins

By Eq. (11.54) with $y_w = 0$

$$h_{o,w} = b_w h_{c,o} / c_{pa}$$

By Fig. 11.18 at $t_s = 40°F$, $b_w = 0.45 \frac{Btu}{lb_{ma} °F}$
(assuming sea level pressure)

By Table A.4E at $40°F$, $W_s = 0.005216 \, lb_{mw}/lb_{ma}$

By Eq. (7.23) and Table 7.3

$$c_{pa} = 0.240 + 0.444(0.005216) = 0.2423 \frac{Btu}{lb_{ma} °F}$$

$$h_{o,w} = \frac{(0.45 \, Btu/lb_{ma} °F)(10 \, Btu/hr \, ft^2 °F)}{0.2423 \, Btu/lb_{ma} °F}$$

$$h_{o,w} = 18.6 \, Btu/hr \, ft^2 \, °F$$

$$y = \frac{h_{o,w}}{k}\left(\frac{L}{0.7}\right)^2$$

$$= \frac{18.6 \, Btu/hr \, ft^2 °F}{120 \, Btu/hr \, ft °F}\left(\frac{1.0 \, in}{0.7}\right)^2\left(\frac{ft}{12 \, in}\right) = 0.0264 \, in.$$

or fin thickness, $2y$, is $0.0527 \, in.$

11.15 cont'd

b) aluminum fins

$$y = \frac{18.6}{223}\left(\frac{1}{0.7}\right)^2\left(\frac{1}{12}\right) = 0.0142 \text{ in.}$$

or fin thickness is 0.0284 in.

11.16

a) copper fins

By Eq. (11.54)

$$h_{o,w} = \left[c_{pa} / (b_w h_{c,o}) + y_w / k_w \right]^{-1}$$

By Fig. 11.18 at $t_s = 10°C$, $b_w = 2.3$ kJ/kg$_a$°C at standard sea level pressure

By Table A.4SI at 10°C, $W_s = 0.007661$ kg$_w$/kg$_a$

By Eq. (7.23) and Table 7.3

$$c_{pa} = 1.00 + 1.86(0.007661) = 1.014 \text{ kJ/kg}_a °C$$

By Table A.6SI at 10°C, $k_w = 0.58$ W/m°C

Thus

$$h_{o,w} = \left[\frac{1.014 \text{ kJ/kg}_a°C}{2.3 \text{ kJ/kg}_a°C \; 50 W/m^2°C} + \frac{10^{-3} m}{0.58 \text{ W/m}°C} \right]^{-1}$$

$$h_{o,w} = 94.9 \text{ W/m}^2°C$$

Let $r_2/r_1 = 4.0$, $L\sqrt{h_{o,w}/ky} = 0.75$

$$L = r_2 - r_1 = 0.75 / \sqrt{\frac{94.9 \text{ W/m}^2°C}{386 \text{ W/m}°C \; 5 \times 10^{-5} m}} = 0.0107 m$$

$r_2 = 4mm + 10.7mm = 14.7 mm$

$r_2/r_1 = 14.7mm / 4mm = 3.68$

11.16 cont'd

Let $r_2/r_1 = 3.75$, $L\sqrt{h_{o,w}/ky} = 0.76$

$L = 0.76/70.12 = 10.8$ mm
$r_2 = 4 + 10.8 = 14.8$ mm
$r_2/r_1 = 14.8/4 = 3.70$

Use $r_2 = 14.8$ mm, $r_2/r_1 = 14.8/4 = 3.70$

By Eq. (11.32)

$$a = \frac{(14.8\text{ mm})^2 \pi}{2\text{ cm}} \left(\frac{\text{cm}}{10\text{ mm}}\right)^2 = 3.44 \text{ cm}$$

b) aluminum fins

let $r_2/r_1 = 3.0$, $L\sqrt{h_{o,w}/ky} = 0.85$

$$L = 0.85 / \sqrt{\frac{94.9}{208(5\times10^{-5})}} = \frac{0.8}{95.5} = 8.9 \text{ mm}$$

$r_2 = 8.9$ mm $+ 4$ mm $= 12.9$ mm
$r_2/r_1 = 12.9/4 = 3.23$

Use $r_2/r_1 = 3.1$, $r_2 = 12.4$ mm

$$a = \frac{(12.4)^2 \pi}{2(100)} = 2.42 \text{ cm}$$

11.17

Procedure follows that outlined for Prob. 11.16

at $-10°C$, $b_w = 1.4$ $kJ/kg_a\ °C$ (sea level Pressure)

$W_s = 0.0016062\ kg_w/kg_a$

$c_{pa} = 1.00 + 1.86(0.0016062) = 1.003\ kJ/kg_a\ °C$

By Eq. (11.54)

$$h_{o,w} = \left[\frac{1.003}{1.4(50)} + \frac{10^{-3}}{1}\right]^{-1} = 65.2\ W/m^2\ °C$$

a) copper

Let $r_2/r_1 = 4.0$, $L\sqrt{h_{o,w}/ky} = 0.70$

$$L = 0.70\bigg/\sqrt{\frac{65.2}{386(5\times10^{-5})}} = \frac{0.70}{58.1} = 12.0\ mm$$

$r_2 = 12.0 + 4 = 16.0$, $r_2/r_1 = 16.0/4 = 4.0$

Use $r_2/r_1 = 4.0$, $r_2 = 4.0 \times 4 = 16.0\ mm$

$$a = \frac{(16)^2 \pi}{2(100)} = 4.0\ cm$$

b) aluminum

Let $r_2/r_1 = 3.5$, $L\sqrt{h_{o,w}/ky} = 0.79$

$$L = 0.79\bigg/\sqrt{\frac{65.2}{208(5\times10^{-5})}} = \frac{0.79}{79.2} = 10.0\ mm$$

$r_2 = 10.0 + 4 = 14.0$, $r_2/r_1 = 14.0/4 = 3.5$

Use $r_2/r_1 = 3.5$, $r_2 = (4\ mm)3.5 = 14.0\ mm$

$$a = \frac{(14.0)^2 \pi}{2(100)} = 3.1\ cm$$

11.18

a) We must solve Eq. (11.63).
Consider areas based on one row of tubes and one ft^2 face area
From the solution of Prob. 11.10

$A_F = 29.91 \ ft^2 / ft^2_{face}$ row   for 8 fins/in.

$A_F = 14.95 \ ft^2 / ft^2_{face}$ row   for 4 fins/in.

$$A_{P,o} = \frac{8 \ tubes \ \pi \ (0.5 in)}{12 \ in/ft} \left(1 - \frac{(4 \ fins/in)(12 \ in)(0.01 in)}{12 \ in/ft}\right)$$

$A_{P,o} = 1.005 \ ft^2 / ft^2_{face}$ row

$A_{P,i} = 0.92 \ ft^2 / ft^2_{face}$ row   (same as Prob. 11.10)

$A_o = A_F + A_{P,o} = 15.96 \ ft^2 / ft^2_{face}$ row

$A_o / A_{P,i} = 17.35$, $A_{P,o} / A_F = 0.0672$

We will assume $t_p = -9°F$ and $t_{w,m} = -6°F$

By Eq. (11.59) and Table A.4E

$$b_R' = \frac{-1.644 + 2.443}{-9°F - (-12°F)} = 0.266 \ \frac{Btu}{lb_{m_a} \ °F}$$

By Table A.4E, $b_{w,m}$ at $-6°F$

$$b_{w,m} \cong \frac{h_s(-5°F) - h_s(-10°F)}{-5°F - (-10°F)}$$

$$= \frac{-0.561 - (-1.915)}{-5 - (-10)} = 0.271 \ \frac{Btu}{lb_{m_a} \ °F}$$

11.18 con't

We may assume $c_{pa} = 0.240 \text{ Btu/lbm}_a \, ^\circ F$

By Eq. (11.54)

$$h_{o,w} = \left[\frac{0.240 \text{ Btu/lbm}_a \, ^\circ F}{0.271 \text{ Btu/lbm}_a \, ^\circ F \cdot 8 \text{ Btu/hr ft}^2 \, ^\circ F} + \frac{0.05 \text{ in} (ft/12 \text{in})}{0.27 \text{ Btu/hr ft} \, ^\circ F} \right]^{-1}$$

$$h_{o,w} = 7.93 \text{ Btu/hr ft}^2 \, ^\circ F$$

From Prob. 11.10, $r_2 - r_1 = 0.727 \text{ in}$, and $r_2/r_1 = 3.91$

$$L \sqrt{\frac{h_{o,w}}{k_F \, y_F}} = \frac{0.727 \text{ in}}{(12 \text{in}/ft)} \sqrt{\frac{7.93 \text{ Btu/hr ft}^2 \, ^\circ F \,(12\text{in}/ft)}{120 \text{ Btu/hr ft} \, ^\circ F \,(0.01 \text{ in})/2}} = 0.763$$

By Fig. 11.11, $\phi_w = 0.73$

Evaluating the terms in Eq. (11.63)

$$R_{i,w} = \frac{b_R' \, A_o}{A_{P,i} \, h_i} = \frac{(0.266 \text{ Btu/lbm}_a \, ^\circ F)\, 17.35}{500 \text{ Btu/hr ft}^2 \, ^\circ F} = 0.00923 \, \frac{\text{hr ft}^2}{\text{lbm}_a}$$

$$R_{F,w} = \frac{b_{w,m} (1-\phi_w)}{h_{o,w} (A_{P,o}/A_F + \phi_w)}$$

$$= \frac{(0.271 \text{ Btu/lbm}_a \, ^\circ F)(1-0.73)}{(7.93 \text{ Btu/hr ft}^2 \, ^\circ F)(0.0672 + 0.73)} = 0.01157 \, \frac{\text{hr ft}^2}{\text{lbm}_a}$$

$$R_{o,w} = \frac{b_{w,m}}{h_{o,w}} = \frac{0.271 \text{ Btu/lbm}_a \, ^\circ F}{7.93 \text{ Btu/hr ft}^2 \, ^\circ F} = 0.03417 \, \frac{\text{hr ft}^2}{\text{lbm}_a}$$

11.18 con't.

$$R_{t,w} = R_{i,w} + R_{F,w} + R_{o,w} = 0.05497 \frac{hr\ ft^2}{lbm_a}$$

$$U_{o,w} = R_{t,w}^{-1} = \frac{18.2\ Btu}{hr\ ft^2\ (Btu/lbm_a)}$$

We now check our assumed values for the pipe temperature and mean frost temperature

By Eq. (11.64) and Table A.4E

$$t_p = -12°F + \frac{\frac{18.2\ Btu}{hr\ ft^2(Btu/lbm_a)}\ 17.35\ (0.835 + 2.443)\ Btu/lbm_a}{500\ Btu/hr\ ft^2\ °F}$$

$t_p = -9.9°F$ which is close to our assumed value of $-9°F$

By Eq. (11.65) and Table A.4E

$$h_{s,w,m} = \frac{0.835\ Btu}{lbm_a} - \left[\frac{\frac{0.24\ Btu}{lbm_a\ °F}\left(\frac{7.93\ Btu}{hr\ ft^2\ °F}\right) 0.73}{\frac{0.271\ Btu}{lbm_a\ °F}\left(\frac{8.0\ Btu}{hr\ ft^2\ °F}\right)} \right] \times$$

$$\left[1 - \frac{\frac{0.266\ Btu}{lbm_a\ °F}\left(\frac{18.2\ Btu}{hr\ ft^2(Btu/lbm_a)}\right)(17.35)}{500\ Btu/hr\ ft^2\ °F} \right] \frac{(0.835 + 2.443)}{lbm_a\ °F}$$

11.18

$h_{s,w,m} = -0.913$ Btu/lb$_{ma}$

This gives a value for $t_{w,m}$ of $-6.3°F$ which closely approximates the assumed value of $-6°F$.

b) With negligible frost layer thickness (frost just beginning to form) the procedure is identical to that given for part a) except $Y_w = 0$. The results are summarized below.

Values remain as determined in part a) except where noted below.

$h_{o,w} = 9.03$

$L\sqrt{\dfrac{h_{o,w}}{k_F Y_F}} = 0.814$, $\phi_w = 0.70$

$R_{F,w} = 0.01174$, $R_{o,w} = 0.03001$

$R_{t,w} = 0.05098$, $U_{o,w} = 19.6$

11.18

Summary of results

Component	Part (a)	Part (b)
$R_{i,w}$ hr ft²/lb$_{m_a}$	0.00923	0.009
$R_{F,w}$ "	0.01157	0.0117
$R_{o,w}$ "	0.03417	0.030
$R_{t,w}$ "	0.05497	0.050
$R_{i,w} / R_{t,w}$	16.8%	18.1%
$R_{F,w} / R_{t,w}$	21.0%	23.0%
$R_{o,w} / R_{t,w}$	62.2%	58.9%
$U_{o,w}$ Btu/hr ft²(Btu/lb$_{m_a}$)	18.2	19.6

11.19

We must solve Eq. (11.63)
Area ratios remain as determined in Prob. 11.8.

We will assume $t_p = 8°C$ and $t_{w,m} = 10°C$

By Eq. (11.59) and Table A.4SI

$$b'_R = \frac{24.852 - 20.644}{8-6} = 2.10 \text{ kJ/kg}_a \cdot °C$$

$$b_{w,m} \cong \frac{h_s(12°C) - h_s(8°C)}{12°C - 8°C} = \frac{34.179 - 24.852}{4} = 2.33 \text{ kJ/kg}_a \cdot °C$$

By Eq. (7.23), Table 7.3 and Table A.4SI
$W_s = 0.010712 \text{ kg}_w/\text{kg}_a$
$c_p = 1.00 + 1.86(0.010712) = 1.0199 \text{ kJ/kg}_a \cdot °C$

By Eq. (11.54), assuming water film 1mm thick,
$k_w = 0.58 \text{ W/m} \cdot °C$ at 10°C from Table A.6SI

$$h_{o,w} = \left[\frac{1.0199 \text{ kJ/kg}_a \cdot °C}{2.33 \text{ kJ/kg}_a \cdot °C \cdot 50 \text{W/m}^2 °C} + \frac{1 \times 10^{-3} \text{ m}}{0.58 \text{ W/m} \cdot °C} \right]^{-1}$$

$$h_{o,w} = 95.4 \text{ W/m}^2 °C$$

From Prob. 11.8, $r_2/r_1 = 2.52$, $L = 0.76 \text{ cm}$

$$L\sqrt{\frac{h_{o,w}}{k_F \, y_F}} = \frac{0.76 \text{ cm}}{(100 \text{ cm/m})} \sqrt{\frac{95.4 \text{ W/m}^2°C \, (10^3 \text{mm/m})}{208 \text{ W/m} \cdot °C (0.05 \text{mm})}} = 0.73$$

From Fig. 11.11, $\phi_w = 0.77$

11.19 Cont'd

By Table A.6 SI at 6°C
$c_p = 4.20 \ kJ/kg°C$
$\mu = 1500 \ kg/m \cdot s \times 10^{-6}$
$k = 0.5724 \ W/m°C$
$Pr = 11.1$

By Table A.1 SI at 6°C
$\rho = 1000.0 \ kg/m^3$

$\nu = \mu/\rho = \dfrac{1.5 \times 10^{-3} \ kg/m \cdot s}{10^3 \ kg/m^3} = 1.5 \times 10^{-6} \ m^2/s$

$Re_{D_i} = \dfrac{\vec{V} D_i}{\nu} = \dfrac{1 \ m/s \ (0.9 cm)}{1.5 \times 10^{-6} \ m^2/s} \left(\dfrac{1 \ m}{100 \ cm}\right) = 6000$

By Eq. (2.52)

$h_i = \dfrac{0.5724 \ W/m°C}{0.9 \times 10^{-2} \ m} (0.023)(6000)^{0.8} (11.1)^{0.4}$

$h_i = 4035 \ W/m^2°C$

Substituting values into Eq. (11.63)

$U_{o,w} = \left\{ \left[\dfrac{2.10 \ kJ/kg_a °C \ (0.3661 \ m^2)}{0.0283 \ m^2 \ (4035 \ W/m^2°C)}\right] + \right.$

$\left. \left[\dfrac{2.33 \ kJ/kg_a °C \ (1-0.77)}{95.4 \ W/m^2°C \ (0.0289/0.3372 + 0.77)}\right] + \dfrac{2.33 \ kJ/kg_a °C}{95.4 \ W/m^2°C} \right\}^{-}$

11.19 Cont'd

$$U_{o,w} = \{R_{i,w} + R_{F,w} + R_{o,w}\}^{-1}$$

$$= \{0.00673 + 0.00656 + 0.02442\}^{-1}$$

$$= \{0.03771\}^{-1}$$

$$U_{o,w} = 26.51 \ W/m^2 \ (kJ/kg_a)$$

We now check our assumed values for the pipe temperature and the mean water temperature

By Eq. (11.64) and Table A.4SI

$$t_p = 6°C + \frac{26.51 \ W \ (0.3661 \ m^2)(42.17 - 20.644) \ kJ/kg_a}{m^2(kJ/kg_a)(4035 \ W/m^2°C)(0.0283 \ m^2)}$$

$$t_p = 6°C + 1.83°C = 7.83°C$$

This is close to our assumed value of 8°C

By Eq. (11.65) and Table A.4SI

$$h_{s,w,m} = \frac{42.17 \ W}{m^2 °C} - \left[\frac{1.0199 \ kJ/kg_a °C \ (95.4 \ W/m^2°C)(0.77)}{2.33 \ kJ/kg_a °C \ (50 \ W/m^2°C)}\right] \times$$

$$\left[1 - \frac{\frac{2.10 \ kJ}{kg_a °C}\left(\frac{26.51 \ W}{m^2(kJ/kg_a)}\right) 0.3661 \ m^2}{(4035 \ W/m^2°C)(0.0283 \ m^2)}\right] (42.17 - 20.644) \ kJ/kg_a$$

11.19 Cont'd

$h_{s,w,m} = 30.80 \text{ kJ/kg}_a$

This gives a value for $t_{w,m}$ of $10.6°C$ which closely approximates the assumed value of $10°C$.

11.20. We will assign the symbols $\dot{m}$, h, t to the moist air side and the symbols M, C_p, T to the refrigerant side. For any differential section, we may write

$$-\dot{m}_a \, dh = U_{o,w} \, dA_o \, (h - h_{s,R}) \tag{1}$$

$$\dot{M}C_p(T - T_1) = \dot{m}_a(h - h_2) \tag{2}$$

We assume

$$T = \frac{h_{s,R} - a_R}{b_R}$$

and

$$T - T_1 = \frac{h_{s,R} - h_{s,R,1}}{b_R} \tag{3}$$

By Eqs. (2) and (3)

$$h_{s,R} = h_{s,R,1} + \frac{\dot{m}_a b_R}{MC_p}(h - h_2) \tag{4}$$

By Eqs. (1) and (4), we obtain

$$\frac{dh}{K_1 h - K_2} = -\frac{U_{o,w} \, dA_o}{\dot{m}_a} \tag{5}$$

where

$$K_1 = 1 - \frac{\dot{m}_a b_R}{MC_p} \tag{6}$$

$$K_2 = h_{s,R,1} - \frac{\dot{m}_a b_R}{MC_p} h_2 \tag{7}$$

Integration of Eq. (5) yields

$$\ln\left[\frac{K_1 h_2 - K_2}{K_1 h_1 - K_2}\right] = -\frac{K_1 U_{o,w} A_o}{\dot{m}_a} \tag{8}$$

By Eqs. (6) and (7)

$$K_1 h_2 - K_2 = h_2 - h_{s,R,1} \tag{9}$$

11.20 Cont'd

We may also write

$$\frac{\dot{m}_a b_R}{MC_P} = \frac{h_{s,R,2} - h_{s,R,1}}{h_1 - h_2} \quad (10)$$

By Eqs. (6), (7) and (10)

$$K_1 h_1 - K_2 = h_1 - h_{s,R,2} \quad (11)$$

$$K_1 = \frac{(h_1 - h_{s,R,2}) - h_2 - h_{s,R,1})}{h_1 - h_2} \quad (12)$$

By Eqs. (8), (9), (11), and (12) and noting that $\dot{Q} = \dot{m}_a (h_1 - h_2)$, we obtain

$$\ln\left[\frac{h_1 - h_{s,R,2}}{h_2 - h_{s,R,1}}\right] = \frac{U_{o,w} A_o}{\dot{Q}} \left[(h_1 - h_{s,R,2}) - (h_2 - h_{s,R,1})\right] \quad (13)$$

but by definition of Δh_m

$$\dot{Q} = U_{o,w} A_o \Delta h_m \quad (14)$$

By Eqs. (13) and (14), we obtain Eq. (11.67)

11.21. For a differential section at an arbitrary location and ignoring energy of condensed water, we may write

$$U_{o,w} \, dA_o \, (h - h_{s,R}) = -\dot{m}_a \, dh \qquad (1)$$

$$\dot{M} C_p \, (T - T_1) = \dot{m}_a \, (h - h_2) \qquad (2)$$

In Eq. (2), substitute

$$\frac{h_{s,R} - h_{s,R,1}}{b_R} = T - T_1$$

Equation (2) may then be written in the form

$$h_{s,R} - h_{s,R,1} = \frac{\dot{m}_a \, b_R}{\dot{M} C_p} \, (h - h_2) = c_1 \, (h - h_2)$$

and thus

$$h - h_{s,R} = (1 - c_1) \, h - (h_{s,R,1} - c_1 \, h_2) \qquad (3)$$

By Eqs. (1) and (3)

$$\int_{h_1}^{h_2} \frac{dh}{(1 - c_1) \, h - (h_{s,R,1} - c_1 \, h_2)} = \frac{U_{o,w}}{\dot{m}_a} \int_0^{A_o} dA_o$$

Integration leads directly to Eq. (11.72)

11.22
a) From Example 11.7, $\dot{Q} = 618,000$ Btu/hr. Thus

$$\dot{m}_w = \frac{\dot{Q}}{c_{p,w} \Delta t_w} = \frac{618,000 \text{ Btu/hr}}{1.00 \frac{\text{Btu}}{\text{lbm}_w \cdot °F} (8°F)} = 77,250 \text{ lbm}_w/\text{hr}$$

By Table A.1E at the mean water temperature of 44°F

$$v_w = 0.01602 \text{ ft}^3/\text{lbm}_w$$

The average water volume flow rate is

$$\dot{V}_w = \dot{m}_w v_w = \frac{77,250 \text{ lbm}_w (0.01602 \text{ ft}^3/\text{lbm}_w)}{\text{hr} (3600 \text{ sec}/\text{hr})} = 0.344 \frac{\text{ft}^3}{\text{sec}}$$

With a mean water velocity of 7 ft/sec, the total flow cross sectional area is

$$A_w = \frac{\dot{V}_w}{\vec{V}_w} = \frac{0.344 \text{ ft}^3/\text{sec}}{7 \text{ ft/sec}} = 0.0491 \text{ ft}^2$$

From Example 11.5 the inner diameter of the tubes is $d_i = 0.565$ in. and the cross sectional area of each tube is 0.00174 ft². Thus the number of tubes needed in each row to handle all the water flowing in parallel tubes is

$$\text{\# Tubes} = \frac{0.0491 \text{ ft}^2}{0.00174 \text{ ft}^2/\text{tube}} = 28.2 \text{ tubes}$$

We will round this off to 28 tubes. The actual water velocity is then

$$\vec{V}_w = \frac{7 \text{ ft}}{\text{sec}} \left(\frac{28.2}{28}\right) = 7.05 \text{ ft/sec}$$

By Table 11.5 the coil is 28 × 1.5 in = 42 in. high The face area is 24 ft² so the width of coil must be $(24 \text{ ft}^2/42 \text{ in})(144 \text{ in}^2/\text{ft}^2) = 82.3$ in.

11.22 cont'd

b) From Example 11.7, $h_1 = 33.98$ Btu/lbma,
   $h_2 = 22.01$ Btu/lbma

By Table A.4E, $h_{s,R,1} = 15.233$ Btu/lbma,

$h_{s,R,2} = 19.24$ Btu/lbma

By Eq. (11.67)

$$\Delta h_m = \frac{(33.98 - 19.24) - (22.01 - 15.233)}{\ln\left[\frac{33.98 - 19.24}{22.01 - 15.233}\right]} = 10.25 \frac{\text{Btu}}{\text{lbma}}$$

We will evaluate the quantities needed in Eq. (11.68). We will assume a mean pipe temperature of 48°F and a mean water film temperature, $t_{w,m}$ of 54°F.

By Eq. (11.59) and Table A.4E with a mean water temperature of 44°F

$$b'_R = \frac{19.24 - 17.17}{48 - 44} = 0.518 \text{ Btu/lbma °F}$$

By Table A.4E we find $b_{w,m} = 0.616$ Btu/lbma °F

Inspection shows that $h_{c,o}$, $h_{o,w}$ and ϕ_w should be closely equal to the values computed in Example 11.7.

11.22 cont'd

By Eq. (2.52), $h_i = \dfrac{k_w}{d_i} 0.023 \, Re_{d_i}^{0.8} \, Pr_i^{0.4}$

By Table A.6E at 44°F
$c_p = 1.00$ Btu/lbm °F
$\mu = 3.53$ lbm/hr ft
$k = 0.331$ Btu/hr ft °F
$Pr = 10.72$

By Table A.1E at 44°F
$\rho = 62.42$ lbm/ft³

$\nu = \mu/\rho = \dfrac{3.53 \text{ lbm/hr ft}}{62.42 \text{ lbm/ft}^3} = 0.0566 \text{ ft}^2/\text{hr}$

$Re_{d_i} = \dfrac{7.02 \text{ ft/sec} (0.565 \text{ in})(3600 \text{ sec/hr})}{0.0566 \text{ ft}^2/\text{hr} (12 \text{ in/ft})} = 21{,}020$

$h_i = \dfrac{0.331 \text{ Btu/hr ft °F}}{0.565 \text{ in} (1 \text{ ft}/12 \text{ in})} \, 0.023 \, (21{,}020)^{0.8} \, (10.72)^{0.4}$

$h_i = 1200$ Btu/hr ft² °F

By Table 2.3, use $1/h_{d,i} = 0.0005$ hr ft² °F/Btu

Substituting values into Eq. (11.68)

$U_{o,w} = \left[\dfrac{0.518 \text{ Btu/lbm}_a \text{°F}(19.31)}{1200 \text{ Btu/hr ft}^2\text{°F}} + \dfrac{0.518 \text{ Btu/lbm}_a\text{°F}(19.31)}{(0.0005 \text{ hr ft}^2\text{°F/Btu})^{-1}} \right.$

$\left. + \dfrac{0.616 \text{ Btu/lbm}_a\text{°F}(0.32)}{\dfrac{27.7 \text{ Btu}}{\text{hr ft}^2\text{°F}}} (0.105 + 0.68) + \dfrac{0.616 \text{ Btu/lbm}_a\text{°F}}{\dfrac{27.7 \text{ Btu}}{\text{hr ft}^2\text{°F}}} \right]^{-1}$

11.22 Cont'd

$$U_{o,w} = [0.00834 + 0.00500 + 0.00009 + 0.02224]^{-1}$$

$$U_{o,w} = 28.0 \text{ Btu/hr ft}^2 (\text{Btu/lb}_{ma})$$

We will now check the assumed values for $t_{p,m}$ and $t_{w,m}$. By Eq. (11.64)

$$t_{p,m} = 44°F + \frac{28.0 \text{ Btu}}{\text{hr ft}^2(\text{Btu/lb}_{ma})} \frac{(19.31)(28.00 - 17.24) \text{ Btu/lb}_{ma}}{1200 \text{ Btu/hr ft}^2 °F}$$

$t_{p,m} = 48.85°F$ which is sufficiently close to the assumed value of 48°F

At the mean water temperature of 44°F, $h_{s,w,m} = 17.17$ Btu/lb$_{ma}$. The mean moist air enthalpy is $h_{s,R,m} + \Delta h_m = 27.5$ Btu/lb$_{ma}$

By Eq. (11.65)

$$h_{s,w,m} = \frac{27.5 \text{ Btu}}{\text{lb}_{ma}} - \left[\frac{\frac{0.245 \text{ Btu}}{\text{lb}_{ma} °F} \left(\frac{27.7 \text{ Btu}}{\text{hr ft}^2(\text{Btu/lb}_{ma})} \right) 0.68}{\frac{0.616 \text{ Btu}}{\text{lb}_{ma} °F} \frac{11.7 \text{ Btu}}{\text{hr ft}^2 °F}} \right] \times$$

$$\left[1 - \frac{\frac{0.513 \text{ Btu}}{\text{lb}_{ma} °F} \left(\frac{28.0 \text{ Btu}}{\text{hr ft}^2(\text{Btu/lb}_{ma})} \right) 19.31}{1200 \frac{\text{Btu}}{\text{hr ft}^2 °F}} \right] (27.5 - 17.24) \frac{\text{Btu}}{\text{lb}_{ma}}$$

11.22 Cont'd

$h_{s,w,m} = 22.5$ Btu/lb$_{m_a}$

and by Table A.4E, $t_{w,m}$ is approximately 54°F

By Eq. (11.66)

$$A_o = \frac{618,000 \text{ Btu/hr}}{\frac{28.0 \text{ Btu}}{\text{hr ft}^2(\text{Btu/lb}_{m_a})}\left(\frac{10.25 \text{ Btu}}{\text{lb}_{m_a}}\right)} = 2153 \text{ ft}^2$$

$$\text{Number of rows} = \frac{2153 \text{ ft}^2}{\left(\frac{22.86 \text{ ft}^2}{\text{ft}^2_{\text{face}} \cdot \text{row}}\right) 24 \text{ ft}^2_{\text{face}}} = 3.92$$

Use 4 rows

11.23
a) We will neglect the thermal resistance of the pipe. Let t_w be the temperature of the outer edge of the frost layer and t_p the pipe temperature. For a differential length of dL we may write

$$dq = \frac{h_o \, dA_o \, (h - h_{s,w})}{c_{p,a}} \quad (1)$$

$$dq = \frac{2\pi k_{frost}(t_w - t_p) dL}{\ln(D_o/D_i)_{frost}} = \frac{2\pi k_{frost}(h_{s,w} - h_{s,p}) dL}{b_w \ln(D_o/D_i)_{frost}} \quad (2)$$

$$dq = h_i \, dA_{P,i} (t_p - t_R) = \frac{h_i \, dA_{P,i}}{b'_R}(h_{s,P} - h_{s,R}) \quad (3)$$

$$dq = U_{o,w} \, dA_o \, (h - h_{s,R}) \quad (4)$$

Combining Eqns. (1-4) and noting that
$dA_{P,i} = \pi d_{P,i} \, dL$ and $dA_o = \pi D_o \, dL$

$$U_{o,w} = \left[\frac{b'_R D_o}{h_i D_{P,i}} + \frac{b_w D_o \ln(D_o/D_i)_{frost}}{2 k_{frost}} + \frac{c_{p,a}}{h_o} \right]^{-1} \quad (5)$$

We will assume that $t_p = -19°F$ and $t_w = -10°F$
By Eqn. (11.59) and Table A.4E

$$b'_R = \frac{-4.269 - (-4.527)}{-19 - (-20)} = 0.258 \text{ Btu/lbm}_a \, °F$$

$$b_w = 0.267 \text{ Btu/lbm}_a \, °F$$

11.23 Cont'd

Substituting values into Eqn. (5)

$$U_{o,w} = \left[\frac{0.258 \text{ Btu/lb}_{m_a}°F (4.375\text{in})}{500 \text{ Btu/hr ft}^2°F (2.059\text{in})} + \frac{0.267 \text{ Btu/lb}_{m_a}°F (4.375\text{in}) \ln}{2(0.27 \text{ Btu/hr ft}°F)(12\text{in})} \right.$$

$$\left. + \frac{0.240 \text{ Btu/lb}_{m_a}°F}{3.0 \text{ Btu/hr ft}^2°F} \right]^{-1}$$

$$U_{o,w} = \left[0.0011 + 0.1101 + 0.0800 \right]^{-1}$$

$$U_{o,w} = 5.23 \text{ Btu/hr ft}^2 (\text{Btu/lb}_{m_a})$$

Inspection shows that values for b_R' and b_w are not sensitive to the estimates made for t_P and t_w so the assumed temperatures should be close enough.

By Eq. (4) using the total length of pipe

$$\dot{Q} = U_{o,w} A_o (h - h_{s,R})$$

$$= \frac{5.23 \text{ Btu} (1000 \text{ ft}) \pi (4.375\text{in}) (0.835 + 4.527) \text{ Btu/lb}_{m_a}}{\text{hr ft}^2 (\text{Btu/lb}_{m_a})(12\text{in/ft})}$$

$$\dot{Q} = 32,100 \text{ Btu/hr}$$

b) When the frost layer thickness is zero, the second term in Eq. (5) vanishes, thus

$$U_{o,w} = \left[\frac{0.258 (2.375)}{500 (2.059)} + \frac{0.240}{3.0} \right]^{-1} = \frac{12.4 \text{ Btu}}{\text{hr ft}^2 (\text{Btu/lb}_{m_a})}$$

11.23 Cont'd

Then

$$\dot{Q} = \frac{12.4(1000)\pi(2.375)}{12}(0.835 + 4.527) = 41,300 \text{ Btu/hr}$$

11.24

a) Because of the relatively high inlet dry bulb temperature ($100°F$) and the relatively low dew point temperature ($53.7°F$) we should check to determine whether the initial portion of the coil is wet or dry. For a Type II surface, we know from Example 11.4 that

$$A_o/A_{p,i} = 19.31, \quad A_{p,o}/A_F = 0.105$$
$$r_2/r_1 = 2.70$$

$$L\sqrt{\frac{h_{c,o}}{k_F Y_F}} = \frac{0.576}{12}\sqrt{\frac{(10)(2)(12)}{(120)(0.016)}} = 0.537$$

By Fig. 11.11, $\phi = 0.87$
We will assume $1/h_{d,i} = 0$
By Eq. (11.40)

$$U_o = \left[\frac{19.31}{\frac{500 \, Btu}{hr \, ft^2 °F}} + \frac{(1-0.87)}{\frac{10 \, Btu}{hr \, ft^2 °F}}(0.105+0.87) + \frac{1}{\frac{10 \, Btu}{hr \, ft^2 °F}}\right]^{-1}$$

$$U_o = 6.581 \, Btu/hr \, ft^2 °F$$

By Eq. (11.70)

$$t_{F,m} = 100°F - 0.87\left[1 - \frac{\left(\frac{6.581 \, Btu}{hr \, ft^2 °F}\right)(19.31)}{500 \, Btu/hr \, ft^2 °F}\right](100-44)°F = 63.7°F$$

This is higher than the dew point temperature ($53.7°F$) so the initial portion of the coil is dry.

By Eq. (11.71)

11.24 Cont'd

$$t = \frac{53.7°F - 0.87(1 - 6.581(19.31)/500) \, 44°F}{1 - 0.87(1 - 6.581(19.31)/500)} = 71.6°F$$

For air dry bulb temperatures above 71.6°F the condition line follows a line of constant humidity ratio (dew point). For air temperatures below 71.6°F dehumidification will occur.

For the wet portion of the coil we will assume $y_w = 0.005$ in., $t_{p,m} = 48°F$ and $t_{w,m} = 53°F$. We find $b_R' = 0.515$ Btu/lbm$_a$ °F and $b_{w,m} = 0.590$ Btu/lbm$_a$ °F. We may estimate $c_{p,a} = 0.243$ Btu/lbm$_a$ °F.

By Eq. (11.54)

$$h_{o,w} = \left[\frac{0.243 \text{ Btu/lbm}_a °F}{0.590 \text{ Btu/lbm}_a °F \,(10.0 \text{ Btu/hr ft}^2 °F)} + \frac{0.0005 \text{ in}(1 \text{ ft}/12 \text{ in})}{0.33 \text{ Btu/hr ft °F}}\right]$$

$$h_{o,w} = 23.5 \text{ Btu/hr ft}^2 °F$$

Thus

$$L\sqrt{\frac{h_{o,w}}{k_F \, y_F}} = \frac{0.576}{12}\sqrt{\frac{(23.5)(2)(12)}{(120)(0.016)}} = 0.823$$

By Fig. 11.11 with $r_2/r_1 = 2.70$, $\phi_w = 0.73$
By Eq. (11.63)

$$U_{o,w} = \left[\frac{0.515 \text{ Btu/lbm}_a °F (19.31)}{500 \text{ Btu/hr ft}^2 °F} + \frac{0.590 \text{ Btu/lbm}_a °F (1-0.73)}{\frac{23.5 \text{ Btu}}{\text{hr ft}^2 °F}(0.105 + 0.73)} + \frac{0.590 \frac{\text{Btu}}{\text{lbm}_a °F}}{\frac{23.5 \text{ Btu}}{\text{hr ft}^2 °F}}\right]$$

11.24 Cont'd

$U_{o,w} = 18.83$ Btu/hr ft² (Btu/lbm$_a$)

Evaluating Eq. (11.65) at an arbitrary value of h

$$h_{s,w,m} = h - \frac{0.243(23.5)0.73}{0.590(10)}\left(1 - \frac{0.515(18.83)(19.31)}{500}\right)(h - 17.17)$$

$$h_{s,w,m} = 0.558h + 7.59 \quad \text{Btu/lbm}_a \tag{1}$$

On the psychrometric chart we draw a horizontal line from the inlet air state ① to point ⓐ at $t = 71.6°F$.

From point ⓐ we follow the procedure outlined in Example 11.8.

ⓐ From chart C-8E, $h_a = 26.8$ Btu/lbm$_a$
By Eq. (1) above,
$h_{s,w,m,a} = 0.558(26.8) + 7.59 = 22.54$ Btu/lbm$_a$
Locate state s,w,m,a on chart C-8E on the saturation line
Read from the enthalpy-moisture ratio protractor for the line connecting states ⓐ and s,w,m,a
$(\Delta h / \Delta W)_a \cong \infty$
Extend a nearly horizontal line to an arbitrary value of h of 25 Btu/lbm$_a$ and locate point ⓑ, $t_b = 64°F$

ⓑ $h_b = 25.0$ Btu/lbm$_a$
$h_{s,w,m,b} = 0.558(25.0) + 7.59 = 21.54$ Btu/lbm$_a$
$(\Delta h / \Delta W)_b \cong 5500$ Btu/lbm$_w$
at 64°F from Table A.1E, $h_{g,t} = 1089.23$ Btu/lbm

11.24 Cont'd.

By Eq. (9.7) and Table 9.1, we estimate $Le \cong 0.90$
By Eq. (11.45)
$$\frac{dh}{dw} = 0.9(5500) + 1089.23 - 1061(0.9) = 5086 \text{ Btu/lbm}_w$$

Extend a line in this direction from ⓑ to an arbitrary value of $h = 23$ Btu/lbm$_a$ and locate point ⓒ, $t_c = 57.6°F$

ⓒ $h_{s,w,m,c} = 0.558(23.0) + 7.59 = 20.42$ Btu/lbm$_a$
$(\Delta h / \Delta w)_c = 3800$ Btu/lbm$_w$
$$\frac{dh}{dw} = 0.9(3800) + 1086 - 1061(0.9) = 3550$$

Extend a line to $h = 20$ Btu/lbm$_a$ at ⓓ, $t_d = 50°F$

ⓓ as point ⓓ is nearly on the saturation line, the remaining process line to $t = 44°F$ for an infinitely deep coil essentially follows the saturation curve.

b) We will first determine the extent of the dry surface. From the problem statement and Chart C-8E we have $t_1 = 100°F$, $W_1 = 0.00883$ lbm$_w$/lbm$_a$
$v_1 = 14.31$ ft³/lbm$_a$

Thus $\dot{m}_a = \frac{\dot{V}_1}{v_1} = \frac{10,000 \text{ ft}^3/\text{min}}{14.31 \text{ ft}^3/\text{lbm}_a}\left(\frac{60 \text{ min}}{hr}\right) = 41,929$ lbm$_a$/hr

$c_{p,a} = 0.240 + 0.444(0.00883) = 0.244$ Btu/lbm$_a$·°F

By Eq. (11.17)
$$K_1 = \ln\left(\frac{t_1 - t_R}{t_2 - t_R}\right) = \ln\left(\frac{100 - 44}{71.5 - 44}\right) = 0.711$$

11.24 cont'd

$$A_o = \frac{\dot{m}_a c_{p,a} K_1}{U_o} = \frac{41{,}929 \text{ lbm}/\text{hr} (0.244 \text{ Btu/lbm}_a \,°F) \, 0.711}{6.581 \text{ Btu/hr ft}^2 °F}$$

$$A_o = 1105 \text{ ft}^2 \text{ of dry surface}$$

2 rows

$$A_o = 2(22.86 \text{ ft}^2/\text{ft}^2_{face} \text{ row})(20 \text{ ft}^2_{face}) = 914.4 \text{ ft}^2$$

(All of this is dry)

$$K_1 = \frac{U_o A_o}{\dot{m}_a c_{p,a}} = \frac{(6.581)(914.4)}{(41{,}929)(0.244)} = 0.588$$

By Eq. (11.17)

$$t_2 = 100°F - 56°F (1 - e^{-0.588}) = 75.1°F$$

From the condition line drawn on chart C-8E,

$$t^*_{2 \text{ rows}} = 61.9°F$$

4 rows

$$A_o = 2(914.4 \text{ ft}^2) = 1828.8 \text{ ft}^2$$
$$A_{o,w} = 1828.8 - 1105 = 723.8 \text{ ft}^2 \text{ of wet surface}$$

At the point where condensation begins, ⓐ, $h = 26.9$ Btu/l

$$\frac{U_{o,w} A_{o,w}}{\dot{m}_a} = \frac{18.83 \text{ Btu/hr ft}^2 (\text{Btu/lbm}_a) \, 723.8 \text{ ft}^2}{41{,}929 \text{ lbm}_a/\text{hr}} = 0.324$$

By Eq. (11.73)

$$h_2 = 17.17 \frac{\text{Btu}}{\text{lbm}_a} + (26.9 - 17.17) \frac{\text{Btu}}{\text{lbm}_a} e^{-0.324} = 24.21 \text{ Btu/lbm}_a$$

At the intersection of this enthalpy with the condition line we read $t_2 \underset{318}{=} 61.7°F$, $t_2^* = 56.6°F$

11.24 Cont'd.

8 rows

$A_o = 2(1828.8 \text{ ft}^2) = 3657.6 \text{ ft}^2$

$A_{o,w} = 3657.6 - 1105 = 2552.6 \text{ ft}^2$ wetted surface

$$\frac{U_{o,w} A_{o,w}}{\dot{m}_a} = \frac{18.83(2552.6)}{41,929} = 1.146$$

By Eq. (11.73)

$h_2 = 17.17 + (26.90 - 17.17) e^{-1.146} = 20.26 \text{ Btu/lb}_{m_a}$

At the intersection of this enthalpy with the condition line, we read $t_2 = 50.2°F$, $t_2^* = 49.8°F$

12.1

The increase in air velocity near the person will increase the convective heat transfer coefficient on the surface of the skin and clothing. The water vapor mass transfer coefficient will also increase. This will result in a decrease in skin surface temperature which will be felt as a cool sensation. The person will be more comfortable thermally when the surrounding air conditions are warm and humid by sitting in front of a fan.

12.2

Convective sensible heat transfer: no
  the air dry bulb temperature is higher than the skin temperature

Convective latent heat transfer (moisture transfer): no
  the air humidity ratio from a psychrometric chart (Chart C-8E) is $0.0279\ lbm_w/lbm_a$. The saturated humidity ratio at the skin temperature of 85°F is $0.0264\ lbm_w/lbm_a$. Therefore moisture will travel from the very humid air toward the skin surface.

Infrared radiative heat transfer: yes
  The skin temperature is higher than the surrounding surfaces temperature so thermal radiation will occur from the skin to the surrounding surfaces.

12.3

$$\dot{Q}_{sen} = \left[h_c(t_{cl} - t) + h_R(t_{cl} - \bar{t}_r) \right] A_{cl} \quad (12.3)$$

let $h = h_c + h_R$ and $t_o = \dfrac{h_c t + h_R \bar{t}_r}{h}$

then $\bar{t}_r = \dfrac{h t_o - h_c t}{h_R}$

substituting into Eq. (12.3)

$$\dot{Q}_{sen} = \left[h_c(t_{cl} - t) + h_R\left(t_{cl} - \dfrac{h t_o - h_c t}{h_R}\right) \right] A_{cl}$$

$$\dot{Q}_{sen} = \left[(h_c + h_R) t_{cl} - h t_o \right] A_{cl}$$

$$\dot{Q}_{sen} = h(t_{cl} - t_o) A_{cl} \quad \text{or} \quad R_o = 1/h A_{cl} \quad (12.6)$$

let $\dot{Q}_{sen} = \dfrac{(t_{sk} - t_o)}{R}$

$R = \dfrac{R_{cl}}{A_{sk}} + \dfrac{1}{h A_{cl}}$ (see Figure 12.1)

$$\dot{Q}_{sen} = \dfrac{(t_{sk} - t_o)}{\dfrac{R_{cl}}{A_{sk}} + \dfrac{1}{h A_{cl}}}$$

$$\dot{Q}_{sen} = \dfrac{A_{sk}(t_{sk} - t_o)}{R_{cl} + \dfrac{A_{sk}}{h A_{cl}}} \quad (12.7)$$

12.4

a) decrease (need higher heat loss rate)

b) decrease (heat transfer by convection is proportional to $h\Delta t$; when h is reduced Δt should be increased which usually results in a lower air temperature)

c) decrease (additional thermal resistance should be compensated by an increase in temperature difference that usually results in a lower air temperature)

d) increase (more heat is lost by radiation, less heat loss by convection)

e) increase (drier air will allow more latent cooling, the dry bulb temperature should be raised to reduce the amount of sensible heat loss)

12.5

By Eq. (12.5)

$$t_o = \frac{h_c t + h_R \bar{t}_r}{h}$$

$h_R = 4.7 \text{ w/m}^2{}^\circ C \quad (p. 335)$

By Eq. (12.10b)

$h_c = 8.3 (0.5)^{0.6} = 5.48 \text{ w/m}^2{}^\circ C$

$h = 4.7 + 5.48 = 10.2 \text{ w/m}^2{}^\circ C$

$$t_o = \frac{5.48 (25^\circ C) + 4.7 (35^\circ C)}{10.2} = 29.6\,^\circ C$$

12.6

We will follow the procedure used in Ex. 12.2,

$$h_c = 8.3(0.6)^{0.6} = 6.11 \text{ W/m}^2\text{°C}$$

$$h_E^v = \frac{6.11}{0.895(1.007)} = 6.78 \text{ W kg}_a/\text{m}^2 \text{ kJ}$$

$$\dot{Q}_c = 1.8(6.11)(33.7-25) = 95.68 \text{ W}$$

$$\dot{Q}_R = -44.84 \text{ W}$$

$$\dot{Q}_E = 1.8(6.78)0.3(0.03406-0.014758) \times 2421.79$$

$$= 171.14 \text{ W}$$

12.7

The solution procedure follows that of Ex. 12.3

From Table 12.2,

$$R_T = 1.02 \times 0.88 = 0.90 \text{ hr ft}^2°F/Btu$$

$$A_{cl} = (19.6 \text{ ft}^2) 1.10 = 21.6 \text{ ft}^2$$

$$\dot{Q}_{sen} = 21.6 \text{ ft}^2 (92.7 - 75)°F / 0.90 \text{ hr ft}^2°F/Btu = 425 \text{ Btu/hr}$$

$$A_{cl,w} = 21.6 \text{ ft}^2 (0.1) = 2.16 \text{ ft}^2$$

$$R_T^v = \frac{0.21 (0.90)}{0.42} = 0.45 \text{ hr ft}^2/lb_{ma}$$

$$\dot{Q}_{lat} = (2.16 \text{ ft}^2)(0.03410 - 0.0092) \frac{lb_{mv}}{lb_{ma}} \left(1041.8 \frac{Btu}{lb_{mv}}\right) / \left(0.45 \frac{\text{hr ft}^2}{lb_{ma}}\right)$$

$$\dot{Q}_{lat} = 125 \text{ Btu/hr}$$

$$\dot{Q}_{sk} = 425 + 125 = 550 \text{ Btu/hr}$$

12.8

The plastic will behave as a very good vapor retarder preventing moisture from escaping. This will cause the moisture content of the socks to increase and will eventually create wet conditions. The wet socks will have much less insulating capability than dry socks so the heat loss from the person's feet will increase dramatically. Although the plastic encasement will work well over a short period of time, the longer term moisture condensation will result in a worse situation than if the socks were left uncovered but dry.

12.9

From Ex. 12.4

$$\dot{Q}_{sen} = \dot{m}_{res}\, c_{pa}\, (t_{ex} - t_{inhaled})$$

The regenerative capability of the mask can be described as

$$t_{inhaled} = t + \eta_{sen}(t_{ex} - t)$$

$$= 75°F + 0.5(94.1 - 75)°F = 84.6°F$$

Thus
$$\dot{Q}_{sen} = (2.38)(0.249)(94.1 - 84.6) = 5.7 \text{ Btu/hr}$$

Likewise for the latent regeneration

$$W_{inhaled} = W + \eta_{lat}(W_{ex} - W)$$

$$= 0.0092 + 0.3(0.0317 - 0.0092) = 0.0160 \text{ lbmv/lbma}$$

$$\dot{Q}_{lat} = (2.38)(0.0317 - 0.0160)(1038.2) = 38.9 \text{ Btu/hr}$$

12.10

The procedure and nomenclature follow that used in Ex. 12.5. Assume point P is near the right rear corner of the room shown in Fig. 12.3.

Angle factor to heater to left of point N:
$a = 10m$, $b = 1m$, $c = 2m$
$X = 10m / 1.8(2m) = 2.778$
$Y = 1m / 1.8(2m) = 0.278$
$F_{P-1} = 0.03286$

Angle factor to heater to right of point N:
$a = 2m$, $b = 1m$, $c = 2m$
$X = 0.556$
$Y = 0.278$
$F_{P-2} = 0.01967$

Angle factor to window to left of point N:
$a = 10m$, $b = 2m$, $c = 2m$
$X = 2.778$, $Y = 0.556$, $F_{P-3} = 0.65953$

Angle factor to window to right of point N:
$a = 2m$, $b = 2m$, $c = 2m$
$X = 0.556$, $Y = 0.556$, $F_{P-4} = 0.03494$

Angle factor to remaining surfaces
$F_{P-4} = 1.0 - 0.03286 - 0.01967 - 0.05953 - 0.03494$
$= 0.85300$

$$\bar{T}_r = \left[(0.03286 + 0.01967)(323.15)^4 + (0.05953 + 0.03494)(288.15)^4 + 0.8530(293.15)^4\right]^{1/4}$$

$\bar{T}_r = 294.5 K$ or $21.4°C$

Problem 12.11

Divide plane into 1m × 1m areas. Compute the mean radiant temperature in the center of each area using the procedure given in Problem 12.10. The results are shown here with some contours sketched through the numbers in the table.

15°C Window + Heater (50°C)

24.09	25.03	25.18	25.21	25.22	25.23	25.23	25.22	25.21	25.18	25.03	24.09
21.40	21.81	22.03	22.14	22.19	22.21	22.21	22.19	22.14	22.03	21.81	21.40
20.78	20.95	21.08	21.17	21.22	21.24	21.24	21.22	21.17	21.08	20.95	20.78
20.52	20.61	20.69	20.74	20.78	20.80	20.80	20.78	20.74	20.69	20.61	20.52
20.38	20.43	20.48	20.51	20.54	20.55	20.55	20.54	20.51	20.48	20.43	20.38
20.29	20.33	20.35	20.38	20.39	20.40	20.40	20.39	20.38	20.35	20.33	20.29
20.23	20.25	20.27	20.29	20.30	20.30	20.30	20.30	20.29	20.27	20.25	20.23
20.19	20.20	20.21	20.22	20.22	20.23	20.23	20.22	20.22	20.21	20.20	20.19

Contours: 25°C, 22°C, 21°C, 20.5°C; surrounding walls/floor at 20°C.

Mean radiant temperature distribution at a height of 1 m above the floor.

12.13

By Eq. (12.20)

$$\bar{T}_r = \left[\frac{0.9(21+273.15)^4 - 0.15(20+273.15)^4 + \frac{5W}{m^2 °C}(21-20)°C \frac{m^2 K^4}{5.670 \times 10^{-8} W}}{0.9 - 0.15}\right]$$

$\bar{T}_r = 295.5 K$ or $\bar{t}_r = 22.34 °C$

By Eq. (12.21)

$$t = \frac{0.9(5.670 \times 10^{-8} W/m^2 K^4)}{5 W/m^2 °C}\left[(21+273.15)^4 - \left(\frac{0.9(21+273.15)^4 - 0.15(20+273.15)^4 + 5W/m^2 °C(21-20)°C/5.670 \times 10^{-8} W/m^2 K^4}{0.9 - 0.15}\right)\right] + 21°C$$

$t = 19.59 °C$

12.14

We will use Eq. (12.22) to determine $\bar{T}_r$.
We will estimate the convective heat transfer coefficient from Eq. (2.54) with the addition of the conduction limit of $Nu_D = 2.0$. The viscosity ratio term is neglected for gases.

$$Re_D = \rho \vec{V} D / \mu$$

From Table A.5E at 70°F
- $\rho = 0.07493 \; lb_m / ft^3$
- $\mu = 0.04400 \; lb_m / hr \, ft$
- $k = 0.01491 \; Btu / hr \, ft \, °F$
- $Pr = 0.709$

$$Re_D = \frac{(0.07493 \; lb_m / ft^3)(25 \; ft/min)(0.5 \; ft)(60 \; min/hr)}{0.04400 \; lb_m / hr \, ft}$$

$$Re_D = 1277$$

By Eq. (2.54)

$$h = \frac{0.01491 \; Btu/hr \, ft \, °F}{0.5 \; ft} \left[2 + \left(0.4(1277)^{0.5} + 0.06(1277)^{2/3} \right) 0.709^{0.4} \right]$$

$$h = 0.60 \; Btu / hr \, ft^2 \, °F$$

By Eq. (12.22)

$$\bar{T}_r = \left[(68 + 459.67)^4 + \frac{0.6 \; Btu \, (68-72)°F}{hr \, ft^2 \, °F \, (0.95)(0.1714 \times 10^{-8} \; Btu/hr \, ft^2 \, °R^4)} \right]^{1/4}$$

$$\bar{T}_r = 525.14 \, °R \quad \text{or} \quad \bar{t}_r = 65.47 \, °F$$

12.15

We will set Eq. (12.24) to equal values at $t = 25°C$ and $\phi = 50\%$ ($ET^* = 25°C$) and a point where $t = 26°C$. Solving for W will provide the necessary second point from which the line may be constructed.

$t_{sk} = 33.7°C$

By Table A.4SI, $W_{s,sk} = 0.034080 \text{ kg}_v/\text{kg}_a$

set $t_o = 25°C$, By Table A.4SI, $W_{s,to} = 0.020206 \text{ kg}_v/\text{kg}$

$C_{sen}(33.7-25)°C + C_{lat}(0.034080 - 0.020206/2) \text{ kg}_v/\text{kg}_a =$

$C_{sen}(33.7-26)°C + C_{lat}(0.034080 - W) \text{ kg}_v/\text{kg}_a$

Dividing through by C_{lat},

$\dfrac{C_{sen}}{C_{lat}}(8.70)°C + 0.023977 = \dfrac{C_{sen}}{C_{lat}}(7.70)°C + (0.034080 - W) \text{ kg}_v/\text{kg}_a$

Substitute the given value for C_{sen}/C_{lat} and solve for W

$0.005(8.70) + 0.023977 = 0.005(7.70) + 0.034080 - W$

$W = 0.00510 \text{ kg}_v/\text{kg}_a$

Therefore the line of constant ET^* passes through the points ($t = 25°C$, $\phi = 50\%$) and ($t = 26°C$, $W = 0.00510 \text{ kg}_v/\text{kg}_a$).

Problem 12.15

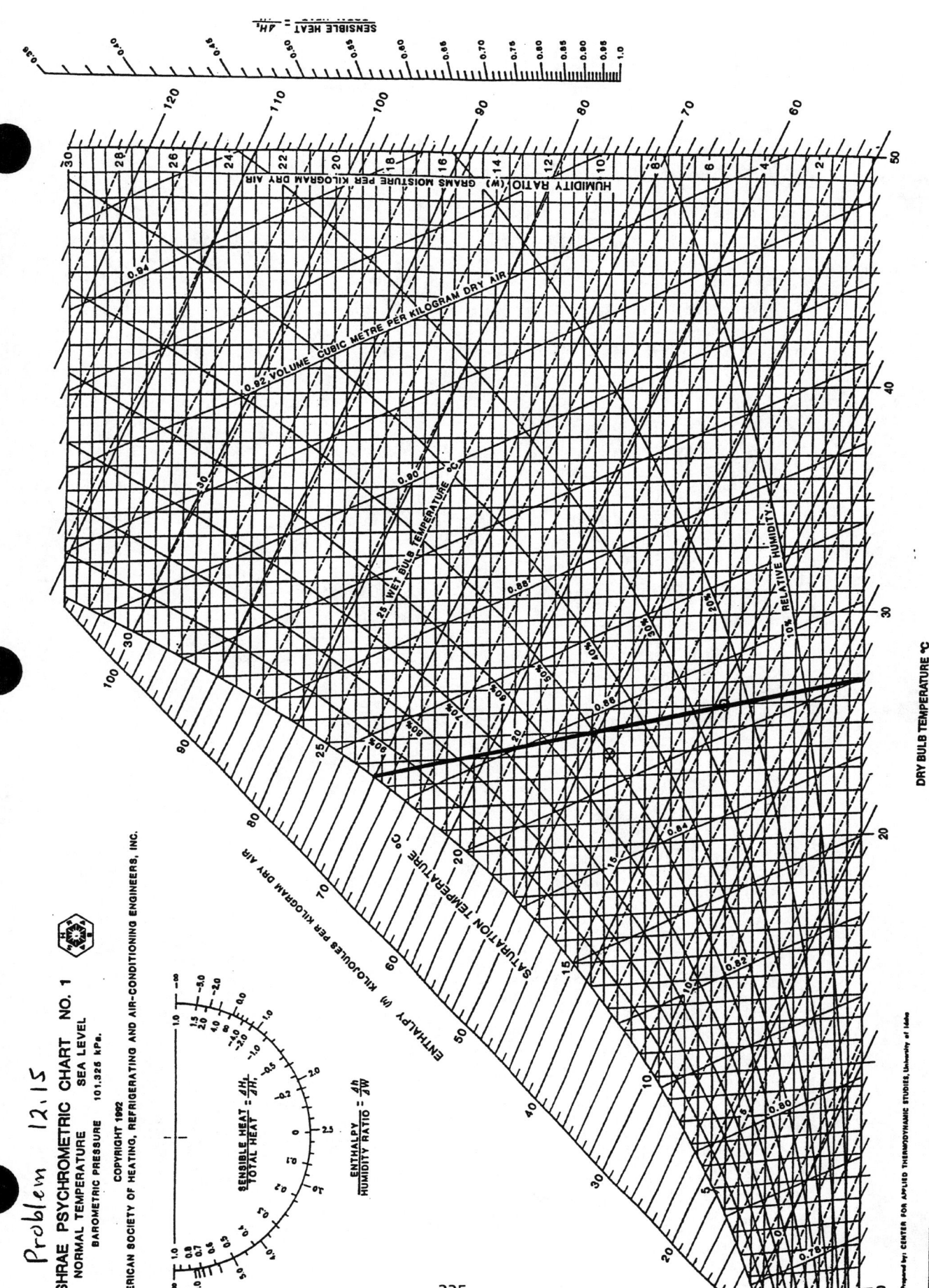

Chart C-8SI

12.16

a) refer to Eq. (12.24), C_{sen}/C_{lat} decreases so a larger variation in t_o is associated with a fixed change in ΔW, slope of ET^* line decreases

b) just the opposite of part a), slope increases

c) C_{sen} will change less than C_{lat} as the air velocity has no effect on the radiative heat exchange. The ratio C_{sen}/C_{lat} will decrease so the slope of the ET^* line will decrease.

d) The change in metabolic rate does not affect the sensible heat transfer. However the skin wetter area increases so C_{lat} increases. This causes the ratio C_{sen}/C_{lat} to decrease so the slope of the constant ET^* line decreases.

12.17

Assume standard sea level pressure.
From chart C-8E, $W = 0.013 \; lbm_w/lbm_a$

By Eq. (12.26a)

$$\dot{Q}_E = 700 \frac{Btu}{hr} + 2(\vec{V})^{0.5}(92-95)°F + 22(90-95)°F$$

$$\dot{Q}_E = 590 - 6(\vec{V})^{0.5} \quad Btu/hr \qquad (1)$$

By Eq. (12.27a)

$$\dot{Q}_{E,max} = 11,700(\vec{V})^{0.4}(0.03676 - 0.013)$$

$$\dot{Q}_{E,max} = 278(\vec{V})^{0.4} \qquad (2)$$

By Eq. (12.25)

$$\dot{Q}_E = 0.3 \; \dot{Q}_{E,max} \qquad (3)$$

By Eqns. (1) - (3)

$$590 - 6(\vec{V})^{0.5} = 0.3\left[278(\vec{V})^{0.4}\right]$$

$$\vec{V}^{0.5} + 13.9(\vec{V})^{0.4} - 98.3 = 0$$

By trial we find $\vec{V} = 101.5 \; ft/min$

12.18

$$\dot{Q}_M = 3.0 \text{ met} \left(\frac{58.2 \text{ W}}{m^2 \text{ met}}\right)(1.8 \text{ m}^2) = 0.314 \text{ kW}$$

From chart C-8SI, $W = 0.0164 \text{ kg}_v/\text{kg}_a$

By Eq. (12.26b)

$$\dot{Q}_E = 0.314 \text{ kW} + 0.015(0.5)^{0.5}(30-35) + 0.012(27-35)$$

$$\dot{Q}_E = 0.165 \text{ kW}$$

By Eq. (12.27b)

$$\dot{Q}_{E,max} = 28.4(0.5)^{0.4}(0.03676 - 0.0164)$$

$$\dot{Q}_{E,max} = 0.438 \text{ kW}$$

By Eq. (12.25)

H.S.I. = $(100)(0.165)/(0.438) = 38$

12.19

We will evaluate the heat stress index to make our recommendation.

As the absolute barometric pressure is not standard sea level pressure, we cannot use the sea level psychrometric chart or moist air tables.

By Eqs. (7.18) & (7.29) and Table A.1E

$$W = \frac{0.622(0.43005)}{14.15 - 0.43005} = 0.01950 \ lbm_v/lbm_a$$

$$W = \frac{0.01950(1094.3 - 42.67) - 0.245(90-75)}{1100.53 - 42.67} = 0.01591 \ lbm_v/lbm_a$$

From Table 12.1, let met = 1.8

$$\dot{Q}_M = \frac{33 \ Btu}{hr \ ft^2}(19.6 \ ft^2) = 647 \ Btu/hr$$

By Eq. (12.26a)

$$\dot{Q}_E = 647 + 2(150)^{0.5}(90-95) + 22(90-95) = 415 \ Btu/hr$$

By Eq. (12.27a)

$$\dot{Q}_{E,max} = 11,700(150)^{0.4}(0.03676 - 0.01591) = 1810 \ Btu/hr$$

By Eq. (12.25), H.S.I. = 100(415)/(1810) = 23

According to Table 12.3, these conditions should not be detrimental to healthy, physically fit workers.

12.20
Assume the mean radiant temperature is equal to the dry-bulb temperature. Therefore, by Eqn. 12.5 the operative temperature is equal to the dry-bulb temperature. For the conditions of the problem we can use Fig. 12.7 to establish if people will be comfortable.

a.) For these conditions use t and t* in Chart C-8E to get ϕ and use t and ϕ in Fig. 12.7.

    a.1 t = 75 °F, t* = 65 °F therefore ϕ = 59% and from Fig. 12.7 it is in the comfort zone.
    a.2 t = 80 °F, t* = 70 °F therefore ϕ = 61% and from Fig. 12.7 it is not in the comfort zone.
    a.3 t = 95 °F, t* = 75 °F therefore ϕ = 40% and from Fig. 12.7 it is not in the comfort zone.

b.) Use the temperatures and relative humidities given directly in Fig. 12.7

    b.1 t = 70 °F, ϕ = 20% Out of the comfort zone
    b.2 t = 75 °F, ϕ = 30% In the comfort zone
    b.1 t = 80 °F, ϕ = 40% Out of the comfort zone

12.21
Assume the mean radiant temperature is equal to the dry-bulb temperature. Therefore, by Eqn. 12.5 the operative temperature is equal to the dry-bulb temperature. Initial conditions: $t_1 = 20$ °C, $\phi_1 = 30\%$ therefore from Fig. 12.7 ET* = 19.5 °C (approx.). For constant ET* the operative temperature at $\phi_2 = 50\%$ is then $t_2 = 19.5$ °C.

Denote the outdoor conditions with subscript 3. $t_3 = 0$ °C, $\phi_3 = 100\%$. Then the ratio of the energy requirements needed to temper the outdoor air can be written as

$$\frac{\dot{Q}_2}{\dot{Q}_1} = \frac{h_2 - h_3}{h_1 - h_3} = \frac{38 - 9.5}{31.5 - 9.5} = 1.3$$ (Where the h values are taken from Chart C-8SI.)

Therefore there is a 30% <u>increase</u> in the energy needed to temper the air if the change is made.

12.22

We will evaluate the Heat Stress Index for the measured conditions. Assume standard sea level barometric pressure.

From Chart C-8E, $W = 0.0162 \; lb_{m\nu}/lb_{ma}$

By Eq. (12.26a)

$$\dot{Q}_E = 500 + 2(40)^{0.5}(90-95) + 22(90-95)$$

$$\dot{Q}_E = 327 \; Btu/hr$$

By Eq. (12.27a)

$$\dot{Q}_{E,max} = 11,700 \, (40)^{0.4}(0.03676 - 0.0162)$$

$$\dot{Q}_{E,max} = 1052 \; Btu/hr$$

By Eq. (12.25)

$$H.S.I. = 100(327)/(1052) = 31$$

According to Table 12.3, this is a moderate to severe heat strain involving a substantial decrement in mental performance. Thus unfavorable conditions for good mental effort.

12.23

We will use Eq. (12.29) to determine the Predicted Mean Vote for these conditions.

From Table A.1SI, $P_v = P_{sat}(10°C) = 1.228$ kPa

By Eq. (12.29) and Table 12.5

$PMV = 0.220(19) + 0.233(1.228) - 5.673$

$PMV = -1.21$

The men will be slightly cool.

12.24

a) By Eq. (12.29) and Table 12.5

$0 = 0.243(20) + 0.278(P_v) - 6.802$

$P_v = 6.99 \text{ kPa}$

By Eq. (7.27) and Table A.1SI

$\phi = \dfrac{6.99 \text{ kPa}}{2.339 \text{ kPa}} = 3.0$

This is three times the saturated value. Therefore, a PMV of zero is not possible under these conditions.

b) $0 = 0.243(25) + 0.278(P_v) - 6.802$

$P_v = 2.62 \text{ kPa}$

$\phi = \dfrac{2.62 \text{ kPa}}{3.169 \text{ kPa}} = 0.83$

or $\phi = 83\%$

12.25

We will determine the PMV and then use Figure 12.8 to estimate the number of persons dissatisfied.

By Eq. (12.29), Table 12.5 and Table A.1E

$$PMV = 0.135(74) + 1.92(0.27514) - 11.122$$

$$PMV = -0.60$$

From Figure 12.8, PPD = 12%

Therefore approximately 12 persons out of the cohort of 100 should feel slightly cool.

12.26

4 Occupants	(0.46 lbm$_w$/hr)(24 hrs)	11.04 lbm$_w$
2 Showers	(2 showers)(0.54 lbm$_w$/shower)	1.08 lbm$_w$
1 Lunch cooked on an electric range		0.55 lbm$_w$
1 Dinner cooked on an electric range		1.27 lbm$_w$
6 house plants (interpolate between 5 & 7 plants)		0.95 lbm$_w$
1 load of laundry dried in a vented electric dryer		0 lbm$_w$
	Total (rounded to the nearest lbm$_w$) =	15 lbm$_w$

Or rounded to the nearest whole number the total is approximately 15 lbm$_w$.

12.27

$$\dot{V}_o = \left(\frac{7 \text{ people}}{1000 \text{ ft}^2}\right)(10,000 \text{ ft}^2)\left(20 \text{ ft}^3/\text{min - person}\right) = 1400 \frac{\text{ft}^3}{\text{min}}$$

12.28

500 m^2 of office space	(7 people/100 m^2)(500 m^2)(10 L/s-person)	350 L/s
50 m^2 of corridors	(0.25 L/s-m^2)((50 m^2)	12.5 L/s
5 restrooms, total 20 wc's	(25 L/s-wc)(20 wc's)	500 L/s
30 m^2 reception area	(60 people/100 m^2)(30 m^2)(10 L/s-person)	144 L/s
50 m^2 of conference rooms	(50 people/100 m^2)(50 m^2)(10 L/s-person)	250 L/s
	Total (to the nearest L/s) =	1257 L/s

12.29

By Eqn. 12.39 with $\dot{G} = 0$ (no internal generation)
$$\dot{R} = \dot{V}_o(C_o - C_s)$$
and from the definition of the filter efficiency
$$\dot{R} = \varepsilon_{ac} C_o \dot{V}_o$$
$$\therefore \varepsilon_{ac} = \frac{C_o - C_s}{C_o} = \frac{0.3 - 0.1}{0.3} = 0.67$$

12.30

Let $\dot{V}$ = total supply air $\therefore \dot{V}_o = 0.15\dot{V}$ and $\dot{V}_r = 0.85\dot{V}$ Thus by Eqn 12.43 with $\dot{G} = 0$,

$$C_s = \frac{C_o(0.15)(1 - \varepsilon_{ac})}{0.15 + 0.85\varepsilon_{ac}}$$

Substituting $C_o = 0.3$ mg/m^3 and $C_s = 0.1$ mg/m^3 and solving, we get $\varepsilon_{ac} = 0.23$.

12.31

By Eq. (12.40)

$$C_o \dot{V}_o + C_s \dot{V}_r - \dot{R} + \dot{G} = (\dot{V}_o + \dot{V}_r) C_s$$

Solve for $C_s \dot{V}_o$

$$C_s \dot{V}_o = C_o \dot{V}_r - \dot{R} + \dot{G} \qquad (1)$$

By Eq. (12.41)

$$\dot{R} = \varepsilon_{ac} (\dot{V}_o + \dot{V}_r) C_m$$

By Eq. (12.42)

$$\dot{R} = \varepsilon_{ac} (\dot{V}_o + \dot{V}_r) \left(\frac{C_o \dot{V}_o + C_s \dot{V}_r}{\dot{V}_o + \dot{V}_r} \right) = \varepsilon_{ac} \left(C_o \dot{V}_o + C_s \dot{V}_r \right)$$

Substitute into Eq. (1)

$$C_s \dot{V}_o = C_o \dot{V}_r - \varepsilon_{ac} (C_o \dot{V}_o + C_s \dot{V}_r) + \dot{G}$$

$$C_s (\dot{V}_o + \dot{V}_r \varepsilon_{ac}) = C_o \dot{V}_o - \varepsilon_{ac} C_o \dot{V}_o + \dot{G}$$

$$C_s = \frac{C_o \dot{V}_o (1 - \varepsilon_{ac}) + \dot{G}}{(\dot{V}_o + \dot{V}_r \varepsilon_{ac})} \qquad (12.43)$$

12.32

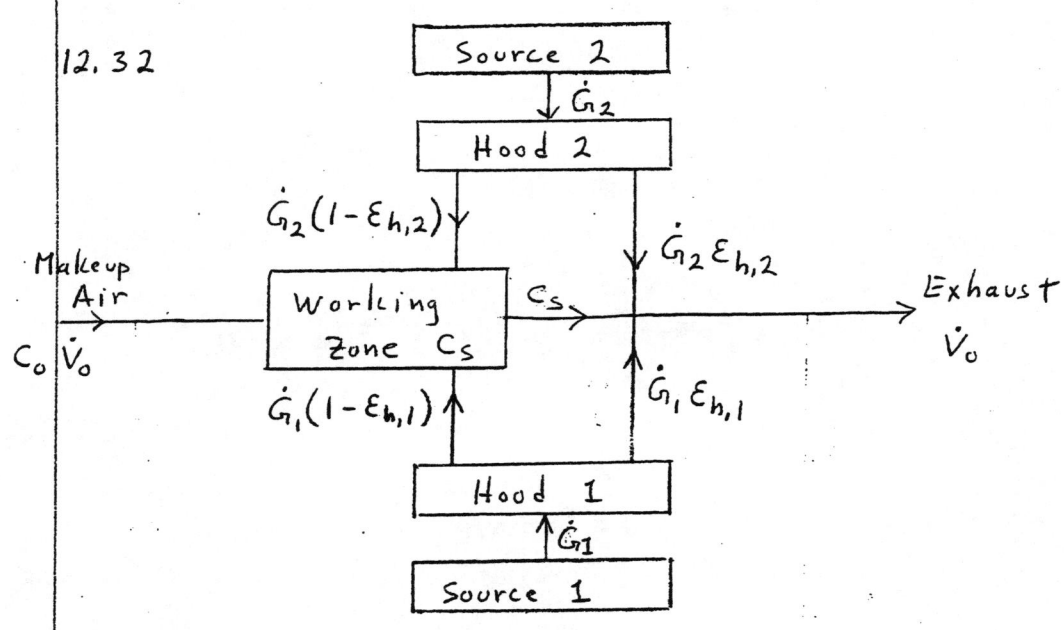

Assume no air cleaner and no recirculated air.
Assume air in working zone is well mixed.
Mass balance around working zone:

$$C_o \dot{V}_o + \dot{G}_1(1-\varepsilon_{h,1}) + \dot{G}_2(1-\varepsilon_{h,2}) = C_s \dot{V}_o$$
↑   ↑   ↑   ↑
makeup   Source #1   Source #2   room air entering exhaust hoods

solve for C_s

$$C_s = C_o + \frac{\dot{G}_1(1-\varepsilon_{h,1}) + \dot{G}_2(1-\varepsilon_{h,2})}{\dot{V}_o}$$

Note: If both exhaust hood capture efficiency values equal 1.0, $C_s = C_o$.

If no exhaust hoods are present,

$$C_s = C_o + \frac{\dot{G}_1 + \dot{G}_2}{\dot{V}_o}$$

12.33

i) $\dot{R} = 0.9(10^4)(10^6) = 9 \times 10^9$

a) $C_s = 10^6 + (10^8 - 9\times10^9)/10^4 = 1.1\times10^5 / ft^3$

b) $C_s = 10^6 + (5\times10^7 - 9\times10^9)/10^4 = 1.05\times10^5 / ft^3$

c) $C_s = 10^6 + (10^8 - 1.8\times10^{10})/(2\times10^4) = 1.05\times10^5 / ft^3$

ii) $\dot{R} = 0.99(10^4)(10^6) = 9.9\times10^9$

a) $C_s = 10^6 + (10^8 - 9.9\times10^9)/10^4 = 2.0\times10^4 / ft^3$

b) $C_s = 10^6 + (5\times10^7 - 9.9\times10^9)/10^4 = 1.5\times10^4 / ft^3$

c) $C_s = 10^6 + (10^8 - 1.98\times10^{10})/(2\times10^4) = 1.5\times10^4 / ft^3$

iii) $\dot{R} = 0.999(10^4)(10^6) = 9.99\times10^9$

a) $C_s = 10^6 + (10^8 - 9.99\times10^9)/10^4 = 1.1\times10^4 / ft^3$

b) $C_s = 10^6 + (5\times10^7 - 9.99\times10^9)/10^4 = 6.0\times10^3 / ft^3$

c) $C_s = 10^6 + (10^8 - 1.998\times10^{10})/(2\times10^4) = 6.0\times10^3 / ft^3$

Chapter 13:

13.1
By Eqn 13.4:
$$h = 15(9-12) = -45°$$

By Eqn 13.1 @ Sept. 1, $n = 244$
$$d = 23.45 \sin\left(\frac{360}{365}(284+244)\right) = 7.72°$$

By Eqn 13.6 @ $l = 42°N$
$$\sin\beta = \cos(42)\cos(-45)\cos(7.72) + \sin(42)\sin(7.72)$$
$$\beta = 37.6°$$

By Eqn 13.7
$$\cos\phi = \frac{1}{\cos(37.6)}\left[\cos(7.72)\sin(42)\cos(-45) - \sin(7.72)\cos(42)\right]$$

Since it is before solar noon we use the $-$ sign
$$\therefore \phi = -62.2°$$

13.2
At sunrise $\beta = 0$. $l = 50°N$ (Given), By Eqn 13.6
$$0 = \cos l \cos h_{sr} \cos d + \sin l \sin d$$
or $\quad \cos h_{sr} = -\tan l \tan d$

a.) June 21st $\quad d = +23.45°$
$$\therefore h_{sr} = -121.1°$$
$h_{sr} = 15(LST - 12) \quad \therefore LST = 3.93$ or $3:55$ a.m.

By Eqn 13.7
$$\cos\phi = \frac{1}{\cos(0)}\left[\cos(23.45)\sin(50)\cos(-121.1) - \sin(23.45)\cos(50)\right]$$

$$\boxed{\phi = -128.2°}$$

b.) By similar calculations with $d = -23.45°$
$$h_{sr} = -58.9°_{49}, \quad LST = 8.1 \ (8:06am), \quad \phi = -51.7°$$

13.3: July 14: By Eqn 13.1 or Table 13.2: $d = 21.7°$
$l = 45°N$. When the horizontal projection of the solar rays are normal to a west facing surface $\phi = 90°$

By Eqn 13.7 with $\cos\phi = 0$
$$\cos(21.7)\sin(45)\cos h = \sin(21.7)\cos(45)$$
$$h = 66.6°$$
By Eqn 13.6 $\underline{\beta = 31.5°}$

13.4: a.) Max. altitude will occur when $d = 23.45°$ and at solar noon ($h = 0$)
∴ By Eqn 13.6 $\underline{\beta_{max} = 68.5°}$ for $l = 45°N$
b) At $23.5°$ North latitude the sun is directly overhead when $d = 23.45°$ ∴ $\underline{\beta_{max} = 90°}$
c.) At $l = 0°$ the sun will be directly overhead at both equinoxes ∴ $\underline{\beta_{max} = 90°}$

13.5: By Eqn 13.7: At Sunrise $\beta = 0$ and for the given condition that $d = 0$
$$\cos\phi = \sin l \cos h$$
By Eqn 13.6
$$\cos h = -\frac{\sin l \sin d}{\cos l \cos d} = 0 \text{ for } d = 0$$
∴ Substituting this into the reduced Eqn 13.7 gives
$$\cos\phi = 0 \quad \text{or} \quad \phi = \pm 90°$$
or $\phi = -90°$ at sunrise and note the solution is independent of latitude, l.

13.6 a.) $L_{loc} = 85°W$, $\ell = 43°N$, July 7, 4 p.m. Central Daylight

By Eqn 13.2 with $E = -0.08$ from Table 13.2

$$LST = 16 + \frac{1}{15}(90-85) + (-0.08) - 1 = 15.25 \text{ hours.}$$

By Eqn 13.4

$$h = 15(15.25 - 12) = 48.75°$$

from Table 13.2 $d = 22.6°$

By Eqn 13.6 we get $\beta = 45°$

For the horizontal $\theta = \theta_H = 90 - \beta = \underline{\underline{45°}}$

b.) For $\beta = 0$ (sunset) $h_{ss} = \cos^{-1}(-\tan(43)\tan(22.6))$

$h_{ss} = 112.8°$; $\Rightarrow$ LST = 19.5 hours.

By Eqn 13.2

$$CT = 19.5 - \frac{1}{15}(90-85) - (-0.08) + 1 = \underline{\underline{20.2 \text{ hrs}}}$$

or 8:12 pm CDST

13.7 July 14 ∴ $d = 21.7°$ (Table 13.2), $l = 36°N$,
$L_{loc} = 80°W$, $E = -0.09$ hrs (Table 13.2)
$L_{std} = 75°W$ (Eastern Time Zone), $DT = 0$ (std. time)

∴ By Eqn 13.2 ; LST = 10.6 hrs.
By Eqn 13.4 $h = 15(10.6-12) = -21°$

a.) For a horizontal surface $\theta = \theta_H$
∴ By Eqn 13.5 $\theta_H = 23.2°$

b.) South Facing Vertical ∴ $\psi = 0$, $\Sigma = 90°$
$\beta = 90 - \theta_H = 90 - 23.2 = 66.8°$
By Eqn 13.7
$\cos\phi = \dfrac{1}{\cos(66.8)}[\cos(21.7)\sin(36)\cos(-21) - \sin(21.7)\cos(36)]$

$\phi = -57.6$ (Note the sign)

By Eqn 13.9 $\gamma = |-57.6 - 0| = 57.6°$
By Eqn 13.10 $\theta = 77.8°$

c.) Given: $\psi = -60°$, $\Sigma = 25°$
From parts a & b: $\beta = 66.8°$, $\phi = -57.6$
By Eqn 13.9: $\gamma = |-57.6 - (-60)| = 2.4°$
By Eqn 13.10: $\theta = 2.0°$

13.8: The surface orientation is seen to be

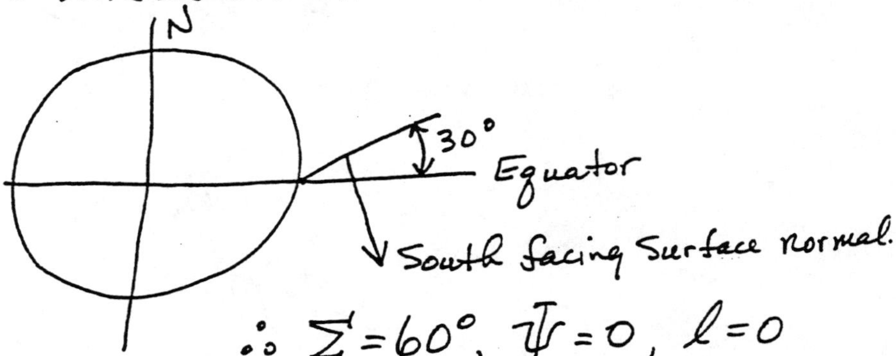

[Diagram showing a circle with N at top, Equator line extending right, with a 30° angle and an arrow labeled "South facing Surface normal"]

$$\therefore \Sigma = 60°, \ \psi = 0, \ \ell = 0$$

At June 22, $d = 23.45°$

The surface just becomes sunlit at $\theta = 90°$

With these substitutions Eqn 13.10 becomes

$$0 = \cos\beta \cos\phi \sin(60) + \sin\beta \cos(60)$$

Eqn 13.7 becomes

$$\cos\phi = \frac{-\sin(23.45)}{\cos\beta}$$

Eqn 13.6 becomes

$$\sin\beta = \cos h \cos(23.45)$$

Solving the equations gives

$$\cos h = \tan(23.45) \tan(60)$$

$$h = \pm 41.3° \quad \text{or} \quad \frac{41.3}{15} = 2.75 \text{ hours on either side of solar noon}$$

∴ Surface is exposed to direct solar radiation <u>5.5 hours</u>

13.9: From page 382 the earth's mean distance from the sun is $R_S = 1.5 \times 10^8$ km

$$\pi D_S^2 \sigma T_S^4 = 4\pi R_S^2 I_{N,0}$$

or

$$T_S = \left(\frac{4 I_{N,0}}{\sigma}\right)^{1/4} \left(\frac{R_S}{D_S}\right)^{1/2} = \left[\frac{4(1367)}{5.67 \times 10^{-8}}\right]^{1/4} \left[\frac{1.5 \times 10^8}{1.39 \times 10^6}\right]^{1/2}$$

$$T_S = 5,790 \text{ K}$$

13.10: Let $dQ_{H,0}$ = solar flux striking a horizontal surface at the outer edge of the atmosphere in the differential time period dt

Then:

$$dQ_{H,0} = I_{DN} \cos\theta_H \, dt$$

or:

$$dQ_{H,0} = \frac{I_{DN}(\cos l \cos h \cos d + \sin l \sin d) \, dh}{w}$$

where dh is the change in hour angle in radians in the time period dt hr, and w is the earth's angular velocity, $\pi/12$ radians per hr. For one day, I_{DN}, d and l may be considered as constants. Since hour angles are symmetrical with respect to solar noon, we have

$$Q_{H,0} = \frac{24 I_{DN}}{\pi} \left[\cos l \cos d \int_0^H \cos h \, dh + \sin l \sin d \int_0^H dh\right]$$

where H = hour angle of sunrise or sunset, radians

Integration gives:

$$Q_{H,0} = \frac{24}{\pi} I_{DN} \sin l \sin d (H - \tan H)$$

By Eqn 13.1 for March 1, $d = -8.3°$
By Eqn 13.6 with $\beta = 0$

$$\cos H = -\tan l \tan d = -\tan(48)\tan(-8.3°) = 0.162$$
$$H = \pm 80.68° = \pm 1.41 \text{ radians}$$

From Fig. 13.16 we see $I_{DN} = 1.017 I_{N,0}$ on March 1

$$\therefore Q_{H,0} = \frac{24}{\pi}(1.017)(432)\sin(42)\sin(-8.3)(1.41 - \tan(80.68))$$

$$\underline{\underline{Q_{H,0} = 1518 \text{ Btu/day-ft}^2}}$$

13.11: Let Q be the energy absorbed in Btu/day-ft^2
We have
$$dQ = 0.5 I_{DN} \cos\theta \, dt$$
$$dh = w \, dt$$

$\therefore$
$$dQ = \frac{0.5 I_{DN}}{w} \cos\theta \, dh \qquad (1.)$$

By Eqn 13.11 and Eqn 13.7 with $\phi = 0$ (South facing)
$$\cos(\gamma) = \cos(\Sigma) = 1.0$$
we have
$$\cos\theta = \cos d \sin l \cos h - \sin d \cos l \qquad (2.)$$

By Eqns (1.) & (2.) above

$$Q = \frac{2(0.5)I_{DN}}{\omega}\left[\cos d \sin \ell \int_0^H \cos h \, dh - \cos \ell \sin d \int_0^H dh\right]$$

$$= \frac{I_{DN}}{\omega}\left[\cos d \sin \ell \sin H - H \cos \ell \sin d\right]$$

where H is the hour angle (in radians) when the plate just becomes sunlit, i.e. at $\theta = \pi/2$. For a vertical south facing surface this occurs when $\phi = \pi/2$ ∴ From Eqn 13.7

$$\cos H = \tan d / \tan \ell$$

By Eqn 13.1: $d = 23.3°$
and $H = 64.5° = 1.13$ radians
From Fig 13.16 $I_{DN} \cong 0.967 \, I_{N,0}$

$$\therefore Q = \frac{0.967(432)}{(\pi/12)}\left[\cos(23.3)\sin(45)\sin(64.5) - 1.13\cos(45)\sin(23.3)\right]$$

$$\underline{Q = 431 \text{ Btu/day-ft}^2}$$

13.12 South Facing ∴ $\Psi = 0$, Vertical ∴ $\Sigma = 90°$
Solar noon ∴ $h = 0$ and $\phi = 0$
$\ell = 42°N$ (Given)

a.) June 21: $d = 23.45$, $A = 345 \frac{\text{Btu}}{\text{hr-ft}^2}$, $B = 0.205$
By Eqn 13.6 $\sin\beta = 0.948$ or $\beta = 71.5°$
By Eqn 13.20
$$I_{DN} = 345 \, e^{-0.205/0.948} = 277.9 \, \frac{\text{Btu}}{\text{hr-ft}^2}$$
By Eqn 13.10 $\theta = 71.5°$

(Note that since $\gamma = 0$ and the surface is vertical we could have observed that $\theta = \beta$.)

By Eqn 13.22
$$I_D = 277.9 \cos(71.5) = 88.2 \text{ Btu/hr-ft}^2$$

b.) Following the same procedure as part a with
$d = -23.45$, $A = 391$ Btu/hr-ft², $B = 0.142$
we calculate $\beta = 24.55°$, $\theta = 24.55°$
$$I_{DN} = 277.8 \text{ Btu/hr-ft}^2 \text{ and } I_D = 252.7 \text{ Btu/hr-ft}^2$$

13.13: a) By Eqn 13.2 with $E = -0.10$ (Table 13.2)
$$LST = 12.00 + \tfrac{1}{15}(90-93) - 0.10 = 11.70 \text{ ; } h = -4.5°$$
For a flat roof $\Sigma = 0$
$\therefore \cos\theta = \cos\theta_H$
From Table 13.3 $d = 20.6°$

By Eqn 13.5
$$\cos\theta_H = \cos(45)\cos(-4.5)\cos(20.6) + \sin(45)\sin(20.6)$$
$$= 0.909$$

For July from Table 13.3
$A = 344$ Btu/hr-ft², $B = 0.207$, $C = 0.136$

By Eqn 13.20
$$I_{DN} = 344 \exp(-0.207/0.909) = 273.9 \text{ Btu/hr-ft}^2$$

By Eqn 13.21
$$I_{dH} = 0.136(273.9) = 37.3 \text{ Btu/hr-ft}^2$$

By Eqn 13.32
$$I_H = 273.9(0.909) + 37.3 = \underline{286} \text{ Btu/hr-ft}^2$$

b.) By the same procedure as part a.)
$LST = 11.45$, $h = -8.25°$, $\theta_H = 14.32°$ $I_H = 307$ Btu/hr-ft²

13.14 Feb. 7 ∴ $d = -15.8°$, $E = -0.24$ hrs (Table 13.2)
$\ell = 33.8°N$, $L_{loc} = 84.4°W$ & $L_{std} = 75°W$ (Eastern time zone)
By Eqn 13.6 with $\beta = 0$ (Sunset) solve
for $h_{ss} = 79.1°$. By Eqn 13.4 LST = 17.27
By Eqn 13.2

a.) $CT = 17.27 - \left(\frac{1}{15}\right)(75 - 84.4) + 0.24 = \underline{18.14 \text{ or } 6:08 \text{ pm E.}}$

b.) West facing ∴ $\Psi = 90°$, Vertical ∴ $\Sigma = 90°$
$\beta = 0$ at sunset, $h = 79.1°$ from part a
∴ By Eqn 13.7 $\phi = 70.84°$
By Eqn 13.9 $\gamma = 19.16°$
By Eqn 13.10 $\theta = 19.16°$

(Note that since the solar rays and the surface normal are horizontal we could have observed that $\theta = \gamma$.)

13.15 $\ell = 42°N$, $L_{loc} = 88°W$, Nov 21.
From Table 13.2, $d = -20.4°$, $E = 0.02$

a.) At sunrise $\beta = 0$, let H = hour angle at sunrise
By Eqn 13.6
$\cos H = -\tan \ell \tan d = -\tan(42°)\tan(-20.4°) = 0.335$
$H = -70.4°$
By Eqn 13.4: $LST = \frac{-70.4}{15} + 12 = 7.31$
By Eqn 13.2
$CT = LST - \frac{1}{15}(L_{std} - L_{loc}) - E$
$= 7.31 - \frac{1}{15}(90 - 88) - 0.02 = \underline{7.16 \text{ hr} = 7:09 \text{ a.m.}}$

b.) By Eqn 13.7:
$\cos \phi = \cos(-20.4)\sin(42)\cos(-70.4) - \sin(-20.4)\cos(42)$
$= 0.469$ ∴ $\underline{\phi = 62.0°}$

c.) By Eqn 13.4 @ LST = 10.5, $h = -22.5°$
By Eqn 13.6: $\beta = 24.2°$

d.) 20° W of S $\Rightarrow \psi = 20$; $\Sigma_i = 65°$
By Eqn 13.7: $\phi = -23.15°$
By Eqn 13.9: $\gamma = 43.15°$
By Eqn 13.10: $\underline{\theta = 39.1°}$

e.) By Eqn 13.13: $\delta = 31.6°$

13.16:
$\ell = 27°$, $L_{loc} = 83°W$, $\psi = -45°$, $\Sigma_i = 55°$
March 26, CT = 13.5 EST, $L_{std} = 75°W$ for EST

By Eqn 13.3
$B = 360(85-81)/364 = 3.96$

and $E = 0.165 \sin(2 \cdot 3.96) - 0.126 \cos(3.96) - 0.025 \sin(3.96)$

$E = -0.10$ hr

By Eqn 13.2: LST = $13.5 + \frac{1}{15}(75-83) - 0.10 = 12.87$
By Eqn 13.4: $h = 13.05°$; By Eqn 13.1: $d = 1.61°$
By Eqn 13.6: $\beta = 61.7°$
By Eqn 13.7: $\phi = 28.4°$
By Eqn 13.9: $\gamma = 73.4°$
By Eqn 13.10 $\underline{\underline{\theta = 52.0°}}$

13.17:
$\ell = 32°N$, $L_{loc} = 111°W$, Mountain Time Zone ∴ $L_{std} = 105°W$
Dec. 15 ∴ $n = 349$

a.) By Eqn 13.3 $B = 265$ and $E = 0.06$ hr
By Eqn 13.2
$12 = CT + \frac{1}{15}(105-111) + 0.06$
$CT = 12.34$ or $12:20$ P.M. MST

b.) $\Psi = 45°$, LST = 10.0 ∴ $h = -30°$, $\Sigma' = 90°$
By Eqn 13.6: $\beta = 27.7$, By Eqn 13.7: $\phi = -31.2$
By Eqn 13.9: $\gamma = 76.2°$
∴ By Eqn 13.10 $\underline{\theta = 77.8°}$

c.) The direct solar rays just begin to strike the window when $\theta = 90°$
By iteration we find $\theta = 90°$ at LST = 8.88
or $\underline{8:53\ a.m.}$

13.18 $\ell = 33°N$, $\Psi = 0$, $\Sigma' = 75°$, June 21 ∴ $d = 23.45°$
Note that direct sunlight will strike a surface anytime $\theta < 90°$. Therefore we need to determine the two hours of the day when $\theta = 90°$ and calculate the time between those hours.
By Eqn 13.10 with $\theta = 90°$
$$0 = \cos\beta \cos(\phi - 0) \sin(75) + \sin\beta \cos(75)$$
By Eqn 13.6
$$\sin\beta = \cos(33)\cos h \cos(23.45) + \sin(33)\sin(23.45)$$
By Eqn 13.7
$$\cos\phi = \frac{1}{\cos\beta}\left[\cos(23.45)\sin(33)\cos h - \sin(23.45)\cos(33)\right]$$

Solving the three equations we get
$h = -67°$, $\beta = 31.15°$ and $\phi = -99.32°$
∴ LST $= \left(\frac{-67°}{15}\right) + 12 = 7.53$ hours
or 4.47 hours before solar noon
From Symmetry, since the surface faces south, the afternoon time at which $\theta = 90$ will be 4.47 hours after solar noon
∴ Total time $= 2(4.47) = \underline{8.94\ hours}$

13.19: 20° North of West ∴ $\Psi = 110°$,

∴ $\Sigma = 30°$

June 15 ∴ By Eqn 13.3, $B = 8308$ & $E = 0$
By Eqn 13.2 $LST = 12.0$ ∴ $h = 0$ & $\phi = 0$
By Eqn 13.1 $d = 23.31°$
By Eqn 13.16 $\beta = 80.31°$, By Eqn 13.9 $\gamma = 110°$
By Eqn 13.10 $\Theta = 34.42°$
From Table 13.3, $A = 345$ Btu/hr-ft², $B = 0.205$, $C = 0.134$

By Eqn 13.20
$$I_{DN} = 345 \exp(-0.205/\sin(80.31)) = 280.2 \text{ Btu/hr-ft}^2$$

By Eqn 13.22
$$I_D = 280.2 \cos(34.42) = 231.2 \text{ Btu/hr-ft}^2$$

By Eqn 13.20
$$I_{dH} = 0.134(280.2) = 37.5 \text{ Btu/hr-ft}^2$$

By Eqn 13.25
$$I_d = 37.5(1+\cos(30))/2 = 35.0 \text{ Btu/hr-ft}^2$$

By Eqn 13.32
$$I_H = 280.2 \cos(9.69) + 37.5 = 313.7 \text{ Btu/hr-ft}^2$$

By Eqn 13.30
$$I_R = 0.3(313.7)(1-\cos(30))/2 = 6.3 \text{ Btu/hr-ft}^2$$

By Eqn 13.31
$$I = 231.2 + 35.0 + 6.3 = \underline{273} \text{ Btu/hr-ft}^2$$

13.20

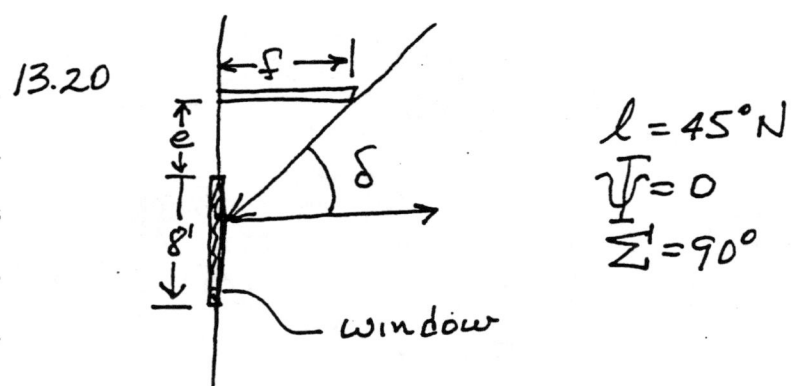

$l = 45°N$
$\Psi = 0$
$\Sigma' = 90°$

Summer Solstice -- completely shaded
$$\therefore e + 8 = f \tan\delta$$
$d = 23.45°$, Solar noon $\therefore h = 0, \phi = 0$
By Eqn 13.6 $\beta = 68.45°$
By Eqn 13.9 $\gamma = 0$ & By Eqn 13.3 $\delta = 68.45$
$$\therefore e + 8 = f(2.53) \qquad \text{Eqn. a)}$$

Winter Solstice -- completely sunlit
$$e = f \tan\delta$$
$d = -23.45$, Solar noon $\therefore h = 0, \phi = 0$
By Eqn 13.6 $\beta = 21.55°$
By Eqn 13.9 $\gamma = 0$ & By Eqn 13.3 $\delta = 21.55$
$$\therefore e = f(0.395) \qquad \text{Eqn. b)}$$

Solving Eqns a.) & b.) above gives
$$e = 1.48 \text{ ft} \quad \& \quad f = 3.75 \text{ ft}$$

13.21 $l = 48°N$, $\Psi = 0$, $\Sigma' = 90°$

Refer to Fig 13.12, Elevation View & note $b = 0$ is given
then for total shading $y = a$ (minimum)
for no shading $y = 0$.
By Eqns 13.6, 13.7 and 13.13 on May 1 at LST = 14.0
$\beta = 48.69°$, $\phi_{362} = 47.05°$, $\delta = 59.1°$

From the geometry for total shading with $a = 2m$

$$\boxed{e + 2m = f \tan(59.1°)}$$

Similarly on Nov 1 at Solar Noon
$\beta = 26.64°$, $\phi = 0$, $\delta = 26.6°$
and for no shading the shadow length can be equal to "e" i.e. $y = 0$

$$\boxed{e = f \tan(26.6)}$$

Solving for "e" & "f"
$e = 0.86 \, m$, $f = 1.71 \, m$.

13.22: $l = 45°N$, $L_{loc} = 93°W$, June 7 @ 3:00 pm CDST
West facing ∴ $\psi = 90°$, Vertical ∴ $\Sigma' = 90°$
By Table 13.2 $d = 22.7°$ and $E = 0.02$
By Eqn 13.2 $LST = 15.0 + \frac{1}{15}(90-93) + 0.02 - 1$
$LST = 13.82$ ∴ By Eqn 13.4 $h = 27.30°$
By Eqn 13.6 $\beta = 58.53°$; By Eqn 13.7 $\phi = 54.11°$
By Eqn 13.9 $\gamma = 35.89°$ By Eqn 13.13 $\delta = 63.62°$
For the window geometry refer to Fig 13.12
∴ $a = 250 \, cm$, $c = 100 \, cm$, $b = 36 \, cm$

$$\therefore F_s = \frac{[250 - 36 \tan(63.62)] \cdot [100 - 36 \tan(35.89)]}{(250)(100)}$$

$\underline{\underline{F_s = 0.52}}$

13.23:

Plan View

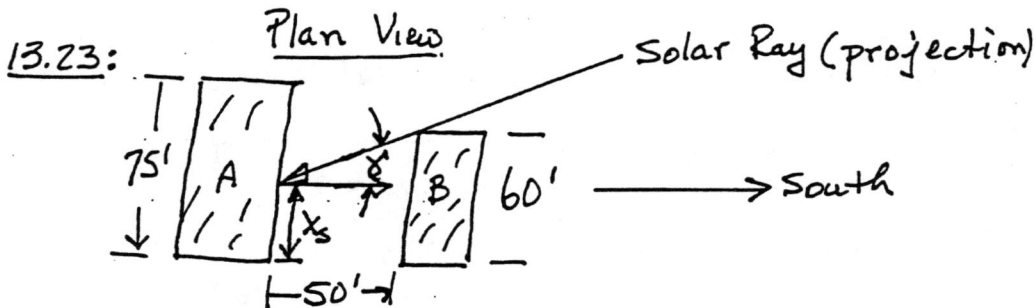

From the geometry: $X_s = 60 - 50 \tan \gamma$

Elevation View

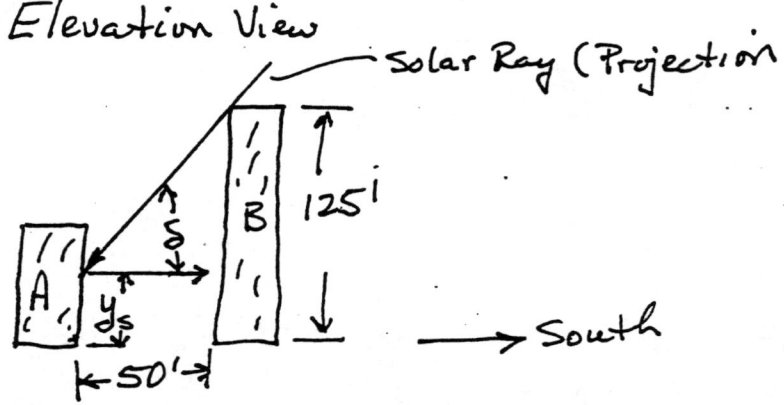

From the geometry: $y_s = 125 - 50 \tan \delta$

$\ell = 42°N$, LST = 11.0, Sept 1, From geometry $Y = 0$
By Eqns 13.1, 13.4, 13.6, 13.7, 13.9 & 13.13
$d = 7.72°$; $h = -15°$; $\beta = 53.25°$; $\phi = -25.38°$, $\gamma = 25.38°$
and $\delta = 55.99°$

∴ $X_s = 60 - 50 \tan(25.38) = 36.3$ ft
$y_s = 125 - 50 \tan(55.99) = 50.9$ ft

Thus, the shaded area is a rectangle of width 36.3 ft and height of 50.9 ft and located at the west end of the south face of building A.

13.24:

$\ell = 45°N$, $L_{loc} = 93°W$, Aug 1 @ 11:00 a.m. CST

By Eqn 13.3: $B = 130.5$ & $E = -0.10$ hr.

By Eqn 13.2: $LST = 10.7$

At any location:

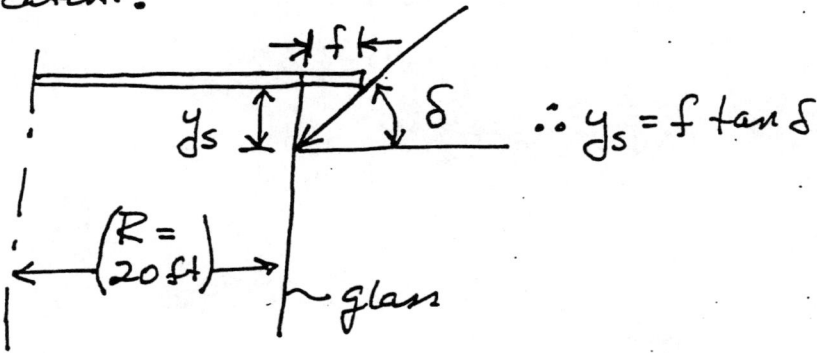

$\therefore y_s = f \tan \delta$

For a section of the building with azimuth angle Ψ and differential angle $d\Psi$ the differential shadow area is

$$dA = y_s R \, d\Psi = f \cdot R \tan \delta \, d\Psi$$

or

$$dA = f \cdot R \tan \beta \sec(\Psi - \phi) \, d\Psi$$

or

$$A = f \cdot R \tan \beta \int_{\Psi = -85°}^{\Psi = -5°} \sec(\Psi - \phi) \, d\Psi$$

$$\therefore A = f \cdot R \tan \beta \ln \left[\tan\left(\frac{\Psi - \phi}{2} + 45°\right) \right]_{\Psi = -85°}^{\Psi = -5°}$$

By Eqns 13.1, 13.4, 13.6 and 13.7:

$d = 17.91°$, $h = -19.5°$, $\beta = 58.4°$ and $\phi = -37.3°$

$$\therefore A = 2(20) \tan(58.4) \ln\left[\frac{1.823}{0.3868}\right] = \underline{\underline{100.8 \text{ ft}^2}}$$

13.25 $\psi = -22°$, $\Sigma' = 35°$, $l = 36°N$, $\rho_g = 0.2$, LST=14.0
July 21, From Table 13.3: $d = 20.6°$, $A = 344$ Btu/hr·ft²
$B = 0.207$ and $C = 0.136$

By Eqns 13.4, 13.6, 13.7, 13.9 and 13.10 we get
$\beta = 59.6°$, $\phi = 67.7°$, $\gamma = 89.7°$, $\theta = 44.9°$

By Eqn 13.20
$$I_{DN} = 344\, e^{-0.207/\sin(59.6)} = 270.6 \text{ Btu/hr·ft}^2$$

By Eqn 13.22
$$I_D = 270.6 \cos(44.9) = 191.6 \text{ Btu/hr·ft}^2$$

By Eqn 13.21
$$I_{dH} = 0.136(270.6) = 36.8 \text{ Btu/hr·ft}^2$$

By Eqn 13.25
$$I_d = 36.8 \left(\frac{1+\cos(35)}{2}\right) = 33.5 \text{ Btu/hr·ft}^2$$

By Eqn 13.32 noting that $\cos\theta_H = \sin\beta$
$$I_H = 270.6 \sin(59.6) + 36.8 = 270.2 \text{ Btu/hr·ft}^2$$

By Eqn 13.30
$$I_R = 0.2(270.2)\left(\frac{1-\cos(35)}{2}\right) = 4.9 \text{ Btu/hr·ft}^2$$

By Eqn 13.31
$$I = 191.6 + 33.5 + 4.9 = 230 \text{ Btu/hr·ft}^2$$

13.26 $\Sigma' = 90°$, $\psi = -25°$, Use the notation of
Fig 13.11 ∴ $a = 6$ ft, $b = 0.75$ ft, $c = 4$ ft
$l = 40°N$, LST = 13.0, January 21 ∴ from
Table 13.3: $d = -20.0°$, $A = 390 \frac{Btu}{hr\cdot ft^2}$, $B = 0.142$, $C = 0.058$

a.) By Eqns. 13.4, 13.6, 13.7, 13.9 and 13.10 we get.
$\beta = 28.4°$, $\phi = 16.1°$, $\gamma = 41.1°$ and $\underline{\underline{\theta = 48.4°}}$

b.) By Eqn 13.13
$$\tan \delta = \tan 28.4 / \cos 41.1 = 0.72 \; ; \; \delta = 35.8°$$
$$y = b \tan \delta = 0.75 \tan(35.8) = 0.54 \text{ ft}$$
$$x = b \tan \gamma = 0.75 \tan(41.1) = 0.65 \text{ ft}$$
$$F_s = \frac{(a-y)(c-x)}{a \cdot c} = \frac{(6-0.54)(4-0.65)}{24} = \underline{0.76}$$

c.) By Eqn 13.31a
$$I = F_s I_D + I_d + I_R$$
By Eqn 13.20
$$I_{DN} = 390 \, e^{-0.142/\sin 28.4} = 289.3 \text{ Btu/hr·ft}^2$$
By Eqn 13.21
$$I_{dH} = 0.058(289.3) = 16.8 \text{ Btu/hr·ft}^2$$
By Eqn 13.22
$$I_D = 289.3 \cos(48.4) = 191.9 \text{ Btu/hr·ft}^2$$

By Eqn 13.24b
$$I_{dV} = 16.8 \left[0.55 + 0.437 \cos(48.4) + 0.313 \cos^2(48.4) \right] = 16.4 \, \frac{\text{Btu}}{\text{hr·ft}^2}$$
By Eqn 13.32 noting that $\cos \theta_H = \sin \beta$
$$I_H = 289.3 \sin(28.4) + 16.8 = 154.4 \text{ Btu/hr·ft}^2$$
By Eqn. 13.30 with $\rho_g = 0.7$
$$I_R = 0.7(154.4)(1-\cos 90)/2 = 54.0$$
∴ By Eqn 13.31a (For the unshaded portion $F_s = 1.0$)
$$I = (1.0)(191.9) + 16.4 + 54 = \underline{262} \text{ Btu/hr·ft}^2$$

13.27: $\Sigma' = 90°$, $\psi = -35°$, Use the notation of Fig 13.11 ∴ $a = 7$ ft, $c = 4$ ft, $b = 1$ ft
$l = 25.8°N$, LST $= 13.0$, September 21 ∴ from Table 13.3 $d = 0$, $A = 365$ Btu/hr-ft², $B = 0.177$, $C = 0.092$

a.) By Eqns 13.4, 13.6, 13.7, 13.9 and 13.10 we get
$h = 15°$, $\beta = 60.24°$, $\phi = -31.43°$, $\gamma = 3.57°$ and
$$\boxed{\theta = 60.3°}$$

b.) Following the calculation procedure in part b of Problem 13.26 we get:
$y = 1.75$ ft, $X = 0.06$ ft ∴ $F_S = 0.74$

Following the calculation procedure in part c of Problem 13.26 we get:
$I_{DN} = 297.7$ Btu/hr-ft², $I_D = 147.5$ Btu/hr-ft²
$I_{dH} = 27.4$ Btu/hr-ft², $I_{dv} = 23.1$ Btu/hr-ft²
$I_H = 285.8$ Btu/hr-ft², $I_R = 57.2$ Btu/hr-ft²

The rate of solar energy striking the window is
$A_w I = A_w (F_S I_D + I_{dv} + I_R) = 28[0.74(147.5) + 23.1 + 57.2]$
$\underline{\underline{A_w I = 5,305 \text{ Btu/hr}}}$

13.28 Use the notation of Fig 13.12 with $b=0$ (no setback). From the geometry we see the minimum width, $(c+2g)$, will occur for total shading when:

$$g = f \tan \gamma'$$

The minimum "f" for total shading for a given value of "e" will be given by

$$a + e = f \tan \delta$$

On June 1 at LST = 10.0, $\ell = 36°N$, $\Sigma' = 90°$, $\Psi = 0$ we calculate:

$$d = 22.0, \ \beta = 60.5, \ \phi = -70.1, \ \gamma' = 70.1, \ \delta = 79.1$$

∴
$$\boxed{\begin{array}{c} g = f \tan(70.1) \\ b + e = f \tan(79.1) \end{array}}$$

If the window is to be totally sunlit, we need the minimum value of "f" to satisfy

$$e = f \tan \delta$$

On Dec. 1 at Solar noon we calculate

$$d = -22.1, \ \beta = 31.9, \ \phi = 0, \ \delta = 31.9°$$

∴
$$\boxed{e = f \tan(31.9°)}$$

Solving the 3 equations in the boxes gives

$$e = 0.82 \text{ ft}, \ f = 1.31 \text{ ft}, \ g = 3.63 \text{ ft}$$

b) Feb 21 at solar noon: From Table 13.3: $d = -10.8°$
$A = 385$ Btu/hr-ft², $B = 0.144$, $C = 0.060$
For a latitude of 36°N we would not expect to find snow on the ground on Feb 21 (Las Vegas, NV, Monterey, CA, and Tulsa, OK are approx. at 36°N)
∴ let's assume $\rho_g \cong 0.3$
Following the calculation procedure described in the solution to Prob. 13.26 we get
$\beta = 43.2°$, $\phi = 0$, $\gamma = 0$, $\delta = 43.2°$ and $\Theta = 43.2°$
(Note from $\gamma = 0$ and $\Sigma = 90$ we could see $\beta = \delta = \Theta$)
$I_{DN} = 312.0 \frac{Btu}{hr \cdot ft^2}$, $I_D = 227.4$ Btu/hr·ft², $I_{dv} = 19.4 \frac{Btu}{hr \cdot ft^2}$
$I_R = 34.8$ Btu/hr·ft²

By the geometry we see
$$y + e = f \tan \delta$$
$$y = 1.31 \tan(43.2) - 0.82 = 0.41 \text{ ft.}$$
$$F_s = \frac{(6 - 0.41)(4)}{24} = 0.93$$

From Eqn 13.31 a

$$I = 0.93(227.4) + 19.4 + 34.8 = \underline{\underline{266}} \text{ Btu/hr·ft²}$$

13.29: $\Sigma = 90°$, $\Psi = -30°$, $\ell = 42°N$
LST = 13.0 ∴ $h = 15°$, July 7 ∴ $d = 22.6°$ (Table 13.2)

By Eqns 13.6, 13.7, 13.9 and 13.11
$\beta = 66.9°$, $\phi = 37.5°$, $\gamma = 67.5°$ and $\theta = 81.4°$

The length of the shadow below the bottom of the overhang is:
$$y = (1 \text{ ft}) \tan \delta = 1 \cdot \frac{\tan(66.9)}{\cos(67.5)} = 6.13 \text{ ft}$$

∴ Shadow extends $y_s = (6.13 - 1.5) = 4.63$ down the window
The overhang is very wide compared to the width of the window
$$\therefore F_S = \frac{(6.5 - 4.63)}{6.5} = 0.29$$

From July from Table 13.3
$A = 344$ Btu/hr-ft², $B = 0.207$, $C = 0.136$

By Eqn 13.20
$$I_{DN} = A e^{-B/\sin\beta} = 344 \exp\left(\frac{-0.207}{\sin(66.9)}\right) = 274.7 \frac{\text{Btu}}{\text{hr-ft}^2}$$

By Eqn 13.21
$$I_{dH} = C I_{DN} = 0.136(274.7) = 37.4 \frac{\text{Btu}}{\text{hr-ft}^2}$$

From Fig 13.20 or Eqn 13.24b with $\cos\theta = 0.150$: $\frac{I_{dV}}{I_{dH}} = 0.62$
∴ $I_{dV} = 0.62(37.4) = 23.2$ Btu/hr-ft²

By Eqn 13.32
$I_H = I_{DN}\cos\theta_H + I_{dH} = I_{DN}\sin\beta + I_{dH} = 274.7\sin(66.9) + 37.4$
$I_H = 290.1$ Btu/hr-ft²

By Eqn 13.30
$$I_R = \rho_g I_H (1 - \cos\Sigma)/2 = 0.3(290.1)/2 = 43.5 \text{ Btu/hr-ft}^2$$

$A \cdot I = A[F_S I_D + I_{dV} + I_R]$
$= (6.5)(4)[0.29(274.7)\cos(81.4)° + 23.2 + 43.4] = 2,044 \frac{\text{Btu}}{\text{hr}}$

13.30 $\Sigma' = 90$, $\Psi = 0$, Solar noon $\therefore \phi = 0$ & $h = 0$
$\ell = 42°N$, $\rho_g = 0.2$, July 1

Using the notation of Fig 13.12
$a = 2$ m, $c = 1.0$ m, $b = 0$, $g = 0.25$ m, $e = 0.15$ m
and $f = 0.46$ m.

Since $\Psi = 0$ and $\phi = 0$; $\gamma = 0$, $\delta = \beta$ and the shadow width is the same width as the overhang. The height of the shadow on the window is then:
$$y = f \tan\beta - e$$

By Eqn 13.1
$\quad d = 23.12°$

By Eqn 13.6 or 13.8
$\quad \beta = 71.12°$ and since $\gamma = 0$ and $\Sigma' = 90$, $\underline{\theta = \beta = 71.12°}$

$\therefore y = 0.46 \tan(71.12) - 0.15 = 1.20$ m

$F_s = \dfrac{2.0 - 1.20}{2} = 0.40$

From Table 13.3 for July use
$A = 1085$ W/m², $B = 0.207$, $C = 0.136$

Following the procedure of part c. of Problem 13.26
$I_{DN} = 871.8$ W/m² $I_D = 282.1$ W/m²
$I_H = 118.6$ W/m² $I_{dv} = 85.9$ W/m²
$I_H = 943.5$ W/m² $I_R = 94.3$ W/m²

The rate of solar energy striking the window is:
$A_w I = A_w [F_s I_D + I_{dv} + I_R] = (2)(1)[0.40(282.1) + 85.9 + 94.3]$
$\quad = \underline{586 \text{ W}}$

13.31: $\Sigma' = 45°$, $\Psi = 0$, Solar Noon $\therefore \phi = 0$, $h = 0$
$l = 45°$ N, Standard day in October $\therefore$ from Table 13.3
$d = -10.5°$, $A = 378 \frac{Btu}{hr\cdot ft^2}$, $B = 0.160$, $C = 0.073$

From geometry we see:
$f = a/\sin(45°)$ and
angle $\xi = 180 - 45 - \delta$

From the law of sines:
$$a_s = \frac{f \sin \delta}{\sin \xi}$$

Since $\Psi = 0$ and $\phi = 0$, $\gamma = 0$ and $\delta = \beta$

$$\therefore F_s = \frac{a - a_s}{a} = 1 - \frac{\sin \beta}{\sin 45 \sin \xi}$$

By Eqn 13.6 or 13.8: $\beta = 34.5°$
$\therefore \xi = 180 - 45 - 34.5 = 100.5°$

$$F_s = 1 - \frac{\sin(34.5)}{\sin(45)\sin(100.5)} = 0.185$$

By Eqn 13.10: $\theta = 10.5°$
By Eqn 13.20: $I_{DN} = 285.0$ Btu/hr-ft²
By Eqn 13.22: $I_D = 280.2$ Btu/hr-ft²
By Eqn 13.21: $I_{dH} = 20.8$ Btu/hr-ft²
By Eqn 13.32: $I_H = 182.2$ Btu/hr-ft²
By Eqn 13.30
$$I_R = 0.3 (182.2) \left(\frac{1 - \cos 45}{2} \right) = 8.0 \text{ Btu/hr-ft}^2$$

By Eqn 13.25:
$$I_d = 20.8 \left(\frac{1 + \cos(45)}{2} \right) = 17.8 \text{ Btu/hr-ft}^2$$

$$I = F_s I_D + I_d + I_R = 0.185(280.2) + 17.8 + 8.0$$
$$I = 77.6 \text{ Btu/hr-ft}^2$$

13.32: $\ell = 45°N$, $L_{loc} = 93°W$, Dec. 15, 9:00 am CST
$L_{std} = 90°$ for CST
By Eqn 13.3: $B = 265.1$, $E = 0.06$ hrs
By Eqn 13.2: $LST = 8.86$ hr $\therefore h = -47.1°$
By Eqns 13.6 and 13.7, $\beta = 9.31°$ $\phi = -43.0°$
($d = -23.34°$ by Eqn 13.1)

For the mirror to reflect the direct solar rays straight down, the surface azimuth Ψ must be equal to ϕ

a.) $\therefore \underline{\underline{\Psi = -43.0°}}$

To determine Σ refer to the sketch below

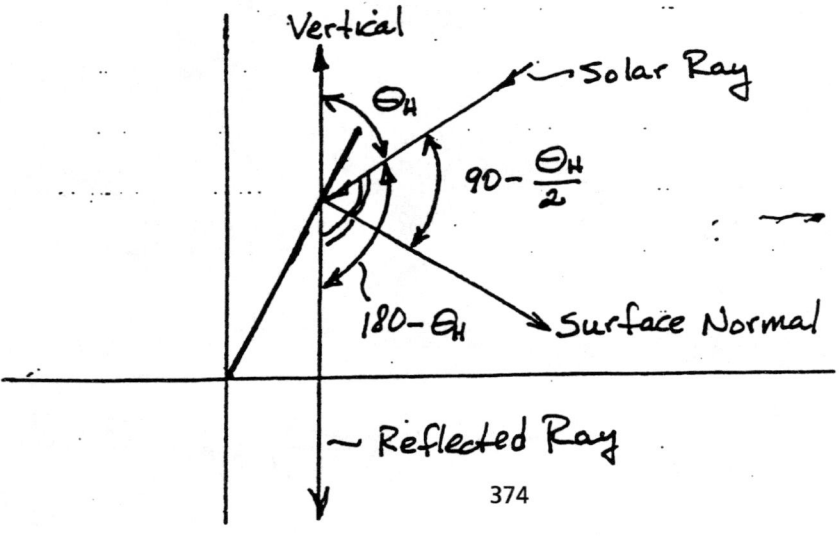

From the sketch we see

$$\Sigma' = \Theta_H + 90 - \frac{\Theta_H}{2} = \frac{\Theta_H}{2} + 90$$

$$\Theta_H = 90 - \beta = 90 - 9.31° = 80.69°$$

$$\therefore \Sigma' = \frac{80.69}{2} + 90 = \underline{130.3°}$$

b.) From the sketch we see $\Theta = 90 - \frac{\Theta_H}{2}$

$$\Theta = 90 - \frac{80.69}{2} = 49.7°$$

(Note Eqn 13.10 could also be used to find Θ)

From Table 13.3 for December
$A = 1233 \, \frac{W}{m^2} \quad B = 0.142$

By Eqn 13.20

$$I_{DN} = 1233 \exp\left(\frac{-0.142}{\sin(9.31)}\right) = 512.6 \, W/m^2$$

$\therefore$ Direct Solar Radiation into the building is:

$$A_m I_D = (15 m^2) \, 512.6 \, \cos(49.7) = 4,970 \, W$$

Problem 14.1

Thermal Resistances, hr-ft^2-°F/Btu	
Resistance Source	Resistance
Inside air film	0.68
3/8 inch plywood (R=1.25x0.375)	0.47
5.5 inch insulating batt of mineral wool	19
8-inch concrete blocks *	0.97
4 inch brick (R = 0.15x4)**	0.6
Outside air film	0.17
Total: R_t	21.89

Notes:
* Arbitrarily Selected lowest R-value
** Arbitrarily Selected lowest R-value/inch

By Eqn 14.5

$$\dot{q} = \frac{t_i - t_o}{R_t} = \frac{72-0}{21.89} = 3.29 \ Btu/hr\text{-}ft^2$$

Problem 14.2

Resistance Source	Thermal Resistances, hr-ft²-°F/Btu	
	Path	
	At studs	Between studs
Outside air film	0.25	0.25
3/8 inch Built up roofing	0.33	0.33
3/4 inch plywood	0.93	0.93
Studs, $R_s = (7.5/0.8)$	9.375	*****
7.5 in. Cellulose Insulation, $R_b = (7.5/0.3)$	*****	25
3/8 inch gypsum wallboard	0.32	0.32
Inside air film	0.92	0.92
Totals:	12.125	27.75

If we include only the 2×8 studs as the framing

$$a_s = \frac{1.5}{12} = 0.125 \text{ and } a_b = \frac{10.5}{12} = 0.875$$

a.) Parallel Path Method: (By Eqn 14.18)

$$R_{t(av)} = \frac{1}{\frac{0.125}{12.125} + \frac{0.875}{27.75}} = 23.90 \; \frac{hr \cdot ft^2 \cdot °F}{Btu}$$

$$\dot{Q}_t/A = \frac{95-78}{23.90} = 0.71 \; Btu/hr \cdot ft^2$$

b) Isothermal Plane Method:

By Eqn 14.22: $R_{s,b} = \left(\frac{0.125}{9.375} + \frac{0.875}{25}\right)^{-1} = 20.69 \; \frac{hr \cdot ft^2 \cdot °F}{Btu}$

By Eqn 14.23 $R_{t(av)} = 2.75 + 20.69 = 23.44 \; \frac{hr \cdot ft^2 \cdot °F}{Btu}$

$$\dot{Q}_t/A = \frac{95-78}{23.44} = 0.73 \; Btu/hr \cdot ft^2$$

Problem 14.3

Thermal Resistances, hr-ft²-°F/Btu		
	Path	
Resistance Source	At studs	Between studs
Inside air film	0.68	0.68
3/8 inch Gypsum Wallboard	0.32	0.32
Studs, R_s = (5.5/0.8)	6.875	******
Insulation, R_b	*****	19
25/32 inch sheathing	2.06	2.06
1.5 inch expanded polystyrene (R=1.5x5)*	7.5	7.5
4 inch brick (R = 0.15x4)**	0.6	0.6
Outside air film	0.25	0.25
Totals:	18.285	30.41

Notes:
* Arbitrarily Selected extruded polystyrene
** Arbitrarily Selected lowest R-value/inch

From the problem statement: $a_s = 0.2$, $a_b = 0.8$

a.) Parallel Path Method. (By Eqn 14.18)

$$R_{t(av)} = \frac{1}{\frac{0.2}{18.285} + \frac{0.8}{30.41}} = 26.85 \; \frac{hr\text{-}ft^2\text{-}°F}{Btu}$$

$$\dot{Q}_t/A = \frac{68-5}{26.85} = 2.35 \; Btu/hr\text{-}ft^2$$

b) Isothermal Plane Method:

By Eqn 14.22 $R_{s,b} = \left(\frac{0.2}{6.875} + \frac{0.8}{19}\right)^{-1} = 14.05 \; \frac{hr\text{-}ft^2\text{-}°F}{Btu}$

By Eqn 14.23 $R_{t(av)} = 11.41 + 14.05 = 25.46 \; \frac{hr\text{-}ft^2\text{-}°F}{Btu}$

$$\dot{Q}_t/A = \frac{68-5}{25.46} = 2.47 \; Btu/hr\text{-}ft^2$$

Problem 14.4

Resistance Source	Thermal Resistances, $m^2 \cdot °C/W$	
	Path	
	At studs	Between studs
Inside air film	0.12	0.12
10-mm plywood (R=8.66x0.01)	0.087	0.087
Studs, (R_s=7.0x0.14)	1.21	******
Insulation, R_b	*****	3.32
200-mm concrete block *	0.17	0.17
10-mm brick (R = 1.02x.010)**	0.01	0.01
Outside air film	0.03	0.03
Totals:	1.627	3.737

Notes:
* Table 14.4 SI lists a range, selected lowest R-value
** Arbitrarily Selected lowest R-value/meter

Given: $a_s = 0.25$ and $a_b = 0.75$

a.) Parallel Path Method:

By Eqn 14.20 with $U = \frac{1}{R}$

$$\boxed{U_{av} = \frac{a_s}{R_{t,s}} + \frac{a_b}{R_{t,b}} = \frac{0.25}{1.627} + \frac{0.75}{3.737} = 0.35 \frac{W}{m^2 \cdot °C}}$$

b.) Isothermal Plane Method:

By Eqn 14.22 $R_{s,b} = \left(\frac{0.25}{1.21} + \frac{0.75}{3.32} \right)^{-1} = 2.31 \frac{m^2 \cdot °C}{W}$

By Eqn 14.23: $R_{t(av)} = 0.417 + 2.31 = 2.727 \frac{m^2 \cdot °C}{W}$

$$\boxed{U_{av} = \frac{1}{2.727} = 0.37 \; W/m^2 \cdot °C}$$

Problem 14.5

Resistance Source	Thermal Resistances, hr-ft²·°F/Btu	
	Path	
	At studs	Between studs
Inside air film	0.68	0.68
3/8 inch plywood (R=1.25x0.375)	0.47	0.47
Studs, R_s = (3.5/0.8)	4.38	******
3.5-inch air space*	*****	0.91
8-inch Concrete Block **	0.97	0.97
4 inch brick (R = 0.15x4)***	0.6	0.6
Outside air film	0.25	0.25
Totals:	7.35	3.88

Notes:
* See selection criteria below
** Arbitrarily Selected lowest R-value
*** Arbitrarily Selected lowest R-value/inch

If we include only the 2x4 studs as framing.
$a_s = \frac{1.5}{16} = 0.094$ & $a_b = \frac{14.5}{16} = 0.906$

* For typical building materials $\epsilon \cong 0.9$ ∴ $\epsilon_{eff} = 0.82$
for the temps. given select mean temp ≈ 50°F & $\Delta T \approx 30°F$
∴ $R_{airspace} = 0.91 \frac{hr \cdot ft^2 \cdot °F}{Btu}$ from Table 14.2 IP.

a.) Parallel Path Method:
$$U_{av} = \frac{0.094}{7.35} + \frac{0.906}{3.88} = 0.246 \text{ Btu/hr·ft}^2°F$$

$$\underline{\frac{\dot{Q}_t}{A} = (0.246 \frac{Btu}{hr \cdot ft^2 °F})(70-10)°F = 14.8 \text{ Btu/hr·ft}^2}$$

b) Isothermal Plane Method
$$R_{s,b} = \left(\frac{0.094}{4.38} + \frac{0.906}{0.91}\right)^{-1} = 0.98 \frac{hr \cdot ft^2 \cdot °F}{Btu}$$

$$R_{t(av)} = 2.97 + 0.98 = 3.95 \frac{hr \cdot ft^2 \cdot °F}{Btu}$$

$$\underline{\dot{Q}_t/A = \frac{70-10}{3.95} = 15.2 \text{ Btu/hr·ft}^2}$$

c.) Replace the resistance of the air space with $R_{ins} = 13 \frac{hr \cdot ft^2}{Btu}$
Then $U_{av} = \frac{0.094}{7.35} + \frac{0.906}{15.97} = 0.070 \text{ Btu/hr·ft}^2°F$

and
$$\underline{\dot{Q}_t/A = (0.070)(70-10) = 4.17 \text{ Btu/hr·ft}^2}$$

14.6: In order to avoid condensation, the surface must be above the dew-point temperature.

From Chart C-8E, the dew-point temperature for moist air at $t_i = 68°F$, $\phi_i = 0.35$ and standard atmos. pressure is $t_d \cong 39.5°F$

From Table 14.5 for a double glass window with a 0.5 inch air space, $U_{cg} = 0.49$ Btu/hr·ft²·°F
$U_{eg} = 0.59$ Btu/hr·ft²·°F

Let the inside surface temperature of the glass be t_s.
Then
$$t_s = t_i - \frac{R_i}{R_t}(t_i - t_o) = t_i - R_i U (t_i - t_o)$$

From Table 14.1 with $\epsilon \cong 0.9$ (typical for glass)
$R_i = 0.68$ hr·ft²·°F/Btu

Consider the edge of glass since the larger U-value results in the warmer allowable outdoor temp.
Therefore with $t_s > 39.5°F$ we get

$$68 - 0.68(0.59)(68 - t_o) > 39.5°F$$

or:
$$t_o > -3°F$$

14.7 Window is double glazed with 9.5 mm argon filled space between sheets of glass

9.5 mm ≅ 0.374 in or 3/8 in gap.

From Table 14.5: $U_{cg} = 0.48$ Btu/hr·ft²·°F
$U_{eg} = 0.59$ Btu/hr·ft²·°F

Consider the edge of glass since that will give the coldest temp at the inside surface of the glass.

$$\therefore R_t = \frac{1}{(0.59) \frac{Btu}{hr\cdot ft^2 \cdot °F} \times 5.68 \frac{W/m^2 \cdot °C}{Btu/hr \cdot ft^2 \cdot °F}} = 0.298 \frac{m^2 \cdot °C}{W}$$

The inside air dew-point temperature must be lower than the inside surface temp of the glass to avoid condensation. ∴ Find inside surface temp of the glass, call this t_s

then:
$$t_s = t_i - \frac{R_i}{R_t}(t_i - t_o)$$

For Minneapolis, the 97.5% design temp = −12°F
or $t_o = -12°F = \underline{-24.4°C}$

From Table 14.1 with $\varepsilon \cong 0.9$ (typical for glass)
$R_i = 0.12 \frac{m^2 \cdot °C}{W}$

$$\therefore t_s = 23 - \left(\frac{0.12}{0.298}\right)(23 - (-24.4)) = 3.9°C$$

From Table A.1SI $P_{w,s}(3.9°C) = 0.808$ kPa

at 23°C $P_{w,s} = 2.81$ kPa

$$\therefore \phi = \frac{0.808}{2.81} = 0.29$$

∴ Humidity must be kept below 29% to avoid condensation at the edge of glass. Condensation at the center of glass would occur at a higher ϕ-value

14.8

$$R_t = R_i + \left(\frac{L}{k}\right)_{glass} + R_{gap,1} + \left(\frac{L}{k}\right)_{glass} + R_{gap,2} + \left(\frac{L}{k}\right)_{glass} + R_o$$

$$\left(\frac{L}{k}\right)_{glass} = \frac{0.190 \text{ inch}}{6.4 \text{ Btu-in/hr-ft}^2\text{°F}} = 0.030 \frac{\text{hr-ft}^2\text{°F}}{\text{Btu}}$$

For both gaps:

$$\varepsilon_{eff} = \frac{1}{\frac{1}{0.9} + \frac{1}{0.02} - 1} = 0.02$$

For a 45° slope with heat flow upward, $\Delta t \approx 10\text{°F}$ & $\bar{t} \cong 50\text{°F}$, $R_{gap} \cong 2.55$ (use value for $\varepsilon_{eff} \approx 0.03$)
From Table 14.1 IP, $R_i = 0.62$, $R_o = 0.17$

$$\therefore R_t \cong 0.62 + 3(0.03) + 2(2.55) + 0.17 = 5.98 \frac{\text{hr-ft}^2\text{°F}}{\text{Btu}}$$

or

a.) $\boxed{U_t \cong 0.17 \text{ Btu/hr-ft}^2\text{°F}}$

b.) The inside surface of the glass must be above the dew-point temp.

at 68°F $P_{w,s} = 0.33917$ psia   Table A1.E

$\therefore P_w = \phi P_{w,s} = 0.36(0.33917) = 0.1221$ psia

or $t_d \cong 40.9\text{°F}$

if t_d = inside surf temp of glass.

$$t_d = t_i - \frac{R_i}{R_t}(t_i - t_o)$$

or

$$t_o = t_i + \frac{R_t}{R_i}(t_i - t_d) = 68 - \frac{5.98}{0.62}(68 - 40.9) =$$

$$t_o = -193\text{°F}$$

(Even in Minneapolis no condensation is going to occur on the inside of this window.)

14.9

$$R_t = R_i + \left(\frac{L}{K}\right)_{glass} + R_{air\,space} + \left(\frac{L}{K}\right)_{glass} + R_o$$

From Table 14.1 with $\epsilon = 0.9$, 45° slope, heat flow up
$R_i = 0.11 \frac{m^2 \cdot °C}{W}$ and at 6.7 m/s $R_o = 0.030 \frac{m^2 \cdot °C}{W}$

$\epsilon_{eff} = \frac{1}{\frac{1}{0.9} + \frac{1}{0.9} - 1} = 0.82$, $t_i = 18°C$, $t_o = 2°C$

∴ For 45°, heat flow up, space = 13 mm, $\epsilon_{eff} = 0.82$,
$\bar{t} \cong 10°C$ & $\Delta t = 5.6°C$ ⟹ $R_{air\,space} = 0.16 \frac{m^2 \cdot °C}{W}$

Since in part b we will compare results to Table 14, use same glass thickness & conductivity as listed in the footnote,

$$\left(\frac{L}{K}\right)_{glass} = \left[\frac{(0.125/12)}{0.53} \frac{hr \cdot ft^2 \cdot °F}{Btu}\right]\left[\frac{1}{5.68} \frac{\left(\frac{m^2 \cdot °C}{W}\right)}{\left(\frac{hr \cdot ft^2 \cdot °F}{Btu}\right)}\right]$$

$\left(\frac{L}{K}\right)_{glass} = 0.003 \frac{m^2 \cdot °C}{W}$

∴ $R_t = 0.11 + 2(0.003) + 0.16 + 0.03 = 0.31 \frac{m^2 \cdot °C}{W}$

(Notice the R of the glass is almost negligible)

or $\boxed{U_t = 3.23 \; W/m^2 \cdot °C}$

b.) Table 14.5 double glazed, vertical, 15 mph wind
$U_{cg} = 0.49 \; Btu/hr \cdot ft^2 \cdot °F$

From Table 14.5B adjust to 45° slope
$U_{cg} = .49 \left(\frac{0.57}{0.5}\right) = 0.56 \; Btu/hr \cdot ft^2 \cdot °F$

From Table 14.6 adjust to 7.5 mph (6.7 m/s)
$U_{cg} = 0.56 \left(\frac{0.56}{0.60}\right) = 0.52 \; Btu/hr \cdot ft^2 \cdot °F = 2.95 \; W/m^2 \cdot °C$

∴ ~3.0 vs ~3.2 or <u>~7% different</u>

14.10

a.) The average U-value for the wall is given by Eqn 14.30. Since the complete height of the wall is insulated, $D_i = D_w$ and Eqn 14.30 becomes

$$U_w = \frac{2k_s}{\pi D_w} \ln\left[1 + \frac{\pi D_w}{2k_s(R_i + R_w + R_{INS})}\right]$$

From Table 14.8 for sandy soil
$$k_s = 15.6 \frac{Btu \cdot in}{hr \cdot ft^2 \cdot °F} = \frac{15.6}{12} \frac{Btu}{hr \cdot ft \cdot °F} = 1.30 \frac{Btu}{hr \cdot ft \cdot °F}$$

From Table 14.1 $R_i = 0.68 \frac{hr \cdot ft^2 \cdot °F}{Btu}$

From Table 14.4 IP & Selecting lowest R-value/inch for concrete
$$R_w + R_{INS} = \underbrace{0.06(8)}_{Concrete} + \underbrace{3.85(2)}_{Polystyrene} + \underbrace{1.25(0.375)}_{Plywood} = 8.65 \frac{hr \cdot ft^2 \cdot °F}{Btu}$$

∴
$$U_w = \frac{2(1.30)}{\pi(6)} \ln\left[1 + \frac{\pi(6)}{2(1.30)(0.68+8.65)}\right]$$

$$\boxed{U_w = 0.079 \; Btu/hr \cdot ft^2 \cdot °F}$$

b.) By Eqn 14.29: $\dot{Q}_w = U_w(L \cdot D_w)(t_i - t_g)$
By Eqn 14.35 $t_g = \bar{t}_a - A$
For Denver from Table B.2, $\bar{t}_a = 38.9°F$ (Avg. Oct-April)
From Table B.1 we see Denver is at 40°N latitude
105°W longitude ∴ from Fig 14.10 $A \cong 22°F$
∴ $t_g = 38.9 - 22 = 16.9°F$; Select $t_i = 70°F$
$\dot{Q}_w = 0.079(300)(6)(70-16.9)$ Btu/hr

$$\boxed{\dot{Q}_w = 7,550 \; Btu/hr}$$

14.11 The heat loss rate from the floor is given by Eqn 14.34. Using the narrowest width, $W_s = 50$ ft

From Table 14.1: $R_i = 0.92 \ \frac{hr\text{-}ft^2\text{-}°F}{Btu}$

From Table 14.4 IP using the lowest R-value/inch for concrete, $R_{concrete} = (0.06)(6) = 0.36 \ \frac{hr\text{-}ft^2\text{-}°F}{Btu}$

$\therefore R_f = 0.36 + 3 = 3.36 \ \frac{hr\text{-}ft^2\text{-}°F}{Btu}$

From Table 14.8 (see problem 14.10 for selection criteria)
$k_s = 1.30 \ Btu/hr\text{-}ft\text{-}°F$

By Eqn 14.34

$$U_f = \frac{2(1.30)}{\pi(50)} \ln\left[\frac{0.92 + 3.36 + \frac{\pi(6)}{2(1.30)} + \frac{\pi(50)}{2(1.30)}}{0.92 + 3.36 + \frac{\pi(6)}{2(1.30)}}\right]$$

$U_f = 0.030 \ Btu/hr\text{-}ft^2\text{-}°F$

By Eqn 14.33
$$\dot{Q}_f = U_f A_f (t_i - t_g)$$

From Problem 14.10 we determined $t_g = 16.9°F$ and we selected $t_i = 70°F$

Therefore:
$$\dot{Q}_f = 0.030(100)(50)(70 - 16.9)$$

$$\dot{Q}_f = 7,970 \ Btu/hr$$

14.12

By Eqn 14.36
$$\dot{Q}_s = F_s P(t_i - t_o)$$

From Table B.2 The annual number of Degree Days in Denver for a base temperature of 65°F is 6016 °F-Days

From Table 14.9 for 100 mm block with brick face, uninsulated and using linear interpolation between the Degree-Days listed.
$$F_s \cong 1.50 \ W/m \cdot °C$$

From Table B.1, The 97.5% Design Temp. for Denver is 1°F = -17 °C. Select an indoor design temp. of 21°C

By Eqn 14.36
$$\boxed{\dot{Q}_s = 1.50(350)(21+17) = 20,000 \ W}$$

14.13. From Table 14.10 determine the effective leakage area.

Component	Area, length or # of components	Leakage area/ area, length or # of components	Leakage Area
Wall Joints			
Ceiling-Wall	64 m	1.5 cm²/lmc	96 cm
Sole Plate	64 m	4 cm²/lmc	256
Windows, 52 total			
Frames	104 m²	0.3 cm²/m²	31.2
Windows	312 m	0.79 cm²/lmc	246.5
Doors			
Frames, total	6.3 m²	0.3 cm²/m²	1.9
Door	3 doors	12 cm²/ea	36.0
Storm Door	3 doors	−6 cm²/ea	−18.0
Common Chimney	1	29 cm²/ea	29.0
Fireplace w/glass door*	0.75 m²	40 cm²/m²	30.0
Bathroom vents w/damper	3	10 cm²/ea	30.0
Dryer Vent w/damper	1	3 cm²/ea	3.0
Kitchen Vent w/damper	1	5 cm²/ea	5.0

$$A_\ell = \underline{\underline{747 \text{ cm}^2}}$$

* Assume fireplace opening ~ 1m × 0.75 m.

For Madison, WI, $t_o = -7°F$ (97.5% from Table B.1)

Assume an indoor design temp of 70°F

∴ $\Delta t = 77°F = 43°C$

From Table 14.11, $a_s = 0.00029 \frac{(L/s)^2}{cm^4 \cdot °C}$;

Table 14.12 $a_w = 0.000137 \ (L/s)^2/[cm^4 \cdot (m/s)^2]$

a.) By Eqn. 14.49 $\dot{V} = 747[0.00029(43) + 0.000137(4)^2]^{1/2} = \underline{\underline{90.5 \text{ L/s}}}$

b) $\dot{Q}_s = \dot{m}_a c_p (t_i - t_o)$ use $\rho_a = \rho_{std} = 1.2 \text{ kga/m}^3$

$\dot{Q}_s = (90.5 \times 10^{-3} \text{ m}^3/\text{s})(1.2 \frac{kg_a}{m^3})(1.00 \frac{kJ}{kg_a \cdot °C})(43°C) = \underline{\underline{4.7 \text{ kW}}}$

14.14 Let subscripts iw = interior wall, ew = exterior walls, r = roof.; $t_i = 65°F$, $t_o = -13°F$

Then: $(UA)_{iw}(t_i - t_u) + (UA)_{ew}(t_o - t_u) + (UA)_r(t_o - t_u)$
$$= \dot{m}_a c_p (t_u - t_o)$$

Inside Wall:
$R_t = R_i + R_{plasterboard} + R_{insl.} + R_{plasterboard} + R_o$
$= 0.68 + 0.32 + 13 + 0.32 + 0.68 = 15 \; \frac{hr\text{-}ft^2\text{-}°F}{Btu}$

∴ $(UA)_{iw} = 90 \; ft^2 / 15 \; \frac{hr\text{-}ft^2\text{-}°F}{Btu} = 6 \; Btu/hr\text{-}°F$

Exterior Walls $U_{av} = 0.6(0.9) + 0.4(0.3) = 0.66 \; Btu/hr\text{-}ft^2\text{-}°F$
∴ $(UA)_{ew} = 0.66(30)(9) = 178.2 \; Btu/hr\text{-}°F$

Roof
$(UA)_r = \left(0.5 \; \frac{Btu}{hr\text{-}ft^2\text{-}°F}\right)(100 \; ft^2) = 50 \; \frac{Btu}{hr\text{-}°F}$

Infiltration: $\dot{V} = (3 \; ACH)(900 \; ft^3) = 2700 \; ft^3/hr$
$\dot{m}_a = \dot{V}/v_a = 2700/11.2 = 241 \; lbm_a/hr$
$c_p \approx c_{pa} = 0.24 \; Btu/lbm_a\text{-}°F$

∴ $6(65 - t_u) + 178.2(-13 - t_u) + 50(-13 - t_u) = 241(0.24)(t_u + 13)$

a) $\boxed{t_u = -11.4°F}$

b) $\dot{Q} = (UA)_{iw}(t_i - t_u) = \left(6 \; \frac{Btu}{hr\text{-}°F}\right)(65 + 11.4) = 458 \; \frac{Btu}{hr}$

14.15

Material	Thermal Res., R		Vapor Res., Z	
Inside Air Film	0.68	$\frac{hr\text{-}ft^2\text{-}°F}{Btu}$	0	Rep.
3/8" Gypsum Wallboard	0.32		0.020	
3.5-inch Mineral wool	13		0.030	
8-inch Block	0.97		0.4	
Outside Air Film	0.17		0	
Total	15.14		0.45	

If condensation occurs it will be near the interface between the insulation and block, call this location x

By Eqn 14.53

$$t_x = t_w - \frac{R_{w\to x}}{R_t}(t_w - t_c) = 72 - \frac{14.97}{15.14}(72+5) = -4.14\,°F$$

$\therefore P_{w,s,x} = 0.01487$ psia (Table A.1E)

$$P_{w,w} = \phi_w P_{w,s,w} = (0.35)(0.38878) = 0.13607 \text{ psia}$$
$$P_{w,c} = \phi_c P_{w,s,c} = (0.60)(0.01420) = 0.00852 \text{ psia}$$

By Eqn 14.55
$$P_{w,x} = P_{w,w} - \frac{Z_{w\to x}}{Z_t}(P_{w,w} - P_{w,c})$$

$$P_{w,x} = 0.13607 - \frac{0.05}{0.45}(0.13607 - 0.00852) = 0.1219 \text{ psia}$$

a) Since this pressure is greater than $P_{w,s,x}$ there would be condensation (and freezing) near the insulation/block interface.

b) Place a vapor retarder on the warm side of the insulation. The most common location will be between the wallboard and insulation.

14.16

Material	Thermal Res, R	Vapor Res, Z Rep
Inside Air Film	0.68 $\frac{hr \cdot ft^2 \cdot °F}{Btu}$	0 Rep
1/2" Plywood	0.63	1.06*
3.5" Insul. Batt	13	0.030
8" Block	0.97	0.4
4" Brick	0.60	1.3
Outside Air Film	0.25	0
Total	16.13	3.07

* Assume interior grade of 1/2" has Z-value equivalent to double that of 1/4" thick plywood.

a.) $\dot{Q}/A = \frac{\Delta t}{R_t} = \frac{72-5}{16.13} = 4.15$ Btu/hr-ft²

b.) Between the insulation and block and between the block and brick.

c.) Find Z_{vr} to eliminate the possibility of condensation at the Block/Brick interface. Locate the vapor retarder at the Plywood/Insulation interface.

$\therefore t_x = t_i - \frac{R_{w \to x}}{R_t}(t_i - t_o) = 72 - \frac{15.28}{16.13}(72-5) = 8.53°F$

$\therefore P_{w,s,x} = 0.02870$ psia (Table A.1E)

$P_{w,w} = \phi_w P_{w,s,w} = 0.3(0.38878) = 0.11663$ psia

$P_{w,c} = \phi_c P_{w,s,c} = 0.5(0.02397) = 0.01199$ psia.

$\therefore$ By Eqn 14.56

$Z_{vr} \geq 1.3 \left(\frac{0.11663 - 0.01199}{0.02870 - 0.01199}\right) - 3.07 = \underline{\underline{5.07}}$ Rep.

14.16 continued

d.) If we use Z_{vr} at 5.07 Rep then at location of insulation/Block interface $Z_{w \to x} = 1.06 + 5.07 + 0.0 = 6.16$ Rep

and $Z_t = 5.07 + 3.07 = 8.14$ Rep.

∴ By eqn 14.55

$$P_{w,x} = 0.11663 - \frac{6.16}{8.14}(0.11663 - 0.01199) = 0.03744 \text{ psia.}$$

Assume $R_{vr} \cong 0$ Then:

$$t_x = 72 - \frac{14.30}{16.13}(72-5) = 12.56°F$$

∴ $P_{w,s,x} = 0.03414$

∴ $P_{w,s,x} < P_{w,x}$ and ∴ condensation would occur here if we use a $Z_{vr} = 5.07$ Rep.

If we had started with the location-x at the insulation/Block interface we would have calculated

$$Z_{vr} > 1.70 \left(\frac{0.11663 - 0.01199}{0.03414 - 0.01199} \right) - 3.07 = \underline{4.97} \text{ Rep}$$

But according to our earlier calculations, $Z_{vr} = 5.07$ was not enough. What is wrong. Nothing. The calculations are ok but notice in the denominator of the eqn 14.56, $(P_{w,s,x} - P_{w,c})$ is the difference between two numbers that are close together. ∴ round-off in these values can easily lead to 10% variation in the number calculated for Z_{vr}.

14.17 If there is a moisture problem, it will occur at the wallpaper/wallboard interface and/or near the wallboard/insulation interface.

We can either compare the calculated vapor pressure using Eqn 14.55 with the saturation vapor pressure at these locations. Or we could go ahead and calculate Z_{vr} for these locations using Eqn 14.56 (Notice if no extra vapor retarder is needed, the value of Z_{vr} from Eqn 14.56 will be less than or equal to zero.)

This second approach not only tells if there is a problem, it also gives a value of Z_{vr} if needed.

		R	Z	
Cool	Inside Air film	0.68 hr-ft²°F/Btu	0	Rep
	Vinyl Wallpaper	0	4.5	
	3/8" Wallboard	0.32	0.02	
	3.5" Insulation	13.0	0.03	
	8" Block	0.97	0.40	
warm	Outside Air film	0.17	0	
	Totals	15.14	4.95	

First look at the interface between wallboard & insul.

$t_x = 89 - \frac{14.14}{15.14}(89-72) = 73.1\ °F$

$P_{w,w} = 0.44462\ psia$ (Sat. Pres. at dew-pt., Table A.1E)

$P_{w,c} = \phi P_{w,s,c} = 0.4(0.38878) = 0.15551\ psia$

$P_{w,s,x} = 0.40352\ psia$

14.17 continued

By Eqn 14.56

$$Z_{vr} = 4.52\left(\frac{0.44462 - 0.15551}{0.40352 - 0.15551}\right) - 4.95 = 0.32 \text{ Rep}$$

At the wallpaper/wallboard interface.

$$t_x = 89 - \frac{14.46}{15.14}(89-72) = 72.8°F$$

$$\therefore P_{w,s,x} = 0.39947 \text{ psia} \quad (\text{Table A.1E})$$

$$Z_{vr} = 4.50\left(\frac{0.44462 - 0.15551}{0.39947 - 0.15551}\right) - 4.95 = 0.38 \text{ Rep}.$$

Therefore if we need to keep the indoor temperature at 72°F, there is a need for a vapor retarder. The most convenient would probably be to put a 0.002 inch (or thicker) sheet of polyethylene ($Z_{vr} \approx 6.3$ Rep or greater) located between the Block and insulation.

14.18

	At framing	Between Framing	
	$R \; \frac{hr \cdot ft^2 \cdot °F}{Btu}$	R	$Z \; perm^{-1}$
Inside Air	0.68	0.68	0
3/8" Gypsum Wallboard	0.32	0.32	0.02
Vapor Retarder	0	0	Z_{vr}
2×6 Framing	6.88	—	—
5.5" Insulation	—	19	0.047 (1)
25/32" sheathing	2.06	2.06	0.022 (2)
1.5" expanded polystyrene	7.5	7.5	1.25 (3)
4" Brick (P/inch = 0.15)	0.6	0.6	1.30
Outside Air	0.25	0.25	0
Totals:	18.29	30.41	$2.639 + Z_{vr}$

× locations

a.) By Eqns 14.19 & 14.12 & $A_s/A_t = 0.15$ (Given)

$$U_{av} = \frac{0.15}{18.29} + \frac{0.85}{30.41} = 0.036 \; Btu/hr \cdot ft^2 \cdot °F$$

$$\dot{Q}/A = (0.036)(68-5) = 2.27 \; Btu/hr \cdot ft^2$$

b.) By Eqn 14.22 $\quad R_{s,b} = \left(\frac{0.15}{6.88} + \frac{0.85}{19}\right)^{-1} = 15.03 \; \frac{hr \cdot ft^2 \cdot °F}{Btu}$

By Eqn 14.23

$$R_{t(av)} = 0.68 + 0.32 + 15.03 + 2.06 + 7.5 + 0.6 + 0.25$$
$$= 26.44 \; \frac{hr \cdot ft^2 \cdot °F}{Btu}$$

$$\dot{Q}/A = (68-5)/(26.44) = 2.38 \; Btu/hr \cdot ft^2$$

c.) $P_{w,w} = \phi_w P_{w,s,w} = 0.3 (0.33917) = 0.10175 \; psia$

$P_{w,c} = P_{w,s,c} = 0.02397 \; psia$

Potential problem locations are at the 3 interfaces:
(1.) Between Insul & Sheathing
(2) Between Sheathing & Polystyrene
(3) Between Polystyrene & Brick

14.18 continued

By Eqn 14.53 find t_x at each location, from Table A.1E find $P_{w,s,x}$ and by Eqn 14.56 Find Z_{vr} for each "x"

Location 1

$$t_{x,1} = 68 - \frac{20}{30.41}(68-5) = 26.6 \,°F$$

$$P_{w,s,1} = 0.06906 \text{ psia}$$

$$Z_{x \to c} = 2.572 \text{ Perm}^{-1}$$

$$Z_{vr} > 2.572 \left(\frac{0.10175 - 0.02397}{0.06906 - 0.02397}\right) - 2.639 = 1.8 \text{ Perm}^{-1}$$

Location 2

$$t_{x,2} = 68 - \frac{22.06}{30.41}(68-5) = 22.3 \,°F$$

$$P_{w,s,2} = 0.05637 \text{ psia}$$

$$Z_{x \to c} = 2.55 \text{ Perm}^{-1}$$

$$Z_{vr} > 2.55 \left(\frac{0.10175 - 0.02397}{0.05637 - 0.02397}\right) - 2.639 = 3.5 \text{ Perm}^{-1}$$

Location 3:

$$t_{x,3} = 68 - \frac{29.56}{30.41}(68-5) = 6.8 \,°F$$

$$P_{w,s,3} = 0.02629 \text{ psia}$$

$$Z_{x \to c} = 1.30 \text{ Perm}^{-1}$$

$$Z_{vr} > 1.30 \left(\frac{0.10175 - 0.02397}{0.02629 - 0.02397}\right) - 2.639 = 40.9 \text{ Perm}^{-1}$$

∴ Use a vapor retarder with $\underline{Z_{vr} > 41 \text{ Perm}^{-1}}$

14.19

a.) If there is a moisture problem it should occur at the polystyrene/concrete interface. The vapor retarder should be placed on the warm side of the polystyrene insulation. In this case, the most common location will be between the plywood paneling and the polystyrene.

b.) Use Eqn. 14.56 to calculate a value for Z_{vr}. If $Z_{vr} < 0$ there was no vapor problem in the wall.

	R	Z
Inside air film	0.68 $\frac{hr\text{-}ft^2 °F}{Btu}$	0 Rep
1/4-inch plywood	0.31	0.53
Vapor Retarder	0	Z_{vr}
1.5" expanded polystyrene*	5.78	0.26
6-inches concrete**	0.36	1.86
Outside Air film	0.17	0
Totals	7.30	2.65 + Z_{vr} Rep

→ (Location)

Notes: * Select the lower limit of Z/inch to get worst case
** Select the lower limit of R/inch to get worst case

By Eqn 14.53: $t_x = 68 - \frac{6.77}{7.30}(68+8) = -2.5°F$

∴ $P_{w,s,x} = 0.01622$ psia (Table A.1E)

$P_{w,w} = (0.35)(0.33917) = 0.11871$ psia, $P_{w,c} = (0.70)(0.01208) = 0.00846$ psia

By Eqn 14.56

$$Z_{vr} > 1.86 \left(\frac{0.11871 - 0.00846}{0.01622 - 0.00846} \right) - 2.65 = 23.8 \text{ Rep}$$

∴ Choose $Z_{vr} > 39.24$ Rep

14.20 The 97.5% design temp. for Chicago is −4°F (Table 8.1)

a.) **Walls**

Thermal Resistances, $\frac{hr\text{-}ft^2\text{-}°F}{Btu}$

Source	At studs	Between Studs
Inside air film	0.68	0.68
3/8-inch gypsum	0.32	0.32
Studs	6.875	—
Insulation	—	19
25/32-inch sheathing	2.06	2.06
1.5-inch polystyrene*	7.5	7.5
4-inch brick **	0.6	0.6
Outside air film	0.17	0.17
Totals	18.21	30.33

* Arbitrarily selected extruded polystyrene
** Arbitrarily selected lowest R-value/inch

Using the parallel path method, By Eqns 14.18 and 14.12

$$(U_{av})_{walls} = \frac{0.25}{18.21} + \frac{0.75}{30.33} = 0.038 \; Btu/hr\text{-}ft^2\text{-}°F$$

Windows: $U_D = 0.33$ from Table 14.5
Roof: $U_R = 1/9.0 = 0.111 \; Btu/hr\text{-}ft^2\text{-}°F$
Doors: $U_D = 0.20$ from Table 14.7

$A_{windows} = 700 \, ft^2$, $A_{doors} = 40.5 \, ft^2$
$A_{walls, frame} = (50)(10)(4) - 700 - 40.5 = 1259.5 \, ft^2$
$A_{Roof} = 2500 \, ft^2$

Prob 14.20 continued – page 2 of 4

By Eqn 14.20 for the above grade frame walls, doors, windows and roof

$$U_{av} = \frac{0.038(1259.5) + 0.2(40.5) + 0.33(700) + 0.111(2500)}{2,000 + 2,500}$$

$$\boxed{U_{av} = 0.125 \text{ Btu/hr-ft}^2\text{-}°F}$$

$$\dot{Q} = U_{av} \cdot A_t (t_i - t_o) = 0.125(4,500)(70+4) = \underline{41,625} \text{ Btu/hr}$$

b.) $R_t = R_i + R_{block} + R_o = 0.68 + 0.97 + 0.17 = 1.82 \frac{\text{hr-ft}^2\text{-}°F}{\text{Btu}}$

Note: Selected lowest R-value for the block
$$\dot{Q} = A(t_i - t_o)/R_t = (4 \times 50 \times 2)(70+4)/1.82$$
$$\dot{Q} = \underline{16,264} \text{ Btu/hr.}$$

c.) By Eqn 14.30 with $D_i = D_w = 6$ ft, $R_i = 0.17 \frac{\text{hr-ft}^2\text{-}°F}{\text{Btu}}$,
$R_w = 0.97 \frac{\text{hr-ft}^2\text{-}°F}{\text{Btu}}$, $R_{INS} = 0$ and $k_s = 15.6$ Btu-in/hr-ft²-°F
or $k_s = 1.30$ Btu/hr-ft-°F (Table 14.8)

$$U_w = \frac{2(1.30)}{\pi(6)} \left\{ \ln\left[1 + \frac{\pi(6)}{2(1.30)(0.17+0.97)}\right]\right\}$$

$U_w = 0.275$ Btu/hr-ft²-°F

From Fig 14.10 $A = 22°F$ for Chicago ∴ By Eqn 14.35
∴ $t_g = 37.4 - 22 = 15.4 °F$

By Eqn 14.29
$$\dot{Q}_w = 0.275(200 \times 6)(70 - 15.4) = 18,018 \text{ Btu/hr.}$$
$$\underline{\dot{Q}_w = 18,018 \text{ Btu/hr.}}$$

Problem 14.20 continued -- page 3 of 4

d.) Use Eqn 14.34 with $R_i = 0.92 \ \frac{hr \cdot ft^2 \cdot °F}{Btu}$, $D_w = 6 \ ft$, $W_s = 50 \ ft$, $R_s = 0.130 \ Btu/hr \cdot ft \cdot °F$
and $R_f = (0.06)(6) = 0.36 \ hr \cdot ft^2 \cdot °F/Btu$ (lowest R/inch)

$$U_f = \frac{2(1.30)}{\pi(50)} \ln\left[\frac{0.92 + 0.36 + \frac{\pi(6)}{2(1.30)} + \frac{\pi(50)}{2(1.30)}}{0.92 + 0.36 + \frac{\pi(6)}{2(1.30)}}\right]$$

$U_f = 0.035 \ Btu/hr \cdot ft^2 \cdot °F$

By Eqn 14.33
$\dot{Q}_f = 0.035(2500)(70 - 15.4) = \underline{4,778 \ Btu/hr}$

e.) From Table 14.10 determine effective leakage area.

Component	Area, length or # of Components	Leakage area per area, length, or # of components	Leakage Area
Wall Joints			
Ceiling-Wall	200 ft	0.071 in²/lftc	14.2 in²
Sole-plate	200 ft	0.19 in²/lftc	38.0
Windows*			
Frames	700 ft²	0.004 in²/ft²	2.8
Windows*	570 lftc	0.011 in²/lftc	6.3
Doors			
Frames, total	40.5 ft²	0.004 in²/ft²	0.2
Door	2 doors	1.9 in²/each	3.8
Chimney **	1	4.5 in²/each	4.5
		Total	69.8 in²

Notes: * The actual size of each window is not given. Based on the window areas for the various walls, assume a typical window area of 25 ft²

Problem 14.20 concluded   page 4 of 4

with dimensions 6 ft high × 4.17 ft wide. Also, assume they are casement windows with weatherstripping. Based on this there would be a total of 28 windows each with 20.34 lftc for a total of 570 lftc.

** Assume the building has a furnace with a chimney.

From Table 14.11 $a_s = 0.0156 \, (ft^3/min)^2 \cdot in^{-4} \cdot °F^{-1}$
From Table 14.12 $a_w = 0.0012 \, (ft^3/min)^2 \cdot in^{-4} \cdot mph^{-2}$

By Eqn 14.49
$$\dot{V} = 69.8 \left[0.0156(70+4) + 0.0012(15)^2 \right]^{0.5} = 83.3 \, cfm$$

By Eqns 14.37a and 14.39 with $c_p = 0.24 \, Btu/lbma\text{-}°F$ and $\rho = 0.075 \, lbma/ft^3$ (standard air). The sensible load due to infiltration is

$$\dot{Q}_s = (0.075)(83.3)(0.24)(70+4) = 111 \, Btu/min$$
$$\dot{Q}_s = 6,660 \, Btu/hr$$

f) The total design load for the building is the sum of the above values.

$$\therefore \dot{Q}_{design} = 41,625 + 16,264 + 18,018 + 4,778 + 6,660$$
$$= 87,000 \, Btu/hr.$$

Note -- This is the sensible load rounded to the nearest 1000 Btu/hr.

15.1

By Eq. 15.10
$$t_e = t_o + \frac{I_T \alpha_s}{h_o} - \frac{\varepsilon \Delta R}{h_o}$$

Assume standard conditions for the dry bulb temp.
$t_{o\,max} = 35°C$, daily range = 11.7°C
By Table 15.2, $t_o = 35°C - 0.23(11.7°C) = 32.3°C$

By Eq. 13.32
$$I_T = I_H = I_{DN} \cos\theta_H + I_{dH}$$

By Eq. 13.20
$$I_{DN} = A e^{-B/\sin\beta}$$

By Table 13.3, $A = 1085\ W/m^2$, $B = 0.207$, $C = 0.136$
By Eq. 13.8, $\beta_{noon} = 90° - |\ell - d|$
By Table 13.2, $d = 20.4°$
$\beta_{noon} = 90° - |35° - 20.4°| = 75.4°$
$$I_{DN} = 1085\ \tfrac{W}{m^2}\ \exp\left[-0.207/\sin(75.4°)\right] = 876\ W/m^2$$

For horizontal surface, $\theta_H = 90° - \beta = 90° - 75.4° = 14.6°$

By Eq. 13.21
$$I_{dH} = C I_{DN} = 0.136(876\ W/m^2) = 119\ W/m^2$$

$$I_T = 876\ \tfrac{W}{m^2} \cos(14.6°) + 119\ \tfrac{W}{m^2} = 967\ \tfrac{W}{m^2}$$

For dark-colored surface, $\alpha_s/h_o = 0.052\ m^2°C/W$
For horizontal surface, $\dfrac{\varepsilon \Delta R}{h_o} = 4°C$

$$t_e = 32.3°C + 967\ \tfrac{W}{m^2} \cdot 0.052\ \tfrac{m^2 °C}{W} - 4°C$$

$$t_e = 78.6°C$$

15.2

Consider the following thermal network:

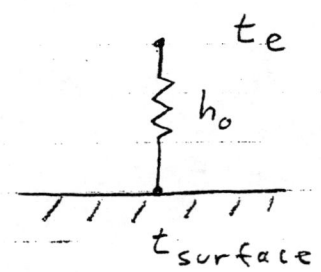

If the surface is perfectly insulating (adiabatic), there is no heat flux to or from the surface so the surface temperature must equal the sol-air temperature

$t_{surface} = t_e = 78.6\,°C$

15.3

By Eq. 15.10
$$t_e = t_o + I_T \frac{d_s}{h_o} - \frac{\varepsilon \Delta R}{h_o}$$

For a dark-colored surface, $d_s/h_o = 0.3 \, hr \, ft^2 \, °F/Btu$
For a vertical surface, $\varepsilon \Delta R / h_o = 0$
Thus $t_e = 10°F + I_T \, 0.3 \, hr \, ft^2 °F/Btu$

By Eq. 13.31
$$I_T = I_D + I_d + I_R$$

By Eq. 13.22
$$I_D = I_{DN} \cos \theta$$

By Eq. 13.20
$$I_{DN} = A \exp[-B/\sin \beta]$$

By Table 13.3, $A = 391 \, \frac{Btu}{hr \, ft^2}$, $B = 0.142$, $C = 0.057$, $d = -23.45°$

By Eq. 13.8
$$\beta_{noon} = 90° - |\ell - d| = 90° - |45° - (-23.45°)| = 21.6°$$

For south-facing surface at solar noon, $\theta = \beta_{noon} = 21.6°$

$$I_{DN} = \frac{391 \, Btu}{hr \, ft^2} \exp\left[\frac{-0.142}{\sin(21.6°)}\right] = 266 \, \frac{Btu}{hr \, ft^2}$$

$$I_D = 266 \, \frac{Btu}{hr \, ft^2} \cos(21.6°) = 247 \, \frac{Btu}{hr \, ft^2}$$

By Eq. 13.21
$$I_{dH} = C \, I_{DN} = 0.057 \left(266 \, \frac{Btu}{hr \, ft^2}\right) = 15.2 \, \frac{Btu}{hr \, ft^2}$$

$\cos \theta = \cos(21.6°) = 0.93$

By Eq. 13.24b
$$\frac{I_{dV}}{I_{dH}} = 0.55 + 0.437 \cos \theta + 0.313 \cos^2 \theta$$
$$= 0.55 + 0.437(0.93) + 0.313(0.93)^2 = 1.23$$

$$I_{dV} = 1.23 \, I_{dH} = 1.23 \times (15.2) = 18.7 \, \frac{Btu}{hr \, ft^2}$$

15.3 Cont'd.

By Eq. 13.28
$$I_R = \rho_g I_H F_{Ag}$$

$\rho_g = 0.5$

By Eq. 13.32
$$I_H = I_{DN} \cos \theta_H + I_{dH}$$

$\theta_H = 90° - \beta = 90° - 21.6° = 68.4°$

By Eq. 13.21
$$I_{dH} = C I_{DN} = 0.057 \left(266 \frac{Btu}{hr\, ft^2}\right) = 15.2 \frac{Btu}{hr\, ft^2}$$

$$I_H = 266 \frac{Btu}{hr\, ft^2} (\cos(68.4°)) + 152 \frac{Btu}{hr\, ft^2} = 250 \frac{Btu}{hr\, ft^2}$$

By Eq. 13.29
$$F_{Ag} = \frac{1 - \cos \Sigma}{2} = 0.5$$

$$I_R = 0.5 \left(250 \frac{Btu}{hr\, ft^2}\right) 0.5 = 62.5 \frac{Btu}{hr\, ft^2}$$

$$I_T = 247 + 18.7 + 62.5 = 328 \frac{Btu}{hr\, ft^2}$$

$$t_e = 10°F + 328 \frac{Btu}{hr\, ft^2} \left(\frac{0.3\, hr\, ft^2\, °F}{Btu}\right)$$

$$t_e = 108°F$$

Problem 15.4

1	2	3	4	5	6	7	8	9	10	11	12	13
Solar time	te,°F	w1θ,deg.	cos(w1θ)	(te)cos(w1θ)	sin(w1θ)	(te)sin(w1θ)	w2θ,deg	cos(w2θ)	(te)cos(w2θ)	sin(w2θ)	(te)sin(w2θ)	te calc.,°F
0	77	0	1.000	77.0	0	0	0	1.000	77.0	0.000	0.0	80.1
1	76	15	0.966	73.4	0.259	19.7	30	0.866	65.8	0.500	38.0	78.3
2	76	30	0.866	65.8	0.500	38.0	60	0.500	38.0	0.866	65.8	75.6
3	75	45	0.707	53.0	0.707	53.0	90	0.000	0.0	1.000	75.0	72.9
4	74	60	0.500	37.0	0.866	64.1	120	-0.500	-37.0	0.866	64.1	71.3
5	74	75	0.259	19.2	0.966	71.5	150	-0.866	-64.1	0.500	37.0	72.1
6	77	90	0.000	0.0	1.000	77.0	180	-1.000	-77.0	0.000	0.0	76.2
7	82	105	-0.259	-21.2	0.966	79.2	210	-0.866	-71.0	-0.500	-41.0	83.8
8	89	120	-0.500	-44.5	0.866	77.1	240	-0.500	-44.5	-0.866	-77.1	94.1
9	105	135	-0.707	-74.2	0.707	74.3	270	0.000	0.0	-1.000	-105.0	106.0
10	118	150	-0.866	-102.2	0.500	59.0	300	0.500	59.0	-0.866	-102.2	117.5
11	129	165	-0.966	-124.6	0.259	33.4	330	0.866	111.7	-0.500	-64.5	126.7
12	134	180	-1.000	-134.0	0.000	0.0	360	1.000	134.0	0.000	0.0	132.0
13	135	195	-0.966	-130.4	-0.259	-34.9	30	0.866	116.9	0.500	67.5	132.6
14	129	210	-0.866	-111.7	-0.500	-64.5	60	0.500	64.5	0.866	111.7	128.7
15	120	225	-0.707	-84.9	-0.707	-84.8	90	0.000	0.0	1.000	120.0	121.1
16	106	240	-0.500	-53.0	-0.866	-91.8	120	-0.500	-53.0	0.866	91.8	111.3
17	100	255	-0.259	-25.9	-0.966	-96.6	150	-0.866	-86.6	0.500	50.0	101.3
18	94	270	0.000	0.0	-1.000	-94.0	180	-1.000	-94.0	0.000	0.0	92.5
19	87	285	0.259	22.5	-0.966	-84.0	210	-0.866	-75.3	-0.500	-43.5	86.1
20	85	300	0.500	42.5	-0.866	-73.6	240	-0.500	-42.5	-0.866	-73.6	82.3
21	83	315	0.707	58.7	-0.707	-58.7	270	0.000	0.0	-1.000	-83.0	80.9
22	81	330	0.866	70.1	-0.500	-40.5	300	0.500	40.5	-0.866	-70.2	80.7
23	79	345	0.966	76.3	-0.259	-20.5	330	0.866	68.4	-0.500	-39.5	80.8
summations	2285			-311.1		-97.7			130.8		21.4	
te,m=	95.2		M1=	-25.93	N1=	-8.15		M2=	10.90	N2=	1.78	
te,1=	27.175 °F												
te,2=	11.0 °F												
ψ1=	197.44 deg												
ψ2=	9.28 deg												

15.5

Following the procedure used in Ex. 15.2

$h_0 = 3.0$ Btu/hr ft² °F

By Table 14.1, $h_i = 1.46$ Btu/hr ft² °F

By Table 14.4 IP, $\rho = 120$ lbm/ft³, $k = 6/12 = 0.5$ Btu/hr ft °F

$c = 0.19$ Btu/lbm °F

$\alpha = \dfrac{k}{\rho c} = \dfrac{0.5}{120 (0.19)} = 0.022$ ft²/hr

$U = \left[(1.46)^{-1} + 1/0.5 + (3.0)^{-1} \right]^{-1} = 0.33$ Btu/hr ft² °F

By Eq. 15.25

$\sigma_1 = \sqrt{\dfrac{0.2618}{2(0.022)}} = 2.44$ ft⁻¹, $\sigma_2 = \sqrt{2} \cdot 2.44 = 3.45$ ft⁻¹

$\sigma_1 L = 2.44$ ft⁻¹ (1 ft) $= 2.44$, $\sigma_2 L = 3.45$

By Eq. 15.26

$Y_1 = \left(\dfrac{3.0(1.46)}{2(2.44)^2 (0.5)^2} + 1 \right) \cos(2.44) \sinh(2.44)$
$ 1.471$

$ + \left(\dfrac{3.0(1.46)}{2(2.44)^2 (0.5)^2} - 1 \right) \sin(2.44) \cosh(2.44)$

$ + \dfrac{(3.0 + 1.46)}{2.44 (0.5)} \cos(2.44) \cosh(2.44) = -25.12$

$Z_1 = \left(\dfrac{3.0(1.46)}{2(2.44)^2 (0.5)^2} + 1 \right) \sin(2.44) \cosh(2.44)$

$ - \left(\dfrac{3.0(1.46)}{2(2.44)^2 (0.5)^2} - 1 \right) \cos(2.44) \sinh(2.44)$

$ + \dfrac{(3.0 + 1.46)}{2.44 (0.5)} \sin(2.44) \sinh(2.44) = 29.58$

15.5 cont'd

$$Y_2 = \left(\frac{3.0(1.46)}{2(3.45)^2(0.5)^2} + 1\right)\cos(3.45)\sinh(3.45)$$

$$+ \left(\frac{3.0(1.46)}{2(3.45)^2(0.5)^2} - 1\right)\sin(3.45)\cosh(3.45)$$

$$+ \frac{(3.0+1.46)}{3.45(0.5)}\cos(3.45)\cosh(3.45) = -63.60$$

$$Z_2 = \left(\frac{3.0(1.46)}{2(3.45)^2(0.5)^2} + 1\right)\sin(3.45)\cosh(3.45)$$

$$- \left(\frac{3.0(1.46)}{2(3.45)^2(0.5)^2} - 1\right)\cos(3.45)\sinh(3.45)$$

$$+ \frac{(3.0+1.46)}{3.45(0.5)}\sin(3.45)\sinh(3.45) = -24.61$$

$$\Phi_1 = \tan^{-1}\left(\frac{29.58}{-25.12}\right) = -49.66 \text{ deg}$$

$$\Phi_2 = \tan^{-1}\left(\frac{-24.61}{-63.60}\right) = 21.25 \text{ deg}$$

$$\lambda_1 = \frac{3.0(1.46)(Btu/hr\,ft^2\,°F)^2}{\frac{0.33\,Btu}{hr\,ft\,°F}(2.44)\,ft^{-1}\frac{0.5\,Btu}{hr\,ft\,°F}\sqrt{(-25.12)^2+(29.58)^2}} = 0.280$$

$$\lambda_2 = \frac{3.0(1.46)}{0.33(3.45)(0.5)\sqrt{(-63.60)^2+(-24.61)^2}} = 0.113$$

15.5 Cont'd

a) 10:00 a.m. solar time

From Problem 15.4, $t_{e,1} = 27.175°F$, $t_{e,2} = 11.0°F$
$t_{e,m} = 95.2°F$, $\psi_1 = 197.44°$, $\psi_2 = 9.28°$

By Eq. 15.30

$$q_i = \frac{0.33 \text{ Btu}}{\text{hr ft}^2 °F} \left\{ \left[95.2 + \overset{7.63}{0.280(27.175)} \cos\left(150° - 197.44° + \overset{2.22}{49.66°}\right) \right. \right.$$

$$\left. \left. + 0.113(11.0) \cos\left(\overset{269.47}{300° - 9.28° - 21.25°}\right) \right] - 75 \right\} °F$$

$$-0.611$$

$$q_i = 9.17 \frac{\text{Btu}}{\text{hr ft}^2}$$

b) 8:00 p.m. solar time

$$q_i = 0.33 \left\{ \left[95.2 + \overset{4.590}{0.280(27.175)} \cos(300 - 197.44 - 49.66) \right. \right.$$

$$\left. \left. + 0.113(11.0) \cos(240 - 9.28 - 21.25) \right] - 75 \right\}$$

$$-1.082$$

$$q_i = 7.82 \frac{\text{Btu}}{\text{hr ft}^2}$$

15.6

By Eq. 15.30, at 3:00 pm, $\omega_1 \theta = 225°$, $\omega_2 \theta = 90°$

a) From Table 15.6 for winter, roof #5, assume $t_i = 70°F$

$U = 0.155$ Btu/hr ft² °F

$\lambda_1 = 0.518$, $\Phi_1 = 70.8°$, $\lambda_2 = 0.286$, $\Phi_2 = 97.3°$

From Table 15.5, average Jan. day

$t_{e,m} = 7.9°F$, $t_{e,1} = 14.4°F$, $\psi_1 = 202.2°$,

$t_{e,2} = 8.5°F$, $\psi_2 = 11.9°$

$$q_i = 0.155\left\{\left[7.9 + 0.518(14.4)\cos(225° - 202.2° - 70.8°)\right.\right.$$
$$\left.\left. + 0.286(8.5)\cos(90° - 11.9° - 97.3°)\right] - 70\right\}$$

$q_i = -8.50$ Btu/hr ft²

b) April (use summer month values for April, July, Oct.)

From Table 15.6 for summer, roof #5, assume $t_i = 75°F$

$U = 0.146$ Btu/hr ft² °F

$\lambda_1 = 0.405$, $\Phi_1 = 78.4°$, $\lambda_2 = 0.213$, $\Phi_2 = 102.2°$

From Table 15.5, average April day

$t_{e,m} = 66.0°F$, $t_{e,1} = 36.2°F$, $\psi_1 = 196.0°$

$t_{e,2} = 11.9°F$, $\psi_2 = 0°$

$$q_i = 0.146\left\{\left[66 + 0.405(36.2)\cos(225° - 196° - 78.4°)\right.\right.$$
$$\left.\left. + 0.213(11.9)\cos(90° - 0° - 102.2°)\right] - 75\right\}$$

$q_i = 0.44$ Btu/hr ft²

15.6 cont'd.

c) July
From Table 15.5
$t_{e,m} = 95.7°F$, $t_{e,1} = 39.2°F$, $\psi_1 = 193.0°$
$t_{e,2} = 9.3°F$, $\psi_2 = -1.8°$

$$g_i = 0.146\{[95.7 + 0.405(39.2)\cos(225° - 193° - 78.4°)$$
$$+ 0.213(9.3)\cos(90° + 1.8° - 102.2°)] - 75\}$$

$g_i = 4.91$ Btu/hr ft^2

d) October
From Table 15.5
$t_{e,m} = 63.8°F$, $t_{e,1} = 27.0°F$, $\psi_1 = 199°$
$t_{e,2} = 12.0°F$, $\psi_2 = 10.2°$

$$g_i = 0.146\{[63.8 + 0.405(27.0)\cos(225° - 199° - 78.4°)$$
$$+ 0.213(12.0)\cos(90° - 10.2° - 102.2°)] - 75\}$$

$g_i = -0.32$ Btu/hr ft^2

15.7

By Eq. 15.30
From Table 15.6, summer, roof #2
$U = 0.125$ Btu/hr ft²°F
$\lambda_1 = 0.241$, $\Phi_1 = 88.5°$, $\lambda_2 = 0.121$, $\Phi_2 = 108°$

From Table 15.5
$t_{e,m} = 105.5°F$, $t_{e,1} = 40.3°F$, $\psi_1 = 191°$
$t_{e,2} = 9.2°F$, $\psi_2 = -1.8°$

$$q_i = 0.125 \{ [105.5 + 0.241(40.3)\cos(\omega_1\theta - 191° - 88.5°)$$
$$+ 0.121(9.2)\cos(\omega_2\theta + 1.8° - 108°)] - 75 \}$$

$$q_i = 0.125 \{ 30.5 + 9.71 \cos(\omega_1\theta - 279.5°)$$
$$+ 1.11 \cos(\omega_2\theta - 106.2°) \}$$

see spreadsheet for hourly results

From Table 15.6, summer, roof #3
$U = 0.125$ Btu/hr ft²°F
$\lambda_1 = 0.519$, $\Phi_1 = 72.2°$, $\lambda_2 = 0.286$, $\Phi_2 = 100.2°$

$$q_i = 0.125 \{ [105.5 + 0.519(40.3)\cos(\omega_1\theta - 191° - 72.2°)$$
$$+ 0.286(9.2)\cos(\omega_2\theta + 1.8° - 100.2°)] - 75 \}$$

$$= 0.125 \{ 30.5 + 20.92 \cos(\omega_1\theta - 263.2°)$$
$$+ 2.63 \cos(\omega_2\theta - 98.4°) \}$$

see spreadsheet for hourly results

Roof #2 peaks at 5.05 Btu/hr ft², about one hour after the peak for Roof #3 at 6.60 Btu/hr ft²

Problem 15.7

solar time	ω1θ	ω2θ	qi, roof#2	qi, roof#3
0	0	0	3.97	3.45
1	15	30	3.73	2.96
2	30	60	3.48	2.50
3	45	90	3.24	2.08
4	60	120	3.01	1.71
5	75	150	2.81	1.43
6	90	180	2.65	1.26
7	105	210	2.57	1.26
8	120	240	2.58	1.46
9	135	270	2.69	1.87
10	150	300	2.91	2.48
11	165	330	3.21	3.24
12	180	360	3.57	4.07
13	195	30	3.96	4.90
14	210	60	4.33	5.64
15	225	90	4.65	6.19
16	240	120	4.88	6.52
17	255	150	5.02	6.60
18	270	180	5.05	6.46
19	285	210	4.99	6.12
20	300	240	4.85	5.65
21	315	270	4.67	5.10
22	330	300	4.45	4.54
23	345	330	4.22	3.98

15.8

By Eqns. 15.32 and 15.33

$$q_i = U\left[(t_{e,ave} - t_i) + \lambda_m (t_{e,\delta} - t_{e,ave})\right]$$

From Table 15.8, wall #4
$U = 1.9235$ W/m²°C, $\delta = 7.3$ hrs, $\lambda_m = 0.33$
round off δ to 7 hours

Sol-air temperatures taken from west-facing light-colored wall in Table 15.3

$$q_i = 1.9235\left[6.7 + 0.33(t_{e,\delta} - 32.7)\right]$$

θ, hrs	$t_{e,\theta}$	q_i (W/m²)
1	24.4	21.0
2	24.4	11.5
3	23.8	10.8
4	23.3	10.1
5	23.3	9.4
6	23.8	8.7
7	25.5	8.0
8	27.2	7.6
9	29.4	7.6
10	31.6	7.2
11	33.8	6.9
12	36.1	6.9
13	43.3	7.2
14	49.4	8.3
15	53.8	9.4
16	55.0	10.8
17	52.7	12.2
18	45.5	13.6
19	30.5	15.0
20	29.4	19.6
21	28.3	23.5
22	27.2	26.3
23	26.1	27.614
24	25.0	25.6

Problem 15.9

PROBLEM 15.9 solution 4/98

HEAT GAIN TRANSFER FUNCTION CALCULATIONS

b-VALUES	d-VALUES	i-value	sum of cn	room temp
0.00009	1	0	0.086751	26
0.01125	-1.5166	1		
0.04635	0.64261	2		
0.02654	-0.08382	3		
0.00249	0.00289	4		
0	0	5		
0	0	6		

HEAT GAIN

	SOL-AIR TEMP	qe-theta		qe-theta		qe-theta		qe-theta		qe-theta
Time	te	DAY 1	Time	DAY 2		DAY 3		DAY 4		DAY 5
19	30.5	0		17.022		19.896		20.157		20.181
20	29.4	0		19.657		22.258		22.494		22.516
21	28.3	0		20.810		23.164		23.378		23.397
22	27.2	0		20.668		22.797		22.991		23.009
23	26.1	0		19.791		21.718		21.893		21.909
24	25.0	0		18.546		20.290		20.449		20.463
1	24.4	0.030	25	17.112	49	18.690	73	18.834	97	18.847
2	24.4	-0.014	26	15.573	50	17.001	74	17.131	98	17.143
3	23.8	-0.160	27	14.000	51	15.292	75	15.410	99	15.421
4	23.3	-0.376	28	12.461	52	13.630	76	13.737	100	13.746
5	23.3	-0.649	29	10.977	53	12.036	77	12.132	101	12.141
6	23.8	-0.974	30	9.551	54	10.508	78	10.595	102	10.603
7	25.5	-1.320	31	8.206	55	9.073	79	9.152	103	9.159
8	27.2	-1.616	32	7.006	56	7.790	80	7.861	104	7.868
9	29.4	-1.757	33	6.045	57	6.755	81	6.819	105	6.825
10	31.6	-1.660	34	5.401	58	6.043	82	6.102	106	6.107
11	33.8	-1.268	35	5.121	59	5.702	83	5.755	107	5.760
12	36.1	-0.559	36	5.223	60	5.749	84	5.797	108	5.801
13	43.3	0.466	37	5.699	61	6.175	85	6.218	109	6.222
14	49.4	1.850	38	6.585	62	7.016	86	7.055	110	7.058
15	53.8	3.817	39	8.102	63	8.492	87	8.527	111	8.530
16	55.0	6.524	40	10.402	64	10.755	88	10.787	112	10.790
17	52.7	9.876	41	13.385	65	13.704	89	13.733	113	13.736
18	45.5	13.541	42	16.717	66	17.006	90	17.032	114	17.035
19	30.5	17.022	43	19.896	67	20.157	91	20.181	115	20.183
20	29.4	19.657	44	22.258	68	22.494	92	22.516	116	22.518
21	28.3	20.810	45	23.164	69	23.378	93	23.397	117	23.399
22	27.2	20.668	46	22.797	70	22.991	94	23.009	118	23.010
23	26.1	19.791	47	21.718	71	21.893	95	21.909	119	21.911
24	25.0	18.546	48	20.290	72	20.449	96	20.463	120	20.465

15.10

Procedure follows Ex. 15.6

- t_e
- 1 — A0   outside air film
- 2 — A1   25 mm stucco
- 3
- 4

     C10   200 mm concrete

- 13
- 14 — B1   20.5 mm air space
- 15
- 16 — E1   20 mm gypsum
-      E0   inside air film
- t_i

choose 20.5 mm air space so $15 \times 17.7 \text{ mm}$ = wall thickness. Total of 18 temperature nodes, 16 in the wall plus t_i and t_e.

$R_o = 0.059 \text{ m}^2{}^\circ\text{C/W}$

$R_{12} = 0.0177 \text{ m}/0.692 \text{ W/m}^\circ\text{C} = 0.0256 \text{ m}^2{}^\circ\text{C/W}$

$R_{23} = 0.0073 \text{ m}/0.692 + 0.0104 \text{ m}/1.731 = 0.0166$

$R_{3-4} - R_{12-13} = 0.0177/1.731 = 0.0102$

$R_{13-14} = 0.0126/1.731 + (0.0051/0.0205)0.16 = 0.0471$

$R_{14-15} = (0.0154/0.0205)0.16 + 0.0023/0.727 = 0.1234$

$R_{15-16} = 0.0177/0.727 = 0.0243$

$R_i = 0.121$

15.10 Cont'd.

$C_1 = 0.00885(1858)0.84 = 13.8$
$C_2 = 13.8 + 0.0073(1858)0.84 + 0.00155(2243)0.84 = 28.1$
$C_3 - C_{13} = 0.0177(2243)0.84 = 33.35$

$C_{14} = 0.00375(2243)0.84 = 7.07$
$C_{15} = (0.0023 + 0.00885)(1602)0.84 = 15.0$
$C_{16} = 0.00885(1602)0.84 = 11.91$

Set all wall temperatures initially to the mean of the daily average sol-air temperature and the indoor temperature.
By Table 15.3
$T_{wall} = (32.7 + 26)/2 = 29.4 °C$

See spreadsheet solution for day #3 and corresponding plots

FINITE DIFFERENCE CALCULATION PROBLEM 15.10 — Prob. 15.10

NODE #	RESISTANCE	HEAT CAPACITY	SOLAR TIME	SOL-AIR TEMP	INDOOR TEMP	SOLAR TIME	SOL-AIR TEMP	INDOOR TEMP
e								
1	0.0690		100	24.4	26	1300	43.3	26
2	0.0256	13.8	200	24.4	26	1400	40.4	26
3	0.0166	28.1	300	23.8	26	1500	53.8	26
4	0.0102	33.35	400	23.3	26	1600	55.0	26
5	0.0102	33.35	500	23.3	26	1700	52.7	26
6	0.0102	33.35	600	23.8	26	1800	45.5	26
7	0.0102	33.35	700	25.5	26	1900	30.5	26
8	0.0102	33.35	800	27.2	26	2000	29.4	26
9	0.0102	33.35	900	29.4	26	2100	28.3	26
10	0.0102	33.35	1000	31.6	26	2200	27.2	26
11	0.0102	33.35	1100	33.8	26	2300	26.1	26
12	0.0102	33.35	1200	36.1	26	2400	25.0	26
13	0.0471	33.35						
14	0.1234	7.07						
15	0.0243	15.0						
16	0.1210	11.91						

dt = 900000

RESULTS FROM DAY 1

NODE #	INITIAL TEMP	TEMP HR 0.25	TEMP HR 0.5	TEMP HR 0.75	TEMP HR 1	TEMP HR 1.25	TEMP HR 1.5	TEMP HR 1.75	TEMP HR 2	TEMP HR 2.25	TEMP HR 2.5	TEMP HR 2.75	TEMP HR 3
1	29.4	28.1	27.7	27.5	27.4	27.1	27.0	26.9	26.9	26.8	26.8	26.7	26.7
2	29.4	28.9	28.6	28.4	28.3	28.1	28.0	27.9	27.8	27.8	27.7	27.7	27.6
3	29.4	29.2	29.0	28.9	28.8	28.7	28.6	28.5	28.4	28.3	28.3	28.2	28.1
4	29.4	29.3	29.2	29.1	29.0	28.9	28.8	28.7	28.7	28.6	28.5	28.5	28.4
5	29.4	29.4	29.3	29.2	29.1	29.1	29.0	28.9	28.9	28.8	28.8	28.7	28.7
6	29.4	29.4	29.3	29.3	29.2	29.2	29.1	29.1	29.0	29.0	28.9	28.9	28.8
7	29.4	29.4	29.4	29.3	29.3	29.3	29.2	29.2	29.1	29.1	29.1	29.0	29.0
8	29.4	29.4	29.4	29.4	29.4	29.3	29.3	29.3	29.2	29.2	29.2	29.1	29.1
9	29.4	29.4	29.4	29.4	29.4	29.4	29.3	29.3	29.3	29.3	29.3	29.2	29.2
10	29.4	29.4	29.4	29.4	29.4	29.4	29.4	29.3	29.3	29.3	29.3	29.2	29.2
11	29.4	29.4	29.4	29.4	29.4	29.4	29.4	29.3	29.3	29.3	29.3	29.2	29.2
12	29.4	29.4	29.4	29.4	29.4	29.3	29.3	29.3	29.2	29.2	29.1	29.1	29.1
13	29.4	29.4	29.4	29.3	29.3	29.1	29.0	28.9	28.8	28.8	28.7	28.7	28.6
14	29.4	28.9	28.9	28.6	28.3	28.1	28.0	27.8	27.7	27.7	27.7	27.6	27.5
15	29.4	28.4	28.4	28.1	27.9	27.8	27.6	27.5	27.5	27.4	27.3	27.3	27.3
16													

NODE #	FROM HR 12	TEMP HR 13										TEMP HR 14										TEMP HR 15
1	30.9	31.7	31.9	32.1	32.2	34.5	35.1	35.5	35.8	37.8	38.5	38.9	39.2
2	29.7	30.1	30.3	30.5	30.6	31.6	32.1	32.5	32.8	33.8	34.3	34.8	35.1
3	29.1	29.3	29.5	29.6	29.7	30.2	30.6	30.9	31.1	31.6	32.1	32.5	32.8
4	28.8	28.9	29.1	29.2	29.3	29.6	29.9	30.1	30.3	30.7	31.0	31.4	31.6
5	28.6	28.7	28.8	28.9	29.0	29.2	29.4	29.6	29.7	30.0	30.3	30.5	30.8
6	28.4	28.5	28.6	28.7	28.7	28.9	29.0	29.1	29.3	29.5	29.7	29.9	30.1
7	28.3	28.4	28.4	28.5	28.5	28.6	28.7	28.8	28.9	29.1	29.2	29.4	29.5
8	28.2	28.3	28.3	28.4	28.4	28.5	28.5	28.6	28.7	28.8	28.9	29.0	29.1
9	28.2	28.2	28.2	28.3	28.3	28.3	28.4	28.4	28.5	28.6	28.6	28.7	28.8
10	28.1	28.1	28.1	28.2	28.2	28.2	28.3	28.3	28.3	28.4	28.4	28.5	28.6
11	28.1	28.1	28.1	28.1	28.1	28.1	28.1	28.1	28.2	28.2	28.3	28.3	28.4
12	28.0	28.0	28.0	28.0	28.0	28.0	28.0	28.1	28.1	28.1	28.2	28.2	28.3
13	28.0	27.9	28.0	28.0	28.0	28.0	28.0	28.0	28.0	28.1	28.1	28.1	28.2
14	27.7	27.7	27.7	27.7	27.7	27.7	27.7	27.7	27.7	27.7	27.8	27.8	27.8
15	26.9	26.9	26.9	26.9	26.9	26.9	26.9	26.9	26.9	26.9	26.9	27.0	27.0
16	26.8	26.8	26.8	26.8	26.8	26.8	26.8	26.8	26.8	26.8	26.8	26.8	26.8

Prob. 15.10

TEMP HR 3.25	TEMP HR 3.5	TEMP HR 3.75	TEMP HR 4	TEMP HR 4.25	TEMP HR 4.5	TEMP HR 4.75	TEMP HR 5	TEMP HR 5.25	TEMP HR 5.5	TEMP HR 5.75	TEMP HR 6	TEMP HR 6.25	TEMP HR 6.5
26.4	26.4	26.3	26.3	26.1	26.0	25.9	25.9	25.9	25.8	25.8	25.8	25.9	25.9
27.5	27.4	27.3	27.3	27.2	27.1	27.0	27.0	26.9	26.9	26.8	26.8	26.8	26.8
28.1	28.0	27.9	27.9	27.8	27.7	27.7	27.6	27.5	27.5	27.4	27.4	27.4	27.3
28.4	28.3	28.2	28.2	28.1	28.0	28.0	27.9	27.9	27.8	27.8	27.7	27.7	27.6
28.6	28.5	28.5	28.4	28.4	28.3	28.3	28.2	28.2	28.1	28.0	28.0	28.0	27.9
28.8	28.7	28.7	28.6	28.6	28.5	28.5	28.4	28.4	28.3	28.3	28.2	28.2	28.1
28.9	28.9	28.8	28.8	28.8	28.7	28.7	28.6	28.6	28.5	28.5	28.4	28.4	28.3
29.0	29.0	29.0	28.9	28.9	28.8	28.8	28.7	28.7	28.6	28.6	28.5	28.5	28.5
29.1	29.1	29.0	29.0	29.0	28.9	28.8	28.8	28.8	28.6	28.7	28.6	28.6	28.6
29.1	29.1	29.1	29.0	29.0	28.9	28.9	28.9	28.8	28.8	28.7	28.7	28.7	28.6
29.1	29.1	29.1	29.0	29.0	28.9	28.9	28.8	28.8	28.8	28.7	28.7	28.6	28.6
29.1	29.1	29.0	29.0	28.9	28.9	28.9	28.8	28.8	28.7	28.7	28.7	28.6	28.6
29.0	29.0	28.9	28.9	28.9	28.8	28.8	28.7	28.7	28.7	28.6	28.6	28.6	28.5
28.6	28.6	28.5	28.5	28.5	28.4	28.4	28.4	28.3	28.3	28.3	28.2	28.2	28.2
27.5	27.5	27.4	27.4	27.4	27.4	27.3	27.3	27.3	27.3	27.3	27.2	27.2	27.2
27.2	27.2	27.2	27.2	27.2	27.1	27.1	27.1	27.1	27.1	27.1	27.0	27.0	27.0

TEMP HR 16							TEMP HR 17				TEMP HR 18		
40.8	41.3	41.7	42.0	42.6	42.9	43.1	43.3	42.8	42.8	42.8	42.9	40.8	40.3
35.9	36.4	36.8	37.2	37.6	38.0	38.2	38.5	38.5	38.5	38.6	38.8	38.0	37.7
33.3	33.7	34.1	34.4	34.8	35.1	35.4	35.6	35.8	35.9	36.1	36.2	36.1	36.0
32.0	32.4	32.7	33.0	33.3	33.6	33.9	34.2	34.4	34.5	34.7	34.9	34.9	34.8
31.0	31.3	31.6	31.9	32.1	32.4	32.7	32.9	33.1	33.3	33.5	33.6	33.7	33.8
30.3	30.5	30.7	31.0	31.2	31.4	31.7	31.9	32.1	32.3	32.4	32.6	32.7	32.8
29.7	29.9	30.1	30.3	30.5	30.7	30.8	31.0	31.2	31.4	31.6	31.7	31.9	32.0
29.3	29.4	29.6	29.7	29.9	30.0	30.2	30.4	30.5	30.7	30.8	31.0	31.1	31.2
28.9	29.0	29.2	29.3	29.4	29.5	29.7	29.8	30.0	30.1	30.2	30.4	30.5	30.6
28.7	28.8	28.8	28.9	29.1	29.2	29.3	29.4	29.5	29.7	29.8	29.9	30.0	30.2
28.5	28.5	28.6	28.7	28.8	28.9	29.0	29.1	29.2	29.3	29.4	29.6	29.7	29.8
28.3	28.4	28.5	28.5	28.6	28.7	28.8	28.9	29.0	29.1	29.2	29.3	29.4	29.5
28.2	28.3	28.4	28.4	28.5	28.6	28.7	28.8	28.9	29.0	29.1	29.2	29.3	29.4
27.9	27.9	28.0	28.0	28.1	28.2	28.3	28.3	28.4	28.5	28.6	28.7	28.8	28.9
27.0	27.0	27.0	27.0	27.1	27.1	27.1	27.2	27.2	27.3	27.3	27.3	27.4	27.4
26.8	26.8	26.9	26.9	26.9	26.9	27.0	27.0	27.1	27.0	27.1	27.1	27.2	27.2

Prob. 15.10

TEMP HR 6.75	TEMP HR 7	TEMP HR 7.25	TEMP HR 7.5	TEMP HR 7.75	TEMP HR 8	TEMP HR 8.25	TEMP HR 8.5	TEMP HR 8.75	TEMP HR 9	TEMP HR 9.25	TEMP HR 9.5	TEMP HR 9.75	TEMP HR 10
25.9	25.9	26.4	26.5	26.5	26.6	27.1	27.3	27.3	27.4	28.1	28.3	28.4	28.5
26.8	26.7	26.9	27.0	27.0	27.1	27.3	27.4	27.5	27.5	27.8	28.0	28.1	28.2
27.3	27.3	27.3	27.4	27.4	27.4	27.5	27.5	27.6	27.6	27.8	27.9	28.0	28.0
27.6	27.6	27.6	27.6	27.6	27.6	27.6	27.7	27.7	27.7	27.8	27.9	28.0	28.0
27.9	27.8	27.8	27.8	27.8	27.8	27.8	27.8	27.8	27.8	27.9	27.9	27.9	28.0
28.1	28.1	28.0	28.0	28.0	28.0	28.0	28.0	28.0	28.0	28.0	28.0	28.0	28.0
28.3	28.2	28.2	28.2	28.1	28.1	28.1	28.1	28.1	28.1	28.1	28.1	28.1	28.1
28.4	28.4	28.3	28.3	28.3	28.2	28.2	28.2	28.2	28.1	28.1	28.1	28.1	28.1
28.5	28.5	28.4	28.4	28.4	28.3	28.3	28.3	28.2	28.2	28.2	28.2	28.2	28.2
28.6	28.5	28.5	28.4	28.4	28.4	28.3	28.3	28.3	28.2	28.2	28.2	28.2	28.1
28.5	28.5	28.5	28.5	28.4	28.4	28.3	28.3	28.3	28.2	28.2	28.2	28.1	28.1
28.5	28.4	28.4	28.4	28.4	28.3	28.3	28.2	28.2	28.2	28.1	28.1	28.1	28.0
28.1	28.1	28.1	28.0	28.0	28.0	27.9	27.9	27.9	27.9	27.8	27.8	27.8	27.8
27.2	27.2	27.2	27.1	27.1	27.1	27.1	27.1	27.1	27.0	27.0	27.0	27.0	27.0
27.0	27.0	27.0	26.9	26.9	26.9	26.9	26.9	26.9	26.9	26.9	26.8	26.8	26.8

TEMP HR 19	TEMP HR 20	TEMP HR 21	TEMP HR 22
40.1	40.0	33.5	33.0	32.4	32.1	31.9	31.7	31.2	31.0	30.8	30.7
37.6	37.5	34.1	33.7	33.3	32.9	32.7	32.4	31.9	31.9	31.7	31.5
35.9	35.8	34.2	33.8	33.5	33.2	32.9	32.7	32.5	32.3	32.1	31.9
34.8	34.8	33.9	33.8	33.3	33.1	32.9	32.7	32.5	32.3	32.2	32.0
33.8	33.9	33.4	33.2	33.0	32.8	32.7	32.5	32.4	32.2	32.1	32.0
32.9	33.0	32.8	32.7	32.6	32.5	32.4	32.3	32.2	32.0	31.9	31.9
32.1	32.2	32.2	32.2	32.1	32.0	32.0	31.9	31.9	31.8	31.7	31.7
31.4	31.5	31.6	31.6	31.6	31.6	31.6	31.6	31.5	31.5	31.5	31.4
30.8	30.9	31.0	31.1	31.2	31.2	31.2	31.2	31.2	31.2	31.2	31.2
30.3	30.4	30.6	30.7	30.8	30.8	30.9	30.9	30.9	31.0	31.0	31.0
29.9	30.0	30.3	30.4	30.5	30.5	30.6	30.6	30.7	30.7	30.7	30.8
29.7	29.8	30.1	30.2	30.2	30.3	30.4	30.4	30.5	30.5	30.6	30.6
29.5	29.6	29.9	30.0	30.1	30.2	30.2	30.3	30.3	30.4	30.4	30.4
28.9	29.0	29.3	29.4	29.4	29.5	29.6	29.6	29.6	29.7	29.7	29.7
27.5	27.5	27.7	27.7	27.8	27.8	27.9	27.9	27.9	27.9	28.0	28.0
27.2	27.3	27.4	27.4	27.5	27.5	27.6	27.6	27.6	27.6	27.6	27.6

Prob. 15.10

TEMP HR 10.2	TEMP HR 10.5	TEMP HR 10.7	TEMP HR 11	TEMP HR 11.2	TEMP HR 11.5	TEMP HR 11.7	TEMP HR 12
29.2	29.4	29.6	29.6	30.4	30.6	30.8	30.9
28.5	28.7	28.8	28.9	29.3	29.5	29.6	29.7
28.2	28.3	28.4	28.5	28.7	28.8	29.0	29.1
28.1	28.2	28.3	28.4	28.5	28.6	28.7	28.8
28.1	28.1	28.2	28.3	28.3	28.4	28.5	28.6
28.1	28.1	28.2	28.2	28.3	28.3	28.4	28.4
28.1	28.1	28.1	28.2	28.2	28.2	28.3	28.3
28.1	28.1	28.1	28.1	28.2	28.2	28.2	28.2
28.1	28.1	28.1	28.1	28.1	28.2	28.2	28.2
28.1	28.1	28.1	28.1	28.1	28.1	28.1	28.1
28.1	28.1	28.1	28.1	28.1	28.0	28.0	28.0
28.0	28.0	28.0	28.0	28.0	28.0	28.0	28.0
27.7	27.7	27.7	27.7	27.7	27.7	27.7	27.7
27.0	27.0	26.9	26.9	26.9	26.9	26.9	26.9
26.8	26.8	26.8	26.8	26.8	26.8	26.8	26.8

			TEMP HR 23				TEMP HR 24
30.2	30.0	29.9	29.8	29.3	29.2	29.0	29.0
31.2	31.1	30.9	30.8	30.5	30.3	30.2	30.1
31.7	31.5	31.4	31.3	31.1	30.9	30.8	30.7
31.8	31.7	31.6	31.4	31.3	31.2	31.0	30.9
31.9	31.7	31.6	31.5	31.4	31.3	31.2	31.1
31.8	31.5	31.5	31.4	31.3	31.3	31.2	31.1
31.6	31.5	31.3	31.3	31.2	31.2	31.1	31.1
31.4	31.4	31.3	31.1	31.1	31.1	31.0	31.0
31.2	31.2	31.1	31.1	30.9	30.9	30.9	30.9
31.0	31.0	31.0	31.0	30.8	30.8	30.8	30.8
30.8	30.8	30.8	30.6	30.6	30.6	30.6	30.6
30.4	30.5	30.5	30.5	30.5	30.5	30.5	30.5
29.8	29.8	29.8	29.8	29.8	29.8	29.8	29.8
28.0	28.0	28.0	28.0	28.0	28.0	28.1	28.1
27.7	27.7	27.7	27.7	27.7	27.7	27.7	27.7

Prob. 15.10

RESULTS FROM DAY 3

NODE #	FROM DAY 1	TEMP HR 0.25	TEMP HR 0.5	TEMP HR 0.75	TEMP HR 1	TEMP HR 1.25	TEMP HR 1.5	TEMP HR 1.75	TEMP HR 2	TEMP HR 2.25	TEMP HR 2.5	TEMP HR 2.75	TEMP HR 3
1	29.2	28.8	28.6	28.5	28.4	28.1	28.0	27.9	27.8	27.8	27.7	27.6	27.6
2	30.4	30.2	30.0	29.9	29.8	29.6	29.4	29.3	29.2	29.1	29.1	29.0	28.9
3	31.1	30.9	30.8	30.6	30.5	30.4	30.2	30.1	30.0	29.9	29.8	29.8	29.7
4	31.4	31.2	31.1	31.0	30.9	30.7	30.6	30.5	30.4	30.3	30.2	30.1	30.1
5	31.5	31.4	31.3	31.2	31.1	31.0	30.9	30.8	30.7	30.6	30.5	30.4	30.4
6	31.6	31.5	31.4	31.4	31.3	31.2	31.1	31.0	30.9	30.8	30.8	30.7	30.6
7	31.7	31.6	31.5	31.4	31.4	31.3	31.2	31.1	31.1	31.0	30.9	30.8	30.8
8	31.6	31.6	31.5	31.4	31.4	31.3	31.2	31.2	31.1	31.1	31.0	30.9	30.9
9	31.5	31.5	31.4	31.4	31.3	31.3	31.2	31.2	31.1	31.1	31.0	31.0	30.9
10	31.4	31.4	31.4	31.3	31.3	31.2	31.2	31.2	31.1	31.1	31.0	31.0	30.9
11	31.3	31.3	31.2	31.2	31.2	31.1	31.1	31.1	31.0	31.0	30.9	30.9	30.9
12	31.1	31.1	31.1	31.1	31.1	31.0	31.0	31.0	30.9	30.9	30.8	30.8	30.8
13	31.0	31.0	31.0	30.9	30.9	30.9	30.8	30.8	30.8	30.7	30.7	30.7	30.6
14	30.2	30.2	30.2	30.2	30.2	30.2	30.1	30.1	30.1	30.0	30.0	30.0	29.9
15	28.3	28.3	28.3	28.3	28.3	28.3	28.3	28.2	28.2	28.2	28.2	28.2	28.2
16	27.9	27.9	27.9	27.9	27.9	27.9	27.9	27.9	27.9	27.8	27.8	27.8	27.8

NODE #	FROM HR 12								TEMP HR 14				TEMP HR 15
					TEMP HR 13								
1	31.4	32.2	32.5	32.6	32.7	35.0	35.7	36.0	36.3	38.3	39.0	39.4	39.7
2	30.5	30.9	31.1	31.3	31.4	32.4	32.9	33.3	33.5	34.5	35.1	35.5	35.8
3	30.0	30.2	30.4	30.5	30.6	31.1	31.4	31.7	32.0	32.5	32.9	33.3	33.6
4	29.8	29.9	30.1	30.2	30.3	30.6	30.8	31.1	31.3	31.6	32.0	32.3	32.5
5	29.7	29.8	29.9	29.9	30.0	30.2	30.4	30.6	30.7	31.0	31.2	31.5	31.7
6	29.6	29.6	29.7	29.8	29.8	29.9	30.1	30.2	30.3	30.5	30.7	30.9	31.1
7	29.5	29.5	29.6	29.6	29.7	29.7	29.8	29.9	30.0	30.1	30.3	30.4	30.6
8	29.5	29.5	29.5	29.5	29.6	29.6	29.7	29.7	29.8	29.9	30.0	30.1	30.2
9	29.4	29.4	29.4	29.4	29.5	29.5	29.5	29.6	29.6	29.7	29.7	29.8	29.9
10	29.3	29.3	29.3	29.4	29.4	29.4	29.4	29.4	29.5	29.5	29.5	29.6	29.7
11	29.3	29.3	29.3	29.3	29.3	29.3	29.3	29.3	29.3	29.4	29.4	29.4	29.5
12	29.2	29.2	29.2	29.2	29.2	29.2	29.2	29.2	29.2	29.2	29.3	29.3	29.3
13	29.1	29.1	29.1	29.1	29.1	29.1	29.1	29.1	29.1	29.1	29.1	29.2	29.2
14	28.7	28.6	28.6	28.6	28.6	28.6	28.6	28.6	28.6	28.7	28.7	28.7	28.7
15	27.5	27.5	27.4	27.4	27.4	27.4	27.4	27.4	27.4	27.4	27.4	27.4	27.5
16	27.2	27.2	27.2	27.2	27.2	27.2	27.2	27.2	27.2	27.2	27.2	27.2	27.2

Prob. 15.10

TEMP HR 3.25	TEMP HR 3.5	TEMP HR 3.75	TEMP HR 4	TEMP HR 4.25	TEMP HR 4.5	TEMP HR 4.75	TEMP HR 5	TEMP HR 5.25	TEMP HR 5.5	TEMP HR 5.75	TEMP HR 6	TEMP HR 6.25	TEMP HR 6.5
27.4	27.3	27.2	27.1	26.9	26.8	26.8	26.7	26.7	26.6	26.6	26.6	26.6	26.6
28.8	28.7	28.6	28.5	28.4	28.3	28.2	28.1	28.0	28.0	27.9	27.9	27.9	27.8
29.6	29.5	29.4	29.3	29.2	29.1	29.0	28.9	28.9	28.8	28.7	28.7	28.6	28.6
30.0	29.9	29.8	29.7	29.6	29.5	29.4	29.4	29.3	29.2	29.1	29.1	29.0	29.0
30.3	30.2	30.1	30.0	30.0	29.9	29.8	29.7	29.6	29.6	29.5	29.4	29.4	29.3
30.5	30.4	30.4	30.3	30.2	30.1	30.1	30.0	29.9	29.9	29.8	29.7	29.7	29.6
30.7	30.6	30.6	30.5	30.4	30.3	30.3	30.2	30.1	30.1	30.0	29.9	29.9	29.8
30.8	30.7	30.7	30.6	30.5	30.5	30.4	30.4	30.3	30.2	30.2	30.1	30.0	30.0
30.9	30.8	30.7	30.7	30.6	30.6	30.5	30.4	30.4	30.3	30.3	30.2	30.1	30.1
30.9	30.8	30.7	30.7	30.6	30.6	30.5	30.5	30.4	30.4	30.3	30.2	30.2	30.1
30.8	30.8	30.7	30.7	30.6	30.6	30.5	30.4	30.4	30.3	30.3	30.2	30.2	30.1
30.7	30.7	30.6	30.6	30.5	30.5	30.4	30.4	30.3	30.3	30.2	30.2	30.1	30.1
30.6	30.5	30.5	30.4	30.4	30.3	30.3	30.2	30.2	30.1	30.1	30.0	30.0	29.9
29.9	29.9	29.8	29.8	29.8	29.7	29.7	29.6	29.6	29.6	29.5	29.5	29.4	29.4
28.2	28.1	28.1	28.1	28.1	28.1	28.0	28.0	28.0	28.0	28.0	27.9	27.9	27.9
27.8	27.8	27.8	27.7	27.7	27.7	27.7	27.7	27.7	27.6	27.6	27.6	27.6	27.6

			TEMP HR 16				TEMP HR 17				TEMP HR 18		
41.3	41.8	42.2	42.5	43.0	43.3	43.6	43.8	43.2	43.2	43.3	43.3	41.2	40.7
36.6	37.1	37.5	37.9	38.3	38.6	38.9	39.1	39.1	39.2	39.3	39.4	38.7	38.3
34.1	34.5	34.9	35.2	35.6	35.9	36.1	36.4	36.5	36.7	36.8	37.0	36.8	36.7
32.9	33.2	33.6	33.9	34.2	34.5	34.7	35.0	35.2	35.3	35.5	35.7	35.6	35.6
32.0	32.3	32.5	32.8	33.1	33.3	33.6	33.8	34.0	34.2	34.3	34.5	34.6	34.6
31.3	31.5	31.7	31.9	32.2	32.4	32.6	32.8	33.0	33.2	33.3	33.5	33.6	33.7
30.7	30.9	31.1	31.3	31.4	31.6	31.8	32.0	32.2	32.3	32.5	32.6	32.8	32.9
30.3	30.4	30.6	30.7	30.9	31.0	31.2	31.3	31.5	31.6	31.8	31.9	32.0	32.2
30.0	30.1	30.2	30.3	30.4	30.6	30.7	30.8	30.9	31.1	31.2	31.3	31.5	31.6
29.7	29.8	29.9	30.0	30.1	30.2	30.3	30.4	30.5	30.6	30.8	30.9	31.0	31.1
29.5	29.6	29.7	29.7	29.8	29.9	30.0	30.1	30.2	30.3	30.4	30.5	30.6	30.7
29.4	29.4	29.5	29.5	29.6	29.7	29.8	29.9	30.0	30.1	30.2	30.3	30.4	30.5
29.3	29.3	29.4	29.4	29.5	29.6	29.7	29.7	29.8	29.9	30.0	30.1	30.2	30.3
28.8	28.8	28.8	28.9	28.9	29.0	29.1	29.1	29.2	29.3	29.4	29.4	29.5	29.6
27.5	27.5	27.5	27.5	27.5	27.6	27.6	27.6	27.7	27.7	27.7	27.8	27.8	27.9
27.2	27.2	27.2	27.3	27.3	27.3	27.3	27.4	27.4	27.4	27.4	27.5	27.5	27.5

Prob. 15.10

TEMP HR 6.75	TEMP HR 7	TEMP HR 7.25	TEMP HR 7.5	TEMP HR 7.75	TEMP HR 8	TEMP HR 8.25	TEMP HR 8.5	TEMP HR 8.75	TEMP HR 9	TEMP HR 9.25	TEMP HR 9.5	TEMP HR 9.75	TEMP HR 10
26.6	26.6	27.1	27.2	27.2	27.3	27.8	27.9	28.0	28.0	28.7	28.9	29.0	29.1
27.8	27.8	27.9	28.0	28.0	28.0	28.2	28.3	28.4	28.4	28.7	28.9	29.0	29.0
28.5	28.5	28.5	28.5	28.5	28.5	28.6	28.7	28.7	28.7	28.8	28.9	29.0	29.1
28.9	28.9	28.9	28.8	28.8	28.8	28.8	28.9	28.9	28.9	29.0	29.0	29.1	29.1
29.3	29.2	29.2	29.1	29.1	29.1	29.1	29.1	29.1	29.1	29.1	29.1	29.2	29.2
29.5	29.5	29.4	29.4	29.4	29.3	29.3	29.3	29.3	29.3	29.3	29.3	29.3	29.3
29.8	29.7	29.6	29.6	29.6	29.5	29.5	29.5	29.4	29.4	29.4	29.4	29.4	29.4
29.9	29.9	29.8	29.8	29.7	29.7	29.6	29.6	29.5	29.5	29.5	29.5	29.4	29.4
30.0	30.0	29.9	29.9	29.8	29.8	29.7	29.7	29.6	29.6	29.6	29.5	29.5	29.5
30.1	30.0	30.0	29.9	29.9	29.8	29.8	29.7	29.7	29.6	29.6	29.5	29.5	29.5
30.1	30.0	30.0	29.9	29.9	29.8	29.8	29.7	29.7	29.6	29.6	29.5	29.5	29.5
30.0	30.0	29.9	29.8	29.8	29.8	29.7	29.7	29.6	29.6	29.5	29.5	29.4	29.4
29.9	29.8	29.8	29.7	29.7	29.6	29.6	29.5	29.5	29.5	29.4	29.4	29.3	29.3
29.3	29.3	29.3	29.2	29.2	29.1	29.1	29.0	29.0	29.0	28.9	28.9	28.9	28.8
27.9	27.8	27.8	27.8	27.8	27.7	27.7	27.7	27.7	27.7	27.6	27.6	27.6	27.6
27.5	27.5	27.5	27.5	27.5	27.5	27.4	27.4	27.4	27.4	27.4	27.3	27.3	27.3

TEMP HR 19	TEMP HR 20	TEMP HR 21	TEMP HR 22
40.4	33.4	32.1	31.0
38.1	34.3	33.0	32.0
36.5	34.5	33.4	32.5
35.6	34.3	33.2	32.7
34.7	34.0	33.3	32.7
33.8	33.5	33.0	32.6
33.0	33.0	32.7	32.4
32.4	32.5	32.4	32.2
31.8	32.0	32.1	32.0
31.3	31.6	31.8	31.8
30.9	31.3	31.5	31.6
30.7	31.0	31.3	31.4
30.5	30.8	31.1	31.2
29.8	30.1	30.3	30.4
27.9	28.1	28.2	28.3
27.6	27.8	27.9	27.9

Prob. 15.10

TEMP HR 10.2	TEMP HR 10.5	TEMP HR 10.7	TEMP HR 11	TEMP HR 11.2	TEMP HR 11.5	TEMP HR 11.7	TEMP HR 12
29.8	30.0	30.2	30.2	31.0	31.2	31.3	31.4
29.4	29.5	29.6	29.7	30.1	30.3	30.4	30.5
29.2	29.3	29.4	29.5	29.7	29.8	29.9	30.0
29.2	29.3	29.4	29.4	29.5	29.6	29.7	29.8
29.3	29.3	29.3	29.4	29.4	29.5	29.6	29.7
29.3	29.3	29.4	29.4	29.4	29.5	29.5	29.6
29.4	29.4	29.4	29.4	29.4	29.4	29.5	29.5
29.4	29.4	29.4	29.4	29.4	29.4	29.4	29.4
29.5	29.4	29.4	29.4	29.4	29.4	29.4	29.3
29.5	29.4	29.4	29.4	29.4	29.3	29.3	29.3
29.4	29.4	29.4	29.3	29.3	29.3	29.3	29.2
29.3	29.3	29.3	29.3	29.3	29.2	29.2	29.2
29.3	29.2	29.2	29.2	29.2	29.1	29.1	29.1
28.8	28.8	28.7	28.7	28.7	28.7	28.7	28.7
27.6	27.5	27.5	27.5	27.5	27.5	27.5	27.5
27.3	27.3	27.3	27.3	27.2	27.2	27.2	27.2

TEMP HR 23							TEMP HR 24
30.6	30.4	30.2	30.1	29.7	29.5	29.4	29.3
31.8	31.6	31.4	31.3	31.0	30.8	30.7	30.5
32.3	32.2	32.0	31.8	31.7	31.5	31.4	31.2
32.5	32.4	32.2	32.1	31.9	31.8	31.7	31.5
32.6	32.4	32.3	32.2	32.1	31.9	31.8	31.7
32.5	32.4	32.3	32.2	32.1	32.0	31.9	31.8
32.4	32.3	32.2	32.1	32.1	32.0	31.9	31.8
32.2	32.1	32.1	32.0	32.0	31.9	31.9	31.8
32.0	32.0	31.9	31.9	31.9	31.8	31.8	31.7
31.8	31.8	31.8	31.7	31.7	31.7	31.7	31.6
31.6	31.6	31.6	31.6	31.6	31.5	31.5	31.5
31.4	31.4	31.4	31.4	31.4	31.4	31.4	31.3
31.2	31.2	31.2	31.2	31.2	31.2	31.2	31.2
30.4	30.4	30.4	30.4	30.4	30.4	30.4	30.4
28.4	28.4	28.4	28.4	28.4	28.4	28.4	28.4
28.0	28.0	28.0	28.0	28.0	28.0	28.0	28.0

Problem 15.10

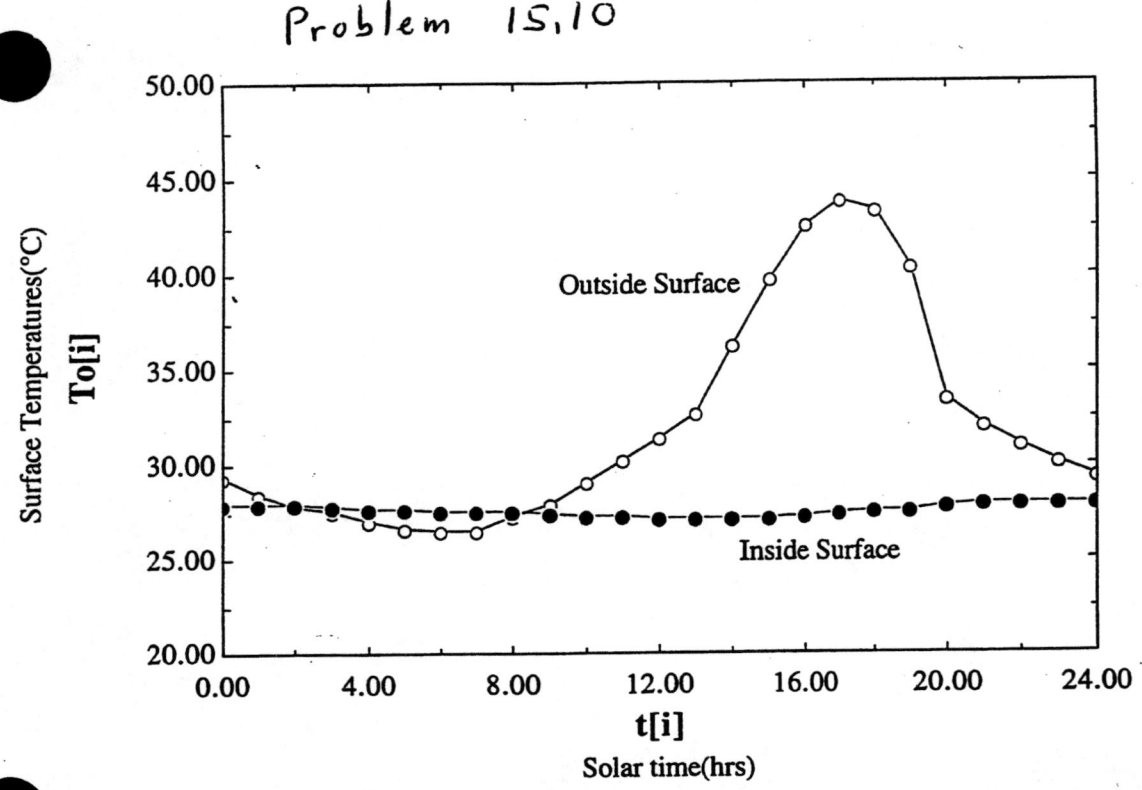

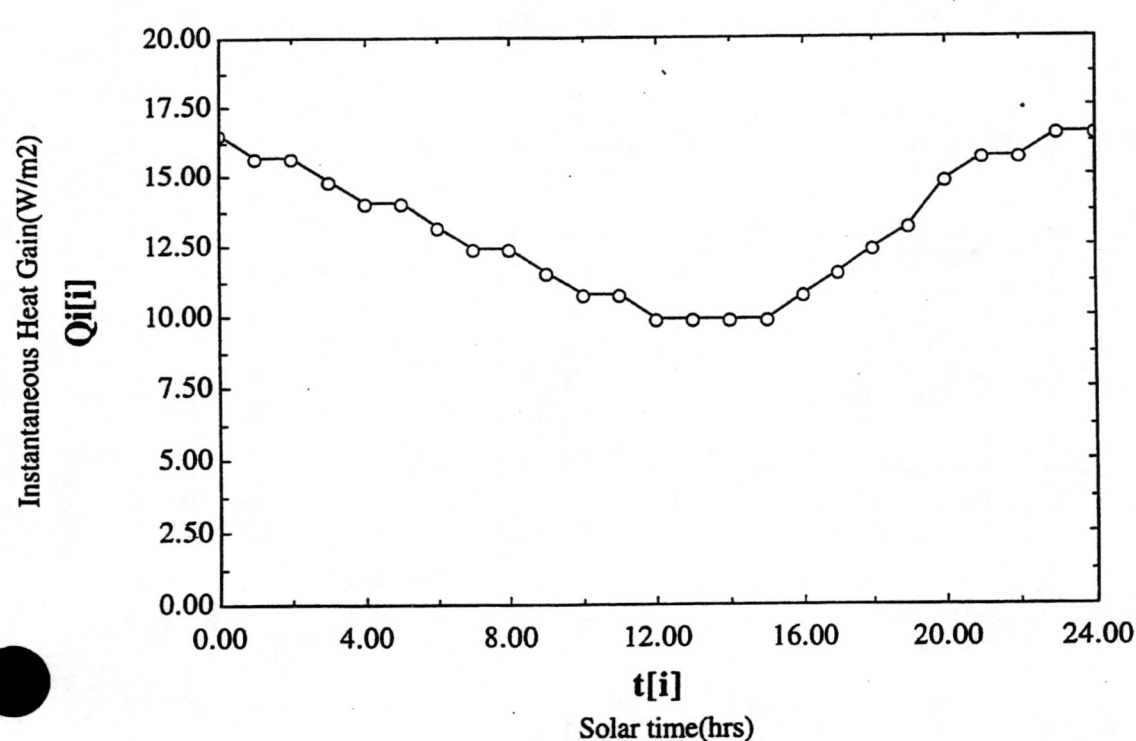

Prob. 15.11

HEAT GAIN TRANSFER FUNCTION CALCULATIONS

Prob. 15.11 soln.

b-VALUES	d-VALUES	i-value	sum of cn	room temp
0.002900	1	0	0.056674	78
0.031432	-0.97905	1		
0.021137	0.13444	2		
0.001201	-0.00272	3		
0.000002	0	4		
0	0	5		
0	0	6		

HEAT GAIN

Time	SOL-AIR TEMP t_e	qe-theta DAY 1	Time	qe-theta DAY 2		qe-theta DAY 3		qe-theta DAY 4		qe-theta DAY 5
19	80	0	0							
20	78	0	0							
21	76	0	0							
22	74	0	0							
23	72	0	0							
24	70	0	0							
1	69	-0.409	25	5.869	49	5.921	73	5.921	97	5.921
2	69	-0.886	26	4.248	50	4.291	74	4.291	98	4.291
3	68	-1.325	27	2.879	51	2.914	75	2.914	99	2.914
4	67	-1.726	28	1.716	52	1.745	76	1.745	100	1.745
5	67	-2.114	29	0.705	53	0.728	77	0.728	101	0.728
6	81	-2.423	30	-0.115	54	-0.095	78	-0.095	102	-0.095
7	102	-2.175	31	-0.284	55	-0.269	79	-0.268	103	-0.268
8	122	-0.877	32	0.671	56	0.684	80	0.684	104	0.684
9	140	1.500	33	2.768	57	2.779	81	2.779	105	2.779
10	155	4.712	34	5.750	58	5.759	82	5.759	106	5.759
11	166	8.448	35	9.298	59	9.305	83	9.305	107	9.305
12	172	12.382	36	13.078	60	13.084	84	13.084	108	13.084
13	172	16.179	37	16.750	61	16.754	85	16.754	109	16.754
14	166	19.501	38	19.968	62	19.972	86	19.972	110	19.972
15	155	22.040	39	22.423	63	22.426	87	22.426	111	22.426
16	139	23.571	40	23.884	64	23.887	88	23.887	112	23.887
17	120	23.939	41	24.196	65	24.198	89	24.198	113	24.198
18	98	23.089	42	23.299	66	23.301	90	23.301	114	23.301
19	80	21.046	43	21.218	67	21.220	91	21.220	115	21.220
20	78	18.103	44	18.243	68	18.245	92	18.245	116	18.245
21	76	15.017	45	15.132	69	15.133	93	15.133	117	15.133
22	74	12.254	46	12.348	70	12.349	94	12.349	118	12.349
23	72	9.842	47	9.919	71	9.920	95	9.920	119	9.920
24	70	7.730	48	7.794	72	7.794	96	7.794	120	7.794

Prob. 15.12

1	2	3	4	5	6
Solar time	$\omega 1\theta$	$\cos(\omega 1\theta - \Psi 1)$	$\omega 2\theta$	$\cos(\omega 2\theta - \Psi 2)$	te, calc(°F)
0	0	-0.926	0	0.979	2.9
1	15	-0.992	30	0.951	1.7
2	30	-0.991	60	0.668	-0.7
3	45	-0.922	90	0.206	-3.6
4	60	-0.790	120	-0.311	-6.1
5	75	-0.605	150	-0.744	-7.1
6	90	-0.378	180	-0.978	-5.9
7	105	-0.125	210	-0.951	-2.0
8	120	0.136	240	-0.668	4.2
9	135	0.388	270	-0.206	11.7
10	150	0.613	300	0.311	19.4
11	165	0.797	330	0.744	25.7
12	180	0.926	0	0.979	29.5
13	195	0.992	30	0.951	30.3
14	210	0.991	60	0.668	27.8
15	225	0.922	90	0.206	22.9
16	240	0.790	120	-0.311	16.6
17	255	0.605	150	-0.744	10.3
18	270	0.378	180	-0.978	5.0
19	285	0.125	210	-0.951	1.6
20	300	-0.136	240	-0.668	0.3
21	315	-0.387	270	-0.206	0.6
22	330	-0.613	300	0.311	1.7
23	345	-0.796	330	0.744	2.8
te,m	7.9				
te,1	14				
$\Psi 1$	202				
te,2	8.5				
$\Psi 2$	12				

Prob. 15.12

HEAT GAIN TRANSFER FUNCTION CALCULATIONS

b-VALUES	d-VALUES	i-value	sum of cn	room temp
0.002900	1	0	0.056674	70
0.031432	-0.97905	1		
0.021137	0.13444	2		
0.001201	-0.00272	3		
0.000002	0	4		
0	0	5		
0	0	6		

Prob. 15.12 soln.

	SOL-AIR TEMP	HEAT GAIN qe-theta		qe-theta		qe-theta		qe-theta		qe-theta
Time	te	DAY 1	Time	DAY 2		DAY 3		DAY 4		DAY 5
19	1.6	0		-19.554		-20.100		-20.104		-20.104
20	0.3	0		-20.466		-20.912		-20.916		-20.916
21	0.6	0		-21.375		-21.741		-21.744		-21.744
22	1.7	0		-22.164		-22.464		-22.466		-22.466
23	2.8	0		-22.775		-23.020		-23.022		-23.022
24	2.9	0		-23.210		-23.411		-23.412		-23.413
1	1.7	-3.810	25	-23.532	49	-23.697	73	-23.698	97	-23.698
2	-0.7	-7.581	26	-23.832	50	-23.966	74	-23.968	98	-23.968
3	-3.6	-10.870	27	-24.192	51	-24.302	75	-24.303	99	-24.303
4	-6.1	-13.745	28	-24.656	52	-24.747	76	-24.747	100	-24.747
5	-7.1	-16.272	29	-25.208	53	-25.282	77	-25.283	101	-25.283
6	-5.9	-18.454	30	-25.772	54	-25.833	78	-25.833	102	-25.833
7	-2.0	-20.233	31	-26.226	55	-26.276	79	-26.276	103	-26.276
8	4.2	-21.523	32	-26.431	56	-26.472	80	-26.472	104	-26.472
9	11.7	-22.253	33	-26.273	57	-26.306	81	-26.306	105	-26.306
10	19.4	-22.405	34	-25.697	58	-25.724	82	-25.724	106	-25.724
11	25.7	-22.033	35	-24.729	59	-24.751	83	-24.751	107	-24.751
12	29.5	-21.270	36	-23.477	60	-23.496	84	-23.496	108	-23.496
13	30.3	-20.308	37	-22.116	61	-22.131	85	-22.131	109	-22.131
14	27.8	-19.363	38	-20.844	62	-20.856	86	-20.856	110	-20.856
15	22.9	-18.636	39	-19.849	63	-19.859	87	-19.859	111	-19.859
16	16.6	-18.273	40	-19.266	64	-19.274	88	-19.274	112	-19.274
17	10.3	-18.335	41	-19.149	65	-19.155	89	-19.155	113	-19.155
18	5.0	-18.796	42	-19.462	66	-19.467	90	-19.467	114	-19.467
19	1.6	-19.554	43	-20.100	67	-20.104	91	-20.104	115	-20.104
20	0.3	-20.466	44	-20.912	68	-20.916	92	-20.916	116	-20.916
21	0.6	-21.375	45	-21.741	69	-21.744	93	-21.744	117	-21.744
22	1.7	-22.164	46	-22.464	70	-22.466	94	-22.466	118	-22.466
23	2.8	-22.775	47	-23.020	71	-23.022	95	-23.022	119	-23.022
24	2.9	-23.210	48	-23.411	72	-23.412	96	-23.413	120	-23.413

Prob 15.13

Solar time	te, North	te, East	te, South	te, West	te, Average(°F)
1	76	76	76	76	76
2	76	76	76	76	76
3	75	75	75	75	75
4	74	74	74	74	74
5	74	74	74	74	74
6	90	121	77	77	89.7
7	90	144	82	82	96.8
8	87	151	89	86	100.2
9	91	148	105	91	106.6
10	95	137	118	95	110.3
11	100	122	129	100	113.1
12	103	104	134	104	112.7
13	106	106	135	128	119.1
14	106	106	129	148	121.3
15	106	106	120	163	121.6
16	104	103	106	168	117.2
17	108	100	100	162	114.8
18	107	94	94	138	106.7
19	87	87	87	87	87
20	85	85	85	85	85
21	83	83	83	83	83
22	81	81	81	81	81
23	79	79	79	79	79
24	77	77	77	77	77

1/3

HEAT GAIN TRANSFER FUNCTION CALCULATIONS

Prob. 15.13 soln.

b-VALUES	d-VALUES	i-value	sum of cn	room temp
0.000157	1	0	0.017425	75
0.005454	-0.93389	1		
0.009608	0.27396	2		
0.002155	-0.02561	3		
0.000051	0	4		
0	0	5		
0	0	6		

HEAT GAIN

Time	SOL-AIR TEMP te	qe-theta DAY 1	Time	qe-theta DAY 2		qe-theta DAY 3		qe-theta DAY 4		qe-theta DAY 5
19	87	0								
20	85	0								
21	83	0								
22	81	0								
23	79	0								
24	77	0								
1	76	0.063	25	2.294						
2	76	0.092	26	2.005						
3	75	0.089	27	1.547						
4	74	0.071	28	1.122						
5	74	0.041	29	0.808						
6	89.7	0.008	30	0.584						
7	96.8	0.070	31	0.416	49	2.294	73	2.294	97	2.294
8	100.2	0.326	32	0.283	50	2.005	74	2.005	98	2.005
9	106.6	0.669	33	0.185	51	1.547	75	1.547	99	1.547
10	110.3	1.005	34	0.117	52	1.122	76	1.122	100	1.122
11	113.1	1.321	35	0.063	53	0.808	77	0.808	101	0.808
12	112.7	1.598	36	0.019	54	0.584	78	0.584	102	0.584
13	119.1	1.812	37	0.075	55	0.416	79	0.416	103	0.416
14	121.3	1.983	38	0.328	56	0.283	80	0.283	104	0.283
15	121.6	2.163	39	0.670	57	0.185	81	0.185	105	0.185
16	117.2	2.326	40	1.006	58	0.117	82	0.117	106	0.117
17	114.8	2.416	41	1.321	59	0.063	83	0.063	107	0.063
18	106.7	2.405	42	1.598	60	0.019	84	0.019	108	0.019
19	87	2.294	43	1.812	61	0.075	85	0.075	109	0.075
20	85	2.005	44	1.983	62	0.328	86	0.328	110	0.328
21	83	1.547	45	2.163	63	0.670	87	0.670	111	0.670
22	81	1.122	46	2.326	64	1.006	88	1.006	112	1.006
23	79	0.808	47	2.416	65	1.321	89	1.321	113	1.321
24	77	0.584	48	2.405	66	1.598	90	1.598	114	1.598
				2.294	67	1.812	91	1.812	115	1.812
				2.005	68	1.983	92	2.005	116	1.983
				1.547	69	2.163	93	1.547	117	2.163
				1.122	70	2.326	94	1.122	118	2.326
				0.808	71	2.416	95	0.808	119	2.416
				0.584	72	2.405	96	0.584	120	2.405

431

15.13 Cont'd

By Day 3 the transfer function solution is periodic to 4 significant figures

At 10:00 p.m. solar time (2200 hrs)

$$\dot{Q}_{walls} = A_{walls} \, \dot{q}_{i\,walls}$$

$$= 2(300 + 200) \text{ ft}^2 \; 1.122 \text{ Btu/hr ft}^2$$

$$\dot{Q}_{walls} = 1122 \text{ Btu/hr}$$

HEAT GAIN TRANSFER FUNCTION CALCULATIONS

Prob. 15.14 soln.

b-VALUES	d-VALUES	i-value	sum of cn	room temp
0.000157	1	0	0.017425	75
0.005454	-0.93389	1		
0.009608	0.27396	2		
0.002155	-0.02561	3		
0.000051	0	4		
0	0	5		
0	0	6		

HEAT GAIN

Time	SOL-AIR TEMP te	qe-theta DAY 1	Time	qe-theta DAY 2		qe-theta DAY 3		qe-theta DAY 4		qe-theta DAY 5
19	87		0							
20	85		0							
21	83		0							
22	81		0							
23	79		0							
24	77		0							
1	76	0.063	25	1.343		1.343		1.343		1.343
2	76	0.092	26	1.276		1.276		1.276		1.276
3	75	0.089	27	1.082		1.082		1.082		1.082
4	74	0.071	28	0.863		0.863		0.863		0.863
5	74	0.041	29	0.675		0.675		0.675		0.675
6	82	0.007	30	0.519		0.519		0.519		0.519
7	82	0.025	31	0.385	49	0.385	73	0.385	97	0.385
8	82	0.127	32	0.268	50	0.268	74	0.268	98	0.268
9	85	0.234	33	0.178	51	0.178	75	0.178	99	0.178
10	89	0.324	34	0.114	52	0.114	76	0.114	100	0.114
11	93	0.432	35	0.061	53	0.061	77	0.061	101	0.061
12	96	0.579	36	0.017	54	0.017	78	0.017	102	0.017
13	99	0.752	37	0.029	55	0.029	79	0.029	103	0.029
14	100	0.931	38	0.129	56	0.129	80	0.129	104	0.129
15	100	1.095	39	0.235	57	0.235	81	0.235	105	0.235
16	99.0	1.220	40	0.324	58	0.324	82	0.324	106	0.324
17	100	1.293	41	0.432	59	0.432	83	0.432	107	0.432
18	99	1.328	42	0.579	60	0.579	84	0.579	108	0.579
19	87	1.343	43	0.752	61	0.752	85	0.752	109	0.752
20	85	1.276	44	0.931	62	0.931	86	0.931	110	0.931
21	83	1.082	45	1.095	63	1.095	87	1.095	111	1.095
22	81	0.863	46	1.220	64	1.220	88	1.220	112	1.220
23	79	0.675	47	1.293	65	1.293	89	1.293	113	1.293
24	77	0.519	48	1.328	66	1.328	90	1.328	114	1.328
				1.343	67	1.343	91	1.343	115	1.343
				1.276	68	1.276	92	1.276	116	1.276
				1.082	69	1.082	93	1.082	117	1.082
				0.863	70	0.863	94	0.863	118	0.863
				0.675	71	0.675	95	0.675	119	0.675
				0.519	72	0.519	96	0.519	120	0.519

15.14

Treat shaded walls similar to North-facing wall so use North-facing wall sol-air temperature values from Table 15.3.

At 10:00 p.m. (2200 hrs)

$\dot{Q}_{walls} = 1000 \text{ ft}^2 (0.863) \text{ Btu/hr ft}^2$

$= 863 \text{ Btu/hr}$

15.15

By Fig. 15.9, $r = 0.043$ at $\theta = 0$.

In Eq. 15.53, the term $r^2 a^2$ should be negligible.

Thus $a = \dfrac{\tau_\lambda}{(1-r)^2} = \dfrac{0.82}{(0.957)^2} = 0.895$

By Eq. (15.56)

$e^{-0.25 K} = 0.895$, $K = 0.444 \text{ in}^{-1}$

At $\theta = 75°$

By Eq. 15.57

$L' = \dfrac{0.25}{\sqrt{1 - \left(\dfrac{\sin 75°}{1.526}\right)^2}} = 0.323 \text{ in.}$

By Eq. 15.56

$a = \exp(-(0.444)(0.323)) = 0.866$

By Fig. 15.6, $r = 0.26$ for $\theta = 75°$

By Eq. 15.53

$\tau_\lambda = \dfrac{(0.74)^2 (0.866)}{1 - [0.26(0.866)]^2} = 0.500$

15.16

By Eq. 13.11, $\cos\theta = \cos\beta \cos\gamma$

By Eq. 13.6, $\sin\beta = \cos l \cosh \cos d + \sin l \sin d$
$l = 42°$
$h = 45°$
$d = 23.0°$ (interpolation from Table 13.2)

Thus $\sin\beta = \cos 42° \cos 45° \cos 23° + \sin 42° \sin 23°$
$\beta = 48.2°$

By Eq. 13.7, $\cos\phi = \dfrac{1}{\cos\beta}\left[\cos d \sin l \cosh - \sin d \cos l\right]$

$= \dfrac{1}{\cos 48.2°}\left[\cos 23° \sin 42° \cos 45° - \sin 23° \cos 42°\right]$

$\phi = 77.4°$, $\gamma = 90° - 77.4° = 12.6°$

$\cos\theta = \cos 48.2° \cos 12.6° = 0.666$, $\theta = 48.3°$

By Eq. 15.57

$L' = \dfrac{0.25}{\sqrt{1 - \left(\dfrac{\sin 48.3°}{1.526}\right)^2}} = 0.287$ in.

By Eq. 15.56

$a = \exp\left[-3.30(0.287)\right] = 0.388$

By Fig. 15.6, $r = 0.06$

By Eq. 15.55

$a_\lambda = 0.94 - \dfrac{(0.94)^2 (0.388)}{1 - (0.06)(0.388)} = 0.589$

15.17

a) By Eq. 13.10, $\cos\theta = \cos\beta \cos\gamma \sin\Sigma + \sin\beta \cos\Sigma$

By Eq. 13.6, $\sin\beta = \cos\ell \cos h \cos d + \sin\ell \sin d$
$\ell = 42°$, $h = 30°$, $d = -22.9°$ (from Table 13.2)

$\sin\beta = \cos 42° \cos 30° \cos(-22.9°) + \sin 42° \sin(-22.9°)$
$\beta = 19.4°$

By Eq. 13.7, $\cos\phi = \dfrac{1}{\cos\beta}\left[\cos d \sin\ell \cos h - \sin d \cos\ell\right]$

$= \dfrac{1}{\cos 19.4°}\left[\cos(-22.9°)\sin 42° \cos 30° - \sin(-22.9°)\cos 42°\right]$

$\phi = 29.2°$
$\psi = 90°$

By Eq. 13.9, $\gamma = |(29.2° - 90°)| = 60.8°$
$\tan\Sigma = 6\,ft/8\,ft$, $\Sigma = 36.9°$
$\cos\theta = \cos 19.4° \cos 60.8° \sin 36.9° + \sin 19.4° \cos 36.9°$
$\theta = 57.2°$

From Fig. 15.8, $\tau_b = 0.83$ (assume $\tau_d = \tau_D$)
By Eq. 13.20, $I_{DN} = A \exp[-B/\sin\beta]$
From Table 13.3, $A = 391$ Btu/hr ft², $B = 0.142$, $C = 0.057$
$I_{DN} = 391 \exp[-0.142/\sin 19.4°] = 255$ Btu/hr ft²

By Eq. 13.22, $I_D = I_{DN} \cos\theta = 255 \cos 57.2° = 138$ Btu/hr ft²

By Eq. 13.21, $I_{dH} = C I_{DN} = 0.057(138) = 7.87$ Btu/hr ft²

By Eq. 13.25, $I_d = I_{dH}(1+\cos\Sigma)/2 = 7.87(1+\cos 36.9°)/2$
$I_d = 7.1$ Btu/hr ft²

By Table 14.1, $h_i = 1.6$ Btu/hr ft² °F (45°, heat flow up)

15.17 Cont'd

Neglecting the thermal resistance of the glass

$$U_{day} = \left[\frac{1}{h_i} + \frac{1}{h_o}\right]^{-1} = \left[\frac{1}{1.6} + \frac{1}{6.0}\right]^{-1} = 1.26 \text{ Btu/hr ft}^2 {}^\circ F$$

Simplifying Eq. 15.74

$$q_i = \tau(I_D + I_d) + U_{day}(t_{o_{day}} - t_i)$$

$$= 0.83(138 + 7.1) + 1.26(0 - 60)$$

$$= 120.4 - 75.6 = 45 \text{ Btu/hr ft}^2 \text{ (gain)}$$

b) No solar gain at night

$$q_i = U_{night}(t_{o_{night}} - t_i)$$

$$U_{night} = \left[\frac{1}{1.6} + \frac{1}{8.0}\right]^{-1} = 1.33 \text{ Btu/hr ft}^2 {}^\circ F$$

$$q_i = 1.33(-10 - 60) = -93 \text{ Btu/hr ft}^2 \text{ (loss)}$$

15.18

By Eq. 13.11, $\cos\theta = \cos\beta \cos\gamma$

By Eq. 13.6, $\sin\beta = \cos l \cos h \cos d + \sin l \sin d$
$l = 36°$, $h = 30°$, $d = 17.9$ (from Table 13.2)

$\sin\beta = \cos 36° \cos 30° \cos 17.9° + \sin 36° \sin 17.9°$
$\beta = 57.9°$

By Eq. 13.7, $\cos\phi = \dfrac{1}{\cos\beta}\left[\cos d \sin l \cos h - \sin d \cos l\right]$

$= \dfrac{1}{\cos 57.9°}\left[\cos 17.9° \sin 36° \cos 30° - \sin 17.9° \cos 36°\right]$

$\phi = 63.7°$

$\gamma = |(\phi - \psi)| = |(63.7° - \psi)|$

We may write, $dA = rH\,d\gamma$
$I_D = I_{DN} \cos\beta \cos\gamma$

$\theta_i = rHI_{DN}\cos\beta \displaystyle\int_{\psi=5°}^{\psi=85°} \tau_D \cos(63.7° - \psi)\,d\psi$

where r is the radius and H is the height of the glass wall. The glass transmissivity may be read from Fig. 15.8

The numerical solution to the integral is 0.91.

By Eq. 13.20, $I_{DN} = A\exp\left[-B/\sin\beta\right]$

From Table 13.3, $A = 341$ Btu/hr ft^2, $B = 0.206$

15.18 cont'd

$$I_{DN} = 341 \exp[-0.206/\sin 57.9°] = 267 \text{ Btu/hr ft}^2$$

$$\dot{q}_i = (20 \text{ ft})(9 \text{ ft}) \frac{267 \text{ Btu}}{\text{hr ft}^2} \cos 57.9° (0.91)$$

$$\dot{q}_i = 23{,}240 \text{ Btu/hr}$$

15.19

Compute τ_c for 2 sheets of DSA glass, $\theta = 60°$, $\rho_2 = 0.3$

By Eq. 15.67, $\tau_c = \dfrac{\tau_i \tau_o}{1 - \rho_2 \rho_3}$

$\rho_3 = \rho$ for single uncoated sheet at $\theta = 60°$.
From Fig. 15.8, $\rho_3 = 0.17$, $\tau_i = 0.8$

By Eq. 15.62, $\tau_o = \dfrac{(1-r_1)(1-r_2)a}{1 - r_1 r_2 a^2}$

From Fig. 15.6, $r_1 = 0.09$
Assume $r_2 \approx \rho_2 = 0.3$

By Eq. 15.56 and Table 15.14

$a = \exp\left[-0.5 \text{ in}^{-1} \; L'\right]$

By Eq. 15.57, $L' = \dfrac{0.125 \text{ in}}{\sqrt{1 - \dfrac{\sin^2 60°}{(1.526)^2}}} = 0.152 \text{ in}$

$a = \exp\left[-0.5 \text{ in}^{-1} (0.152 \text{ in})\right] = 0.927$

$\tau_o = \dfrac{(1 - 0.09)(1 - 0.3) \, 0.927}{1 - 0.09 \, (0.3)} = 0.607$

$\tau_c = \dfrac{0.8 \, (0.607)}{1 - 0.3 \, (0.17)} = 0.51$

15.20

indoors 6   i   5   4 m 3   2 0   1   outdoors

$\tau_c I \leftarrow$ | | $\leftarrow I_5$ | | $\leftarrow I_3$ | | $\leftarrow I$

$$I_3 = I\tau_o \left[1 + \rho_{c_{3-6}} \rho_2 + \left(\rho_{c_{3-6}} \rho_2\right)^2 + \cdots\right]$$

$$= \frac{I\tau_o}{1 - \rho_2 \rho_{c_{3-6}}}$$

$$I_5 = I_3 \tau_m \left[1 + \rho_5 \rho_{c_{1-4}} + \left(\rho_5 \rho_{c_{1-4}}\right)^2 + \cdots\right]$$

$$= \frac{I_3 \tau_m}{1 - \rho_5 \rho_{c_{1-4}}}$$

$$\tau_c I = \tau_i I_5 = \frac{\tau_i \tau_m \tau_o I}{(1 - \rho_5 \rho_{c_{1-4}})(1 - \rho_2 \rho_{c_{3-6}})}$$

or $$\tau_c = \frac{\tau_i \tau_m \tau_o}{(1 - \rho_5 \rho_{c_{1-4}})(1 - \rho_2 \rho_{c_{3-6}})}$$

where ρ_2 and ρ_5 are of the form Eq. 15.63

and $\rho_{c_{1-4}}$ & $\rho_{c_{3-6}}$ are of the form of Eq. 15.69

15.21

By Eq. 15.79, $q_i = SC(SHGF) + U(t_o - t_i)$

By Table 15.7, $SC = 0.58$ (closest value)

By Table 15.16b, $SHGF = 433 \text{ W/m}^2$

$$U \approx \left[\frac{1}{h_o} + \frac{1}{h_i}\right]^{-1}$$

By Table 14.1, $h_i = 8.29 \text{ W/m}^2\text{°C}$

$$U \cong \left[\frac{1}{8} + \frac{1}{8.29}\right]^{-1} = 4.07 \text{ W/m}^2\text{°C}$$

$$q_i = 0.58(433 \text{ W/m}^2) + \frac{4.07 \text{ W}}{\text{m}^2\text{°C}}(33-27)\text{°C}$$

$q_i = 276 \text{ W/m}^2$

$Q_i = 276 \text{ W/m}^2 (1\text{m} \times 2\text{m}) = 552 \text{ W}$

15.22

By Eq. 15.79, $q_i = SC(SHGF_{ave}) + U(t_o - t_i)$

By Eq. 15.86, $SHGF_{ave} = F_s(SHGF_D) + (SHGF_d + SHGF_R)$

Find the sunlit fraction from Section 13.6

$F_s = (a - (y - 1'))/a$
$y = b \tan \delta = b \tan \beta / \cos \gamma$
$\quad = b \tan \beta / \cos \phi$
$F_s = 1 - \dfrac{b}{a} \dfrac{\tan \beta}{\cos \phi} + \dfrac{1'}{a}$

$b = 2 \text{ ft}$
$a = 4 \text{ ft}$

By Eq. 13.6, $\sin \beta = \cos l \cos h \cos d + \sin l \sin d$
$l = 40°$, $h = 30°$, $d = 20.4°$ from Table 13.2

$\sin \beta = \cos 40° \cos 30° \cos 20.4° + \sin 40° \sin 20.4°$
$\beta = 57.8°$

By Eq. 13.7, $\cos \phi = \dfrac{1}{\cos \beta} (\cos d \sin l \cos h - \sin d \cos l)$

$\cos \phi = \dfrac{1}{\cos 57.8°} (\cos 20.4° \sin 40° \cos 30° - \sin 20.4° \cos 40°)$

$\phi = 61.4°$

$F_s = 1 - \dfrac{2' \tan 57.8°}{4' \cos 61.4°} + \dfrac{1'}{4'} = -0.41 \to 0$

($y = 6.63$ ft so entire window is shaded)

use SHGF for N-facing surface
From Table 15.16a, $SHGF_{ave} = SHGF_N = 35 \text{ Btu/hr}$

15.22 cont'd
From Table 15.17, SC = 0.88
From Table 14.5, $U = 0.49$ Btu/hr ft^2 °F

$$q_i = 0.88(35 \text{ Btu/hr ft}^2) + 0.49 \frac{\text{Btu}}{\text{hr ft}^2 \text{°F}} (90-75)\text{°F}$$

$q_i = 38.2$ Btu/hr ft^2

$Q_i = 38.2 (4 \times 3) = 458$ Btu/hr

15.23

let $g_i = 38.2 \text{ Btu/hr ft}^2$

$g_i = SC(SHGF) + U(t_o - t_i)$

Solve for SC

$$SC = \frac{g_i - U(t_o - t_i)}{SHGF}$$

$$= \frac{38.2 \text{ Btu/hr ft}^2 - 0.49 \text{ Btu/hr ft}^2 {}^\circ F (90-75)^\circ F}{81 \text{ Btu/hr ft}^2}$$

$SC = 0.38$

15.24

By Eq. 15.79, $q_i = SC(SHGF_{ave}) + U(t_o - t_i)$

By Eq. 15.86, $SHGF_{ave} = F_s \, SHGF_D + (SHGF_d + SHGF_R)$

From Section 13.6
$$y = b \tan \beta / \cos \phi$$
$$x = b \tan \phi$$

$$F_s = (a-y)(c-x)/ac$$

From Problem 15.22, $\beta = 57.8°$, $\phi = 56.0°$

$y = 0.5 \text{ft} \tan 57.8° / \cos 56° = 1.42 \text{ ft}$
$x = 0.5 \text{ft} \tan 56° = 0.74 \text{ ft}$

$F_s = (4 - 1.42)(3 - 0.74)/(4 \times 3) = 0.49$

$SHGF_{ave} = 0.49(81 - 35) + 35 = 57.5 \text{ Btu/hr ft}^2$

$q_i = 0.88(57.5) + 0.49(90 - 75) = 58.0 \text{ Btu/hr ft}^2$

$Q_i = 58(4 \times 3) = 696 \text{ Btu/hr}$

15.25

a) Assume standard summer design conditions

By Eq. 15.32, $q_i = U(TETD)$

$$\sum_{i=1}^{24} q_i = U \sum_{i=1}^{24} TETD$$

By Eq. 15.33, $TETD_\theta \approx (t_{e_{ave}} - t_i) + \lambda_m (t_{e_\delta} - t_{e_{ave}})$

or $\sum_{i=1}^{24} TETD = 24(t_{e_{ave}} - t_i)$

From Table 15.3, $t_{e_{ave}} = 41.6°C$
From p. 486, $t_i = 25.6°C$
From Table 15.9 for roof #4, $U = 0.45569 \ W/m^2°C$

$$\sum_{i=1}^{24} q_i = 0.45569 \frac{W}{m^2°C} (24 hr)(41.6 - 25.6)°C \left(\frac{3600 \ sec}{hr}\right)$$

$q_i = 630 \ kJ/m^2$

b) By Eq. 15.79, $q_i = SC(SHGF) + U(t_o - t_i)$

$$\sum_{i=1}^{24} q_i = \left[SC \left(\sum_{i=1}^{12} SHGF + \sum_{i=13}^{24} SHGF \right) + U(t_{o_{ave}} - t_i) 24 hrs \right]$$

From Table 15.18, $SC = 0.57$
From Table 15.16b, $\sum_{i=1}^{12} SHGF = \sum_{i=13}^{24} SHGF = 3395 \ W/m^2$

From Table 14.5, assume 1/8 in. acrylic material,
$U = 1.03 \times 5.68 = 5.85 \ W/m^2°C$
(corrections for horizontal orientation and reduced wind speed essentially cancel each other)

From Table 15.1, $t_{o_{ave}} = 91 - 14 = 77°F = 25°C$

15.25 Cont'd

$$\sum_{i=1}^{24} g_i = \left[0.57(3395)2 + 5.85(25-25.6)°C \cdot 24\right]\frac{3600s}{hr}$$

$$g_i = 1.36 \times 10^4 \text{ kJ/m}^2$$

c) Each m^2 of skylight is equivalent to

$$\frac{1.36 \times 10^4}{630} = 21.6 \text{ m}^2 \text{ of roof area in terms}$$

of daily heat gain. Thus for energy saving reasons (assuming the space below the roof is to be cooled) it may be beneficial to minimize the area of skylights. However the tradeoff between daylighting and artificial lighting costs should be considered also.

15.26

150 occupants

From Table 15.19, $IHG_s = 250 \frac{Btu}{hr} \times 150 = 37,500 \frac{Btu}{hr}$ se

$IHG_\ell = 200 \frac{Btu}{hr} \times 150 = 30,000 \frac{Btu}{hr}$ la

3500 W lighting

$IHG_s = 3500 W \times 1.2 \times \left(\frac{3.412\ Btu/hr}{W}\right) = 14,330 \frac{Btu}{hr}$ sen.

4 copy machines

From Table 15.21

$IHG_s = 1000 \frac{Btu}{hr} \times 4 = 4000 \frac{Btu}{hr}$ sen.

2 coffee makers (unhooded)

From Table 15.20

$IHG_s = 3750 \frac{Btu}{hr} \times 2 = 7500 \frac{Btu}{hr}$ sen.

$IHG_\ell = 1910 \frac{Btu}{hr} \times 2 = 3820 \frac{Btu}{hr}$ laT.

100 computer terminals

From Table 15.21

$IHG_s = 1000 \frac{Btu}{hr} \times 100 = 100,000 \frac{Btu}{hr}$ sen.

Totals

Sensible: $IHG_s = 163,330\ Btu/hr$

Latent: $IHG_\ell = 33,820\ Btu/hr$

15.27

a) From Table 15.20

electric: $\text{Input} = \dfrac{820\,W}{kg\ fat} \times 10\,kg\ fat \times 16\,hr \times \dfrac{\$0.08}{kw\ hr}$

$\text{Input} = \$10.50$ electric cost

gas: $\text{Input} = \dfrac{1470\,W \times 10\,kg\ fat}{kg\ fat} \times \left(\dfrac{3.412\ Btu}{hr\ W}\right) \times 16\,hr \times \dfrac{\$0.01}{1000\ Btu}$

$\text{Input} = \$8.03$

b) electric: $IHG = 14\ Btu/hr$

gas: $IHG = 160\ Btu/hr$

c) The gas fryer is less expensive to operate (fuel cost) but has a higher sensible IHG. The IHG is quite small considering other gains in a typical kitchen so this is not a serious issue. On an energy cost basis, the gas appliance is preferred. However, if all other appliances are to be electric, the first cost of running a gas line may negate the savings in energy use. The final recommendation should consider all these factors.

16.1

From Table 16.1, $w_1 = -0.92$

From Table 16.2, $v_0 = 0.681$,

$v_1 = 1 - 0.92 - 0.681 = -0.601$

Problem 16.1

COOLING LOAD TRANSFER FUNCTION CALCULATIONS

NU-VALUES	W-VALUES	I-value
0.681	1	0
-0.601	-0.92	1
0	0	2

COOLING LOAD

Time	HEAT GAIN q-theta	Q-theta DAY 1	Q-theta DAY 2	Q-theta DAY 3	Q-theta DAY 4	Q-theta DAY 5	Q-theta DAY 6	Q-theta DAY 7
	23.01	0	17.094	19.689	20.040	20.087	20.093	20.094
	21.91	0	16.818	19.205	19.528	19.572	19.578	19.578
	20.47	0	16.245	18.441	18.738	18.778	18.784	18.784
1	18.85	0.534	15.480	17.500	17.773	17.810	17.815	17.816
2	17.14	0.835	14.585	16.444	16.695	16.729	16.734	16.734
3	15.42	0.968	13.618	15.328	15.559	15.590	15.595	15.595
4	13.75	0.987	12.625	14.198	14.411	14.440	14.443	14.444
5	12.14	0.912	11.619	13.066	13.262	13.288	13.292	13.292
6	10.60	0.761	10.612	11.943	12.123	12.147	12.151	12.151
7	9.16	0.568	9.630	10.855	11.021	11.043	11.046	11.046
8	7.87	0.377	8.714	9.841	9.993	10.014	10.017	10.017
9	6.83	0.268	7.938	8.975	9.115	9.134	9.137	9.137
10	6.11	0.302	7.359	8.313	8.442	8.459	8.462	8.462
11	5.76	0.529	7.021	7.898	8.017	8.033	8.035	8.036
12	5.80	0.974	6.947	7.755	7.864	7.879	7.881	7.881
13	6.22	1.647	7.141	7.884	7.985	7.998	8.000	8.000
14	7.06	2.584	7.640	8.323	8.416	8.428	8.430	8.430
15	8.53	3.944	8.594	9.223	9.308	9.320	9.321	9.321
16	10.79	5.850	10.128	10.707	10.785	10.796	10.797	10.797
17	13.74	8.254	12.190	12.722	12.794	12.804	12.805	12.806
18	17.04	10.940	14.562	15.051	15.117	15.126	15.127	15.128
19	20.18	13.566	16.898	17.349	17.409	17.418	17.419	17.419
20	22.52	15.689	18.754	19.169	19.225	19.232	19.233	19.233
21	23.40	16.835	19.655	20.036	20.088	20.095	20.095	20.096
22	23.01	17.094	19.689	20.040	20.087	20.093	20.094	20.094
23	21.91	16.818	19.205	19.528	19.572	19.578	19.578	19.578
24	20.47	16.245	18.441	18.738	18.778	18.784	18.784	18.784
		DAY 1	DAY 2	DAY 3	DAY 4	DAY 5	DAY 6	DAY 7

Problem 16.2

COOLING LOAD TRANSFER FUNCTION CALCULATIONS

NU-VALUES	W-VALUES	i-value
0.676	1	0
-0.646	-0.97	1
0	0	2

Time	HEAT GAIN q-theta	COOLING LOAD Q-theta DAY 1	Q-theta DAY 2	Q-theta DAY 3	Q-theta DAY 4	Q-theta DAY 5	Q-theta DAY 6	Q-theta DAY 7	Q-theta DAY 8
	12.349	0	7.596	10.038	11.214	11.780	12.052	12.184	12.247
	9.920	0	6.096	8.465	9.606	10.155	10.419	10.547	10.608
	7.794	0	4.774	7.072	8.178	8.711	8.967	9.091	9.150
1	5.921	-1.032	3.598	5.827	6.900	7.417	7.666	7.786	7.843
2	4.291	-1.926	2.566	4.728	5.769	6.270	6.512	6.628	6.684
3	2.914	-2.670	1.687	3.784	4.794	5.280	5.514	5.627	5.681
4	1.745	-3.293	0.933	2.968	3.947	4.419	4.646	4.755	4.808
5	0.728	-3.829	0.270	2.244	3.194	3.651	3.871	3.977	4.028
6	-0.095	-4.249	-0.272	1.642	2.563	3.007	3.221	3.324	3.373
7	-0.268	-4.241	-0.384	1.473	2.367	2.797	3.004	3.104	3.152
8	0.684	-3.478	0.263	2.064	2.931	3.349	3.550	3.646	3.693
9	2.779	-1.937	1.692	3.439	4.280	4.685	4.880	4.974	5.019
10	5.759	0.219	3.739	5.434	6.250	6.642	6.831	6.922	6.966
11	9.305	2.782	6.197	7.841	8.632	9.013	9.196	9.285	9.327
12	13.084	5.532	8.845	10.439	11.207	11.576	11.754	11.840	11.881
13	16.754	8.240	11.453	12.999	13.744	14.102	14.275	14.358	14.398
14	19.972	10.671	13.787	15.287	16.010	16.357	16.525	16.605	16.644
15	22.426	12.609	15.631	17.087	17.787	18.125	18.287	18.365	18.403
16	23.887	13.891	16.823	18.235	18.914	19.241	19.399	19.475	19.511
17	24.198	14.401	17.245	18.614	19.274	19.591	19.744	19.817	19.853
18	23.301	14.088	16.847	18.176	18.815	19.123	19.271	19.342	19.377
19	21.220	12.958	15.634	16.923	17.543	17.841	17.985	18.054	18.088
20	18.245	11.195	13.791	15.040	15.642	15.932	16.071	16.138	16.171
21	15.133	9.303	11.821	13.033	13.616	13.897	14.033	14.098	14.129
22	12.349	7.596	10.038	11.214	11.780	12.052	12.184	12.247	12.277
23	9.920	6.096	8.465	9.606	10.155	10.419	10.547	10.608	10.637
24	7.794	4.774	7.072	8.178	8.711	8.967	9.091	9.150	9.179
		DAY 1	DAY 2	DAY 3	DAY 4	DAY 5	DAY 6	DAY 7	DAY 8

1/2

Problem 16.2

Q-theta	Q-theta	Q-theta	Q-theta	Q-theta	Q-theta
DAY 9	DAY 10	DAY 11	DAY 12	DAY 13	DAY 14
12.277	12.292	12.299	12.302	12.304	12.305
10.637	10.652	10.658	10.662	10.663	10.664
9.179	9.192	9.199	9.202	9.204	9.204
7.871	7.884	7.891	7.894	7.895	7.896
6.711	6.724	6.730	6.733	6.734	6.735
5.707	5.720	5.726	5.729	5.730	5.731
4.833	4.845	4.851	4.854	4.855	4.856
4.053	4.065	4.070	4.073	4.075	4.075
3.397	3.408	3.414	3.417	3.418	3.418
3.175	3.186	3.192	3.194	3.195	3.196
3.715	3.726	3.731	3.734	3.735	3.736
5.041	5.051	5.056	5.059	5.060	5.060
6.987	6.997	7.002	7.005	7.006	7.006
9.348	9.357	9.362	9.364	9.366	9.366
11.901	11.910	11.915	11.917	11.918	11.919
14.417	14.427	14.431	14.433	14.434	14.435
16.663	16.672	16.676	16.678	16.679	16.680
18.421	18.430	18.434	18.436	18.437	18.437
19.529	19.537	19.541	19.543	19.544	19.545
19.870	19.878	19.882	19.884	19.885	19.885
19.393	19.401	19.405	19.407	19.408	19.408
18.104	18.111	18.115	18.117	18.118	18.118
16.186	16.194	16.197	16.199	16.200	16.200
14.144	14.151	14.155	14.157	14.157	14.158
12.292	12.299	12.302	12.304	12.305	12.305
10.652	10.658	10.662	10.663	10.664	10.664
9.192	9.199	9.202	9.204	9.204	9.205
DAY 9	DAY 10	DAY 11	DAY 12	DAY 13	DAY 14

Problem 16.3

COOLING LOAD TRANSFER FUNCTION CALCULATIONS

NU-VALUES	W-VALUES	I-value
0.681	1	0
-0.641	-0.96	1
0	0	2

COOLING LOAD

Time	HEAT GAIN q-theta	Q-theta DAY 1	Q-theta DAY 2	Q-theta DAY 3	Q-theta DAY 4	Q-theta DAY 5	Q-theta DAY 6	Q-theta DAY 7	Q-theta DAY 8	Q-theta DAY 9	Q-theta DAY 10
	0.305	0	2.302	2.829	3.026	3.100	3.128	3.139	3.143	3.144	3.145
1	-0.383	0	1.754	2.259	2.449	2.520	2.547	2.557	2.561	2.562	2.563
2	-0.936	0	1.292	1.777	1.959	2.027	2.053	2.063	2.066	2.068	2.068
3	-1.437	-0.379	0.862	1.327	1.502	1.568	1.592	1.601	1.605	1.606	1.607
4	-1.808	-0.674	0.517	0.964	1.132	1.195	1.218	1.227	1.231	1.232	1.232
5	-1.991	-0.844	0.299	0.728	0.890	0.950	0.973	0.981	0.984	0.986	0.986
6	-2.164	-1.007	0.090	0.502	0.657	0.715	0.736	0.745	0.748	0.748	0.749
7	-2.378	-1.199	-0.146	0.250	0.398	0.454	0.475	0.482	0.485	0.487	0.487
8	-2.282	-1.181	-0.170	0.210	0.352	0.406	0.426	0.433	0.436	0.437	0.438
9	-0.385	0.067	1.038	1.402	1.539	1.590	1.609	1.617	1.619	1.620	1.621
10	3.488	2.686	3.618	3.968	4.099	4.149	4.167	4.174	4.177	4.178	4.178
11	8.218	5.939	6.834	7.170	7.296	7.343	7.361	7.368	7.370	7.371	7.372
12	12.901	9.220	10.079	10.401	10.522	10.567	10.585	10.591	10.593	10.594	10.595
13	17.058	12.198	13.022	13.332	13.448	13.492	13.508	13.514	13.517	13.517	13.518
14	20.384	14.657	15.449	15.746	15.858	15.899	15.915	15.921	15.923	15.924	15.924
15	22.610	16.402	17.162	17.447	17.554	17.595	17.610	17.615	17.618	17.618	17.619
16	23.501	17.257	17.987	18.261	18.363	18.402	18.417	18.422	18.424	18.425	18.425
17	22.953	17.134	17.834	18.097	18.196	18.233	18.247	18.252	18.254	18.255	18.255
18	21.034	16.060	16.732	16.984	17.079	17.115	17.128	17.133	17.135	17.136	17.136
19	17.838	14.082	14.728	14.970	15.061	15.095	15.108	15.113	15.115	15.115	15.115
20	13.618	11.359	11.978	12.211	12.298	12.331	12.343	12.348	12.350	12.350	12.351
21	8.671	8.080	8.675	8.898	8.982	9.014	9.025	9.030	9.031	9.032	9.032
22	4.043	4.952	5.523	5.738	5.818	5.848	5.860	5.864	5.865	5.866	5.866
23	1.467	3.161	3.710	3.915	3.993	4.022	4.033	4.037	4.038	4.039	4.039
24	0.305	2.302	2.829	3.026	3.100	3.128	3.139	3.143	3.144	3.145	3.145
	-0.383	1.754	2.259	2.449	2.520	2.547	2.557	2.561	2.562	2.563	2.563
	-0.936	1.292	1.777	1.959	2.027	2.053	2.063	2.066	2.068	2.068	2.068
		DAY 1	DAY 2	DAY 3	DAY 4	DAY 5	DAY 6	DAY 7	DAY 8	DAY 9	DAY 10

16.4

Window is shaded before 1000 hrs and after 1400 hrs LST (see Prob. 15.22). Determine shading by overhang at 1100 hrs (1300 hrs) and solar noon.

Following Prob. 15.22

1100 hrs, $h = 15°$

$\sin \beta = \cos 40° \cos 15° \cos 20.4° + \sin 40° \sin 20.4°$

$\beta = 66.58°$

$\cos \phi = \dfrac{1}{\cos 66.58°} \left(\cos 20.4° \sin 40° \cos 15° - \sin 20.4° \cos 40° \right)$

$\phi = 37.6°$

$y = 2 \text{ ft} \tan 66.58° / \cos 37.6° = 5.83 \text{ ft}$

entire window is shaded

1200 hrs, $h = 0°$, $\phi = 0°$, $\gamma = 0°$

$\sin \beta = \cos 40° \cos 20.4° + \sin 40° \sin 20.4°$

$\beta = 70.4°$

$y = 2 \text{ ft} \tan 70.4° = 5.62 \text{ ft}$

entire window is shaded

∴ window is shaded all day, use N-facing values (or shaded surface values) for SHGF

An estimate of hourly outdoor temperatures can be made from Table 15.2 assuming a 20°F daily range.

At 10:00, $90° = t_{o,max} - 0.56 (20°F)$ or $t_{o,max} = 101 °F$

Values tabulated below are obtained using Tables 15.2, 15.16a and Eq. 15.79

The indoor temperature is assumed to remain at 75°F.

16.4 Cont'd.

LST hrs	t_o °F	SHGF Btu/hr ft²	IHG Btu/hr ft²
0100	83.6	0	3.92
0200	82.6	0	3.72
0300	81.8	0	3.33
0400	81.2	0	3.04
0500	81.0	1	3.82
0600	81.4	11	12.82
0700	82.4	20	21.23
0800	84.2	28	29.15
0900	86.8	32	33.94
1000	90.0	35	38.15
1100	93.2	37	41.48
1200	96.4	38	43.93
1300	98.8	37	44.22
1400	100.4	35	43.25
1500	101.0	32	40.90
1600	100.4	28	37.09
1700	99.0	20	29.36
1800	96.8	11	20.36
1900	94.2	1	10.29
2000	91.6	0	8.13
2100	89.4	0	7.06
2200	87.4	0	6.08
2300	85.8	0	5.29
2400	84.6	0	4.70

From Table 16.1, $W_1 = -0.93$
From Table 16.2, $v_o = 0.197$ (no interior shade)
$v_1 = 1 - 0.93 - 0.197 = -0.127$

Problem 16.4

COOLING LOAD TRANSFER FUNCTION CALCULATIONS

NU-VALUES	W-VALUES	I-value
0.197	1	0
-0.127	-0.93	1
0	0	2

COOLING LOAD

Time	HEAT GAIN q-theta	Q-theta DAY 1	Q-theta DAY 2	Q-theta DAY 3	Q-theta DAY 4	Q-theta DAY 5	Q-theta DAY 6	Q-theta DAY 7
0	6.08	0	16.098	19.021	19.534	19.623	19.639	19.642
1	5.29	0	15.241	17.960	18.436	18.520	18.534	18.537
2	4.70	0	14.428	16.957	17.400	17.477	17.491	17.493
3	3.92	0.175	13.594	15.945	16.357	16.429	16.442	16.444
4	3.72	0.398	12.877	15.064	15.447	15.514	15.526	15.528
5	3.33	0.554	12.159	14.193	14.549	14.612	14.623	14.625
6	3.04	0.691	11.484	13.375	13.707	13.765	13.775	13.777
7	3.82	1.009	11.047	12.806	13.114	13.168	13.177	13.179
8	12.82	2.979	12.314	13.950	14.236	14.286	14.295	14.297
9	21.23	5.324	14.006	15.527	15.794	15.841	15.849	15.850
10	29.15	7.998	16.072	17.487	17.735	17.778	17.786	17.787
11	33.94	10.422	17.931	19.247	19.477	19.518	19.525	19.526
12	38.15	12.898	19.881	21.105	21.319	21.357	21.363	21.364
13	41.48	15.322	21.816	22.954	23.153	23.188	23.194	23.195
14	43.93	17.635	23.675	24.733	24.919	24.951	24.957	24.958
15	44.22	19.533	25.150	26.134	26.307	26.337	26.342	26.343
16	43.25	21.070	26.294	27.209	27.370	27.398	27.403	27.403
17	40.90	22.160	27.018	27.869	28.018	28.044	28.049	28.050
18	37.09	22.721	27.239	28.031	28.169	28.194	28.198	28.199
19	29.36	22.204	26.406	27.142	27.271	27.294	27.298	27.298
20	20.36	20.932	24.840	25.524	25.644	25.665	25.669	25.670
21	10.29	18.908	22.542	23.179	23.291	23.310	23.314	23.314
22	8.13	17.879	21.259	21.851	21.955	21.973	21.976	21.977
23	7.06	16.986	20.129	20.680	20.776	20.793	20.796	20.797
24	6.08	16.098	19.021	19.534	19.623	19.639	19.642	19.642
25	5.29	15.241	17.960	18.436	18.520	18.534	18.537	18.537
26	4.70	14.428	16.957	17.400	17.477	17.491	17.493	17.494
		DAY 1	DAY 2	DAY 3	DAY 4	DAY 5	DAY 6	DAY 7

Problem 16.5

COOLING LOAD TRANSFER FUNCTION CALCULATIONS

NU-VALUES	W-VALUES	I-value	
0.75	1	0	
-0.68	-0.93	1	
0	0	2	

	HEAT GAIN	COOLING LOAD, kW						
	kW	Q-theta	Q-theta	Q-theta	Q-theta	Q-theta	Q-theta	Q-theta
Time	q-theta	DAY 1	DAY 2	DAY 3	DAY 4	DAY 5	DAY 6	DAY 7
	0	0	1.795	2.109	2.165	2.174	2.176	2.176
	0	0	1.669	1.962	2.013	2.022	2.024	2.024
	0	0	1.552	1.824	1.872	1.880	1.882	1.882
1	0	0.000	1.444	1.697	1.741	1.749	1.750	1.750
2	0	0.000	1.343	1.578	1.619	1.626	1.628	1.628
3	0	0.000	1.249	1.468	1.506	1.513	1.514	1.514
4	0	0.000	1.161	1.365	1.400	1.407	1.408	1.408
5	0	0.000	1.080	1.269	1.302	1.308	1.309	1.309
6	0	0.000	1.004	1.180	1.211	1.217	1.218	1.218
7	20	15.000	15.934	16.098	16.126	16.131	16.132	16.133
8	20	15.350	16.219	16.371	16.398	16.402	16.403	16.403
9	20	15.676	16.483	16.625	16.650	16.654	16.655	16.655
10	20	15.978	16.730	16.861	16.884	16.888	16.889	16.889
11	20	16.260	16.958	17.081	17.102	17.106	17.107	17.107
12	20	16.522	17.171	17.285	17.305	17.309	17.309	17.309
13	20	16.765	17.369	17.475	17.494	17.497	17.498	17.498
14	20	16.991	17.554	17.652	17.669	17.672	17.673	17.673
15	20	17.202	17.725	17.816	17.832	17.835	17.836	17.836
16	20	17.398	17.884	17.969	17.984	17.987	17.987	17.987
17	0	2.580	3.032	3.111	3.125	3.128	3.128	3.128
18	0	2.399	2.820	2.894	2.907	2.909	2.909	2.909
19	0	2.232	2.623	2.691	2.703	2.705	2.706	2.706
20	0	2.075	2.439	2.503	2.514	2.516	2.516	2.516
21	0	1.930	2.268	2.327	2.338	2.340	2.340	2.340
22	0	1.795	2.109	2.165	2.174	2.176	2.176	2.176
23	0	1.669	1.962	2.013	2.022	2.024	2.024	2.024
24	0	1.552	1.824	1.872	1.880	1.882	1.882	1.882
		DAY 1	DAY 2	DAY 3	DAY 4	DAY 5	DAY 6	DAY 7

16.6

By Eq. 16.7, $\dfrac{ICL_{1800}}{A} = U(CLTD)_{1800}$

From Table 15.8, $U = 1.9235 \dfrac{W}{m^2 {}^\circ C} \left(\dfrac{0.1761\ Btu/hr\ ft^2 {}^\circ F}{W/m^2 {}^\circ C} \right)$

$= 0.3387\ Btu/hr\ ft^2 {}^\circ F$

From Table 16.5, 1800 hrs, W-facing, $CLTD = 24{}^\circ F$
$K = 1.0$

From Table 16.7, $LM = 0$

From Table 15.2, $\bar{t}_o = 94{}^\circ F - 0.5(18{}^\circ F) = 85{}^\circ F$

$CLTD_{corr} = (24{}^\circ F + 0) + (78{}^\circ F - 73{}^\circ F) + (85{}^\circ F - 85{}^\circ F)$

$CLTD_{corr} = 29{}^\circ F$

$\dfrac{ICL}{A}\bigg|_{1800} = 0.3387\ \dfrac{Btu}{hr\ ft^2 {}^\circ F} (29{}^\circ F) = 9.8\ \dfrac{Btu}{hr\ ft^2}$

16.7

(a) From Table 16.5, $K = 0.65$

$$CLTD_{corr} = (24 + 0)0.65 + 5 + 0 = 20.6 °F$$

$$\frac{ICL}{A}\bigg|_{1800} = 0.3387(20.6) = 7.0 \; \frac{Btu}{hr \, ft^2}$$

(b) $K = 0.83$

$$CLTD_{corr} = (24 + 0)0.83 + 5 + 0 = 24.9 °F$$

$$\frac{ICL}{A}\bigg|_{1800} = 0.3387(24.9) = 8.4 \; \frac{Btu}{hr \, ft^2}$$

Light-color reduces ICL by $(9.8 - 7.0)/9.8 = 29\%$

Medium color reduces ICL by $(9.8 - 8.4)/9.8 = 14\%$

16.8

a) From Table 15.8, $U = 0.3724 \text{ W/m}^2\text{°C}$

By Eq. 16.7, $ICL = UA(CLTD)_{corr}$

From Table 16.5, $K = 0.83$
From Table 16.7, $LM_N = -1.1°C$, $LM_E = LM_W = 0$, $LM_S = 2.2°C$

$$CLTD_{corr} = (CLTD_{Table} + LM)0.83 + (25.5 - 23) + (27 - 29.4)$$

$$= (CLTD_{Table} + LM)0.83 + 0.1°C$$

$A = 100 \text{ m}^2$ for each wall orientation

see Tables and plot

b) Maximum ICL for the 4 walls:

Orientation	ICL_{max} (W)	Time of Occurrence
North	432	1800
East	961	1000
South	875	1400
West	1240	1700

16.8

LST hr	N °C	CLTD corr E °C	S °C	W °C
0100	0.8	1.8	3.6	3.4
0200	0.0	0.9	2.8	2.6
0300	-0.8	0.9	2.8	1.8
0400	-0.8	0.1	1.9	0.9
0500	-0.8	0.1	1.9	0.9
0600	0.0	5.1	1.9	0.9
0700	2.5	14.2	2.8	0.9
0800	3.3	21.7	4.4	2.6
0900	3.3	25.0	7.7	4.3
1000	5.0	25.8	11.9	5.1
1100	5.8	23.3	16.0	6.7
1200	7.5	18.4	20.2	8.4
1300	9.1	15.9	22.7	12.6
1400	10.0	14.2	23.5	19.2
1500	10.0	14.2	21.8	25.8
1600	10.8	13.4	19.4	30.8
1700	10.8	12.6	16.0	33.3
1800	11.6	10.9	13.5	30.8
1900	9.1	9.2	11.1	22.5
2000	5.8	6.7	8.6	13.4
2100	4.2	5.9	7.7	9.2
2200	3.3	4.3	6.1	6.7
2300	2.5	3.4	5.2	5.1
2400	1.7	2.6	4.4	4.3

16.8 cont'd

LST	ICL (W)			
hr	N	E	S	W
0100	30	67	134	127
0200	0	34	104	97
0300	-30	34	104	67
0400	-30	4	71	34
0500	-30	4	71	34
0600	0	190	71	34
0700	93	529	104	34
0800	123	808	164	97
0900	123	931	287	160
1000	186	961	443	190
1100	216	868	596	250
1200	279	685	752	313
1300	339	592	845	469
1400	372	529	875	715
1500	372	529	812	961
1600	402	499	722	1147
1700	402	469	596	1240
1800	432	406	503	1147
1900	339	343	413	838
2000	216	250	320	499
2100	156	220	287	343
2200	123	160	227	250
2300	93	127	194	190
2400	63	97	164	160

Problem 16.8

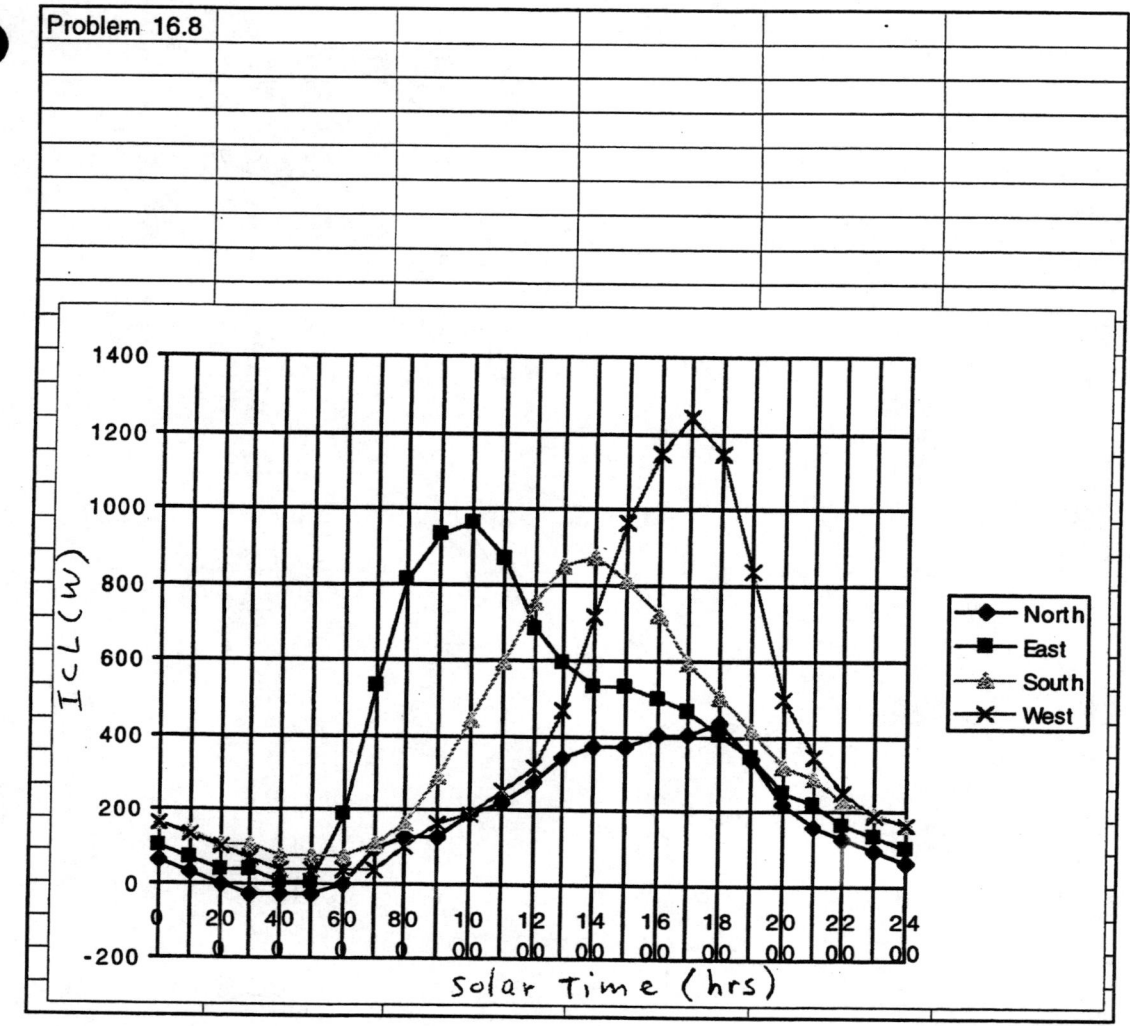

Problem 16.9

Solar time	N CLTD tab.	E CLTD tab.	S CLTD tab.	W CLTD tab.	N CLTD corr	E CLTD corr	S CLTD corr	W CLTD corr
100	9	13	12	17	6.7	10.9	11.9	16.0
200	8	12	11	16	5.8	10.1	11.1	15.2
300	7	11	10	15	5.0	9.2	10.2	14.4
400	7	10	9	14	5.0	8.4	9.4	13.5
500	6	9	8	12	4.2	7.6	8.6	11.9
600	5	8	7	11	3.3	6.7	7.7	11.1
700	5	7	6	10	3.3	5.9	6.9	10.2
800	4	7	6	9	2.5	5.9	6.9	9.4
900	4	8	5	8	2.5	6.7	6.1	8.6
1000	4	9	5	7	2.5	7.6	6.1	7.7
1100	4	11	5	7	2.5	9.2	6.1	7.7
1200	4	13	5	7	2.5	10.9	6.1	7.7
1300	5	14	6	7	3.3	11.7	6.9	7.7
1400	5	15	8	7	3.3	12.6	8.6	7.7
1500	6	16	9	8	4.2	13.4	9.4	8.6
1600	6	16	11	9	4.2	13.4	11.1	9.4
1700	7	17	12	11	5.0	14.2	11.9	11.1
1800	8	17	13	13	5.8	14.2	12.7	12.7
1900	9	16	14	16	6.7	13.4	13.5	15.2
2000	9	16	14	18	6.7	13.4	13.5	16.9
2100	9	16	14	19	6.7	13.4	13.5	17.7
2200	10	15	14	20	7.5	12.6	13.5	18.5
2300	9	14	13	19	6.7	11.7	12.7	17.7
2400	9	13	12	18	6.7	10.9	11.9	16.9

Instantaneous Cooling Load

16.9

N	E	S	W
950	1555	1697	2289
832	1436	1578	2171
713	1318	1460	2052
713	1199	1341	1934
595	1081	1223	1697
476	962	1104	1578
476	844	986	1460
358	844	986	1341
358	962	867	1223
358	1081	867	1104
358	1318	867	1104
358	1555	867	1104
476	1673	986	1104
476	1792	1223	1104
595	1910	1341	1223
595	1910	1578	1341
713	2029	1697	1578
832	2029	1815	1815
950	1910	1934	2171
950	1910	1934	2408
950	1910	1934	2526
1069	1792	1934	2645
950	1673	1815	2526
950	1555	1697	2408

Problem 16.10			
Solar time	CLTD table	CLTD corr	Q gain
hrs	°F	°F	Btu/hr ft2
100	15	-29	-3.16
200	12	-32	-3.49
300	10	-34	-3.71
400	8	-36	-3.93
500	7	-37	-4.04
600	5	-39	-4.25
700	4	-40	-4.36
800	3	-41	-4.47
900	4	-40	-4.36
1000	5	-39	-4.25
1100	9	-35	-3.82
1200	13	-31	-3.38
1300	19	-25	-2.73
1400	24	-20	-2.18
1500	29	-15	-1.64
1600	32	-12	-1.31
1700	34	-10	-1.09
1800	33	-11	-1.20
1900	31	-13	-1.42
2000	29	-15	-1.64
2100	26	-18	-1.96
2200	23	-21	-2.29
2300	20	-24	-2.62
2400	17	-27	-2.95

All gains are negative indicating continuous heat loss. However, the heat loss is lower during the afternoon due to the absorbed solar radiation. Better insulation would reduce the heat loss. There may not be a scenario to achieve a heat gain without changing the wall type.

Problem 16.11		
Solar time	CLTD table	ICL
hrs	°C	kW
100	17	86.6
200	16	81.5
300	15	76.4
400	14	71.3
500	13	66.2
600	13	66.2
700	12	61.1
800	11	56.0
900	11	56.0
1000	11	56.0
1100	12	61.1
1200	13	66.2
1300	15	76.4
1400	16	81.5
1500	18	91.7
1600	19	96.7
1700	20	101.8
1800	21	106.9
1900	21	106.9
2000	21	106.9
2100	21	106.9
2200	20	101.8
2300	19	96.7
2400	18	91.7

16.12

#R5 roof with an additional R-7 insulation
From Table 16.6, see footnote #4
select roof #R1 (max CLTD 2 hrs later, closest match to weight

$$CLTD_{corr} = (CLTD_{Table} + LM)K + (78 - t_i) + (t_o - 85)$$

$CLTD_{Table} = 64°F$ at 1800 hrs

Select July for month
From Table 16.7, $LM = 1°F$

$$CLTD_{corr} = (64 + 1)1.0 + (78 - 78) + (87 - 85) = 67°F$$

By Eq. 16.7

$$ICL_{1800} = UA\, CLTD_{corr}$$

$$R = \frac{1}{U} + 7 = \frac{1}{0.24311(0.1761)} + 7 = 30.36 \frac{hr\, ft^2 °F}{Btu}$$

$$U = 1/R = 1/30.36 = 0.0329\ Btu/hr\, ft^2 °F$$

$$ICL_{1800} = \frac{0.0329\ Btu}{hr\, ft^2 °F}(10,000\ ft^2)\, 67°F$$

$$ICL_{1800} = 22,040\ Btu/hr$$

16.13

From Table 16.6, $CLTD_{max} = CLTD_{1800} = 64°F$

$CLTD_{corr} = (CLTD_{Table} + LM)K + (78 - \bar{t_i}) + (\bar{t_o} - 85)$

From Table 16.7, $LM = -14°F$

$\bar{t_o} = 65°F - 21°F/2 = 54.5°F$

$CLTD_{corr} = (64 - 14)1.0 + (78 - 70) + (54.5 - 85)$

$CLTD_{corr} = 27.5°F$

By Eq. 16.7, $ICL_{max} = UA \, CLTD_{max}$

From Table 15.9, $U = 0.4243 \text{ Btu/hr ft}^2 °F$

$ICL_{max} = 0.4243 \dfrac{\text{Btu}}{\text{hr ft}^2 °F} (10,000 \text{ ft}^2) \, 27.5°F$

$ICL_{max} = 116,700 \text{ Btu/hr}$

16.14

The change in magnitudes is primarily caused by the difference in thermal resistance. Roof #R1 has the smallest thermal resistance (largest U-value) followed by #R3 and #R5.

The time of maximum ICL is determined primarily by the mass per unit area. Roof #R5 has the smallest mass and fastest response whereas #R3 has the largest mass and slowest response.

	#R1	#R3	#R5		
0	16	18	0.7		
100	12	17	0.5		
200	10	16	0.0		
300	7	15	-0.5		
400	5	14	-0.7		
500	3	13	-1.0		
600	2	13	-1.0		
700	1	12	-1.0		
800	0	11	-0.5		
900	2	11	0.7		
1000	4	11	2.2		
1100	8	12	3.6		
1200	13	13	5.3		
1300	18	15	6.6		
1400	24	16	7.8		
1500	29	18	8.5		
1600	33	19	8.8		
1700	35	20	8.5		
1800	36	21	7.8		
1900	35	21	6.6		
2000	32	21	4.9		
2100	28	21	3.4		
2200	24	20	2.4		
2300	19	19	1.5		
2400	16	18	0.7		

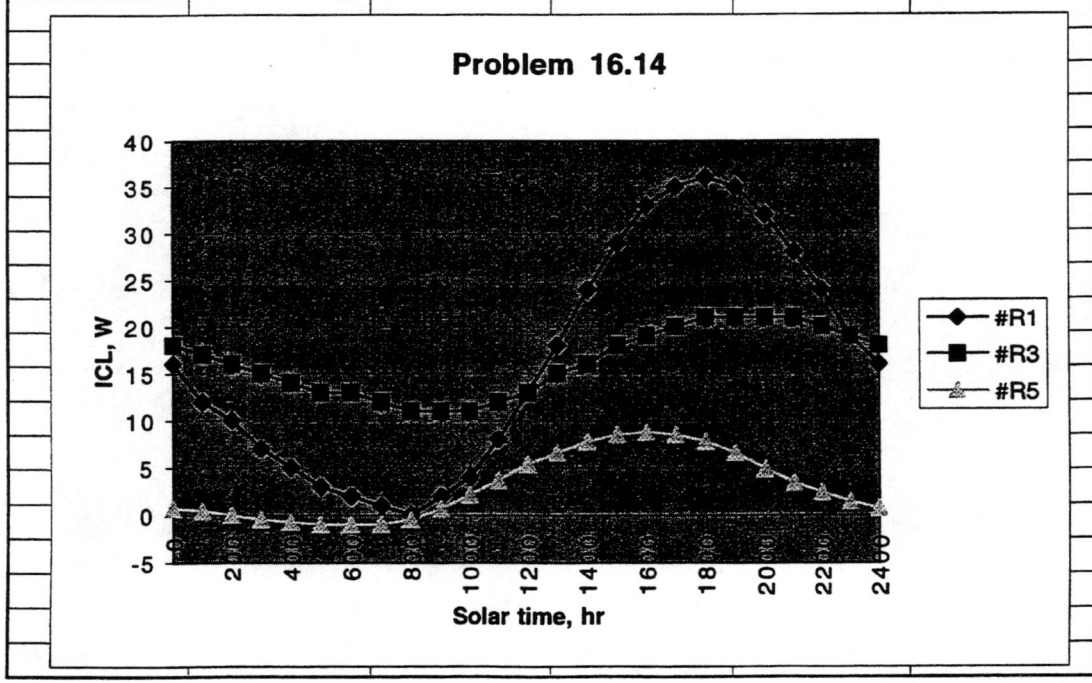

16.15

By Eq. 16.11, $ICL = A_{glass}(SC)SCL + U(A)CLTD$

From Table 15.17, $SC = 0.88$

$A_{glass} = 100 - 10 = 90 \text{ ft}^2$

From Table 16.10, assume gypsum partitions
   Zone Type = C

From Table 16.11, $SCL_{max} = 156$ Btu/hr ft² at 1700 hrs

From Table 14.5, assume aluminum frame with thermal break, 1/4 in. air space, no special coatings
$U = 0.65$ Btu/hr ft² °F

From Table 16.9, $CLTD = 13$°F (1700 hrs)

$ICL_{max} = 90 \text{ ft}^2 (0.88) \dfrac{156 \text{ Btu}}{\text{hr ft}^2} + \dfrac{0.65 \text{ Btu}}{\text{hr ft}^2 \text{°F}} (100 \text{ ft}^2) 13°F$

$ICL_{max} = 12,355 + 845 = 13,200$ Btu/hr

16.16

a) By Eq. 16.11, $ICL = A_{glass}(SC)SCL + U(A)CLTD$

assume $A_{glass} \approx A$

$$\frac{ICL}{A} = (SC)SCL + U(CLTD)$$

From Table 15.17, $SC = 0.88$

From Table 16.11, $SCL_{max} = 492 \ W/m^2$ (1700 hrs)

From Table 14.5, assume aluminum frame with thermal break, 1/4 in. air space, no special coatings or gas, $U = 3.69 \ W/m^2 °C$

From Table 16.9, $CLTD = 7°C$ (1700 hrs)

$$\frac{ICL_{max}}{A} = (0.88) 492 \frac{W}{m^2} + 3.69 \frac{W}{m^2 °C}(7°C)$$

$$= 433 + 26 = 459 \ W/m^2$$

b) Results are nearly the same as for part (a) because the time lag associated with windows is small and the maximum SCL occurs 2 hours after the end of the shading.

16.16 cont.

c) We will assume the U-value does not change appreciably

$$\frac{ICL_{max}}{A} = (0.4)492 + 26 = 223 \; W/m^2$$

16.17

a) N-facing

From Table 16.11, $SCL = 40$ Btu/hr ft^2

$ICL_{1200} = 9.8(0.88)(40) + 82 = 427$ Btu/hr

b) S-facing

$ICL_{1200} = 9.8(0.88)97 + 82 = 919$ Btu/hr

c) W-facing

$ICL_{1200} = 9.8(0.88)40 + 82 = 427$ Btu/hr

16.18

By Eq. 16.11, $ICL = A_{glass}(SC)SCL + U(A)CLTD$

Assume $A_{glass} = A = 10 \text{ m}^2$

From Table 15.17, $SC = 0.55$

From Table 14.5, assume aluminum frame with thermal break, no special coatings, 1/4 in. air space, air-filled, $U = 3.69 \text{ W/m}^2°C$

From Table 16.10, zone type = A

Corrections to tabulated CLTD values from Table 16.9,
$CLTD_{corr} = CLTD_{Table} + (25.5°C - 24°C) + (32°C - 29.4°C)$
$= CLTD_{Table} + 4.1°C$

Problem 16.18					
Area, m2	10				
SC	0.55				
U, W/m2 °C	3.69				
Solar time	SCL	SCL	CLTD table	CLTD corr.	ICL
hrs	Btu/hr(ft2)	W/m2	°C	°C	W
100	0	0	1	5	188
200	0	0	0	4	151
300	0	0	-1	3	114
400	0	0	-1	3	114
500	0	0	-1	3	114
600	9	28	-1	3	271
700	17	54	-1	3	409
800	25	79	0	4	585
900	41	129	1	5	899
1000	64	202	2	6	1335
1100	85	268	4	8	1773
1200	97	306	5	9	2018
1300	96	303	7	11	2075
1400	84	265	7	11	1867
1500	63	199	8	12	1539
1600	42	132	8	12	1175
1700	31	98	7	11	947
1800	20	63	7	11	757
1900	8	25	6	10	511
2000	4	13	4	8	368
2100	2	6	3	7	297
2200	1	3	2	6	242
2300	0	0	2	6	225
2400	0	0	1	5	188

16.19

By Eq. 16.12, $ICL_\gamma = IHG_\gamma \, CLF$ (sensible)

From Table 16.10, zone type = B for internal gains

150 occupants
From Table 16.13, $CLF = 0.97$
$ICL_s = 37,500 \times 0.97 = 36,375$ Btu/hr
$ICL_e = IHG_e = 30,000$ Btu/hr

3500 W lighting
From Table 16.12, $CLF = 0.97$
$ICL_s = 14,330 \times 0.97 = 13,900$ Btu/hr

4 copy machines, 2 coffee makers, 100 computers
From Table 16.13, $CLF = 0.97$
$ICL_s = (4000 + 7500 + 100,000) \, 0.97 = 108,155$ Btu/hr
$ICL_e = IHG_e = 3820$ Btu/hr

Totals:
Sensible: $ICL_s = 158,430$ Btu/hr

Latent: $ICL_e = 33,820$ Btu/hr

16.20

From Table 16.13 (unhooded), $CLF_{max} = 0.93$

From Table 16.14 (hooded), $CLF_{max} = 0.90$

The cooling load factor (and net cooling load) is smaller for the hooded appliance because most of the thermal energy convected into the air should be captured by the hood and exhausted from the building. However, the radiant portion contributes to the cooling load in both cases.

16.21

a) use more energy efficient lamps and ballasts

b) Zone types C and D have smaller maximum values for CLF so the ICL would be reduced. Zone type can be changed by
   i) replacing carpet with tile floor covering
   ii) use concrete block partitions rather than lightweight gypsum board partitions

Problem 17.1				
Time	Outdoor Temp.	#DD, tz=65°F	#DD, tz=60°F	#DD, tz=55°F
hrs	°F	°F*Day	°F*Day	°F*Day
100	42	0.958	0.750	0.542
200	40	1.042	0.833	0.625
300	38	1.125	0.917	0.708
400	36	1.208	1.000	0.792
500	35	1.250	1.042	0.833
600	35	1.250	1.042	0.833
700	39	1.083	0.875	0.667
800	41	1.000	0.792	0.583
900	44	0.875	0.667	0.458
1000	47	0.750	0.542	0.333
1100	50	0.625	0.417	0.208
1200	54	0.458	0.250	0.042
1300	57	0.333	0.125	-0.083
1400	59	0.250	0.042	-0.167
1500	59	0.250	0.042	-0.167
1600	58	0.292	0.083	-0.125
1700	56	0.375	0.167	-0.042
1800	53	0.500	0.292	0.083
1900	50	0.625	0.417	0.208
2000	46	0.792	0.583	0.375
2100	44	0.875	0.667	0.458
2200	41	1.000	0.792	0.583
2300	38	1.125	0.917	0.708
2400	36	1.208	1.000	0.792
Sums		19.250	14.250	9.250

17.2

By Eq. 17.10, $\eta_h = \dfrac{\eta_{ss} \, CF_{p\ell}}{1+\alpha_D}$

By Eq. 17.11, $CF_{p\ell} = 0.7791 + 0.1983\,RLC - 0.0711\,(RLC)^2$

By Eq. 17.12, $RLC = \dfrac{(UA)_{eff}\,(t_z - t_{o,design})(1+\alpha_D)}{\dot{Q}_{output}}$

$\eta_{ss} = 0.92,$

$\dot{Q}_{output} = 1.1\,(UA)_{eff}\,(t_z - t_{o,design})$

$\alpha_D = 0.05$

$RLC = (1+0.05)/1.1 = 0.955$

$CF_{p\ell} = 0.9036$

$\eta_h = \dfrac{0.92\,(0.9036)}{1+0.05} = 0.792$

17.3

From Table B.2, #DD = 4733

$$F_{yr} = \frac{24(1122)4733}{0.85(100,000)} = 1500 \text{ ccf}$$

17.4

$(UA)_{eff} = 1600$ Btu/hr °F, $\dot{Q}_{gains} = 10,000$ Btu/hr

$t_i = 70°F$, $\eta_h = 0.85$

By Eqn. 17.3

$$t_z = 70°F - \frac{10,000 \text{ Btu/hr}}{1600 \text{ Btu/hr°F}} = 63.75°F$$

From Table B.2, $(DD)_{65°F, Jan} = 1017$ °F-days

$(DD)_{60°F, Jan} = 862$ °F-days

By linear interpolation

$(DD)_{63.75°F, Jan} \cong 978$ °F-days

By Eqn 17.8;

$$E_{Jan} = \left(\frac{24 \text{ hrs}}{\text{day}}\right)\left(\frac{1600 \text{ Btu}}{\text{hr °F}}\right)\left(978 °F\text{-days}\right) = 37.56 \times 10^6 \text{ Btu}$$

By Eqn 17.9,

$$F_{Jan} = \frac{37.56 \times 10^6 \text{ Btu}}{(0.85)(100,000 \text{ Btu/ccf})} = 442 \text{ ccf}$$

17.5

By Eq. 17.10, $\eta_h = \dfrac{\eta_{ss} \, CF_{p\ell}}{1+\alpha_D}$

By Eq. 17.11, $CF_{p\ell} = 0.7791 + 0.1983\,RLC - 0.0711\,(RLC)^2$

By Eq. 17.12, $RLC = \dfrac{(UA)_{eff}\,(t_z - t_{o,design})(1+\alpha_D)}{\dot{Q}_{output}}$

$\dot{Q}_{output} = 110{,}000 \text{ Btu/hr}$

$\dot{Q}_{design} = 105{,}000 \dfrac{Btu}{hr} = (UA)_{eff}\,(72-(-12))\,°F$

$(UA)_{eff} = \dfrac{105{,}000}{72+12} = 1250 \text{ Btu/hr·°F}$

$\alpha_D = 0.04$
$\eta_{ss} = 0.93$

Assume $t_z = 65°F$

$RLC = 1250 \text{ Btu/hr·°F}\,(65-(-12))\,°F\,(1+0.04)/110{,}000\,Btu$

$RLC = 0.910$

$CF_{p\ell} = 0.7791 + 0.1983(0.910) - 0.0711(0.910)^2 = 0.901$

$\eta_h = \dfrac{0.93\,(0.901)}{1.04} = 0.805$

17.6

By Eq. 17.13, $F_{yr} = \dfrac{24 \, (UA)_{eff} \, (°DD)_{yr}}{\eta_h \, H}$

By Eq. 17.3, $t_z = t_i - \dfrac{Q_{gains}}{(UA)_{eff}}$

$t_z = 72°F - \dfrac{6000 \; Btu/hr}{1250 \; Btu/hr\,°F} = 67.2°F$

From Table B.2, interpolating between the yearly totals, $°DD_{yr} = 8828$

$\eta_h = 0.805$ and $(UA)_{eff} = 1250 \; \dfrac{Btu}{hr\,°F}$ from Prob. 17.5

$H = 100,000 \; Btu/ccf$ (from Ex. 17.1)

$F_{yr} = \dfrac{24\,(1250)(8828)}{0.805\,(100,000)} = 3290 \; ccf$

17.7

By Eq. 17.14, $\#DD_{Jan} = \frac{1}{24}\left(\sum \|t_z - \bar{t}_{bin}\| N_{bin}\right)$

see attached Table

The Total #DD for January is computed to be 1089.5 which is slightly more than the value of 1017 given in Table B.2.

Problem 17.7							
		Hr. Grp.	Hr. Grp.	Hr. Grp.	Hr. Grp.	Hr. Grp.	Hr. Grp.
Ave. Bin Temp.	tz-tbin	01 to 08	09 to 16	17 to 24	01 to 08	09 to 16	17 to 24
°F	°F	hrs	hrs	hrs	#DD	#DD	#DD
57	8	0	1	1	0	0.3	0.3
52	13	3	3	2	1.6	1.6	1.1
47	18	6	17	7	4.5	12.8	5.3
42	23	18	36	23	17.3	34.5	22.0
37	28	35	57	44	40.8	66.5	51.3
32	33	45	50	54	61.9	68.8	74.3
27	38	42	37	47	66.5	58.6	74.4
22	43	39	27	33	69.9	48.4	59.1
17	48	28	13	23	56.0	26.0	46.0
12	53	18	4	10	39.8	8.8	22.1
7	58	10	2	4	24.2	4.8	9.7
2	63	3	0	1	7.9	0	2.6
				Hr. Grp. Sums	390.3	331.1	368.2
				Jan. sum	1089.5		

17.8

The zero load temperature should be changed to account for the varying internal gains

By Eq. 17.3, $t_z = t_i - \dot{Q}_{gains}/(UA)_{eff}$

Hr. Grp. 01-08
$$t_z = 70°F - 8000 \text{ Btu/hr}/(1600 \text{ Btu/hr·°F}) = 65°F$$

Hr. Grp. 09-16,
$$t_z = 70 - 12,000/1600 = 62.5°F$$

Hr. Grp. 17-24
$$t_z = 70 - 10,000/1600 = 63.75°F$$

see attached table, $\#DD_{Jan} = 1050.8$

By Eqn. 17.13

$$F_{Jan} = \frac{24(1600)\,1050.8}{0.85\,(100,000)} = 475 \text{ ccf}$$

This value is 7% larger than the results given in Prob. 17.4.

1/2

Problem 17.8

Ave. Bin Temp.	Hr. Grp. 01 to 08	Hr. Grp. 09 to 16	Hr. Grp. 17 to 24	Hr. Grp. 01 to 08	Hr. Grp. 09 to 16	Hr. Grp. 17 to 24	Hr. Grp. 01 to 08	Hr. Grp. 09 to 16	Hr. Grp. 17 to 24
°F	tz-tbin, °F	tz-tbin, °F	tz-tbin, °F	hrs	hrs	hrs	#DD	#DD	#DD
57	8	5.5	6.75	0	1	1	0	0.2	0.3
52	13	10.5	11.75	3	3	2	1.6	1.3	1.0
47	18	15.5	16.75	6	17	7	4.5	11.0	4.9
42	23	20.5	21.75	18	36	23	17.3	30.8	20.8
37	28	25.5	26.75	35	57	44	40.8	60.6	49.0
32	33	30.5	31.75	45	50	54	61.9	63.5	71.4
27	38	35.5	36.75	42	37	47	66.5	54.7	72.0
22	43	40.5	41.75	39	27	33	69.9	45.6	57.4
17	48	45.5	46.75	28	13	23	56.0	24.6	44.8
12	53	50.5	51.75	18	4	10	39.8	8.4	21.6
7	58	55.5	56.75	10	2	4	24.2	4.6	9.5
2	63	60.5	61.75	3	0	1	7.9	0.0	2.6
						Hr. Grp. Sums	390.3	305.4	355.2
						Jan. sum	1050.8		

492

Problem 17.9

Ave. Bin Temp. °F	Jan hrs	Feb hrs	Mar hrs	Apr hrs	May hrs	Jun hrs	Jul hrs	Aug hrs	Sep hrs	Oct hrs	Nov hrs	Dec hrs
97	0	0	0	0	0	3	4	1	0	0	0	0
92	0	0	0	0	1	13	16	15	5	0	0	0
87	0	0	0	2	7	27	52	38	9	1	0	0
82	0	0	0	4	20	59	98	76	21	7	0	0
77	0	0	0	9	40	92	128	114	45	14	0	0
72	0	0	0	14	64	124	164	150	72	25	0	0
67	0	0	1	27	89	136	149	159	92	47	2	0
62	0	0	2	43	123	128	92	112	122	74	8	0
57	0	1	6	57	129	75	35	55	134	100	21	1
52	0	1	16	81	111	44	8	19	114	119	36	3
47	0	4	23	96	85	15	0	4	65	123	58	5
42	4	15	54	119	49	2	0	0	34	105	87	18
37	25	44	102	118	18	0	0	0	7	71	118	49
32	52	96	152	88	7	0	0	0	1	40	126	91
27	93	94	133	43	1	0	0	0	0	13	94	120
22	99	87	95	14	0	0	0	0	0	5	69	106
17	86	94	63	2	0	0	0	0	0	0	46	88
12	88	80	40	1	0	0	0	0	0	0	30	74
7	81	56	29	0	0	0	0	0	0	0	12	64
2	71	41	16	0	0	0	0	0	0	0	8	54
-3	57	30	7	0	0	0	0	0	0	0	4	33
-8	41	17	2	0	0	0	0	0	0	0	0	21
-13	29	7	1	0	0	0	0	0	0	0	0	8
-18	11	3	0	0	0	0	0	0	0	0	0	2
-23	6	0	0	0	0	0	0	0	0	0	0	0
-28	1	0	0	0	0	0	0	0	0	0	0	0
hrs/month	744	670	742	718	744	718	746	743	721	744	719	737
Ave. Temp. °F	12.7	19.0	28.0	45.1	58.8	68.4	72.6	70.7	60.6	50.7	33.1	19.0

Problem 17.9 plot

	Ave. Temp.
Jan	12.7
Feb	19.0
mar	28.0
Apr	45.1
May	58.8
Jun	68.4
Jul	72.6
Aug	70.7
Sep	60.6
Oct	50.7
Nov	33.1
Dec	19.0

Ave. Monthly Temps. for Mpls.

Problem 17.10 Bin Calculation, Minneapolis Weather Data, Constant Output Furnace

(UA)effective=	1122	Btu/hr	This is the effective (UA)-value for the building
ti, °F=	68		Indoor temperature
internal gain=	9000	Btu/hr	This is the sum of internal gains expected
Sys. Capacity=	98700	Btu/hr	This is the constant output of the furnace at steady-state
Dc=	0.25		This is the part load degredation coefficient
Sys. efficiency=	0.85		This is the steady-state furnace efficiency
tz, °F=	60.0		This is the calcuated Zero-load temp.

A	B	C	D	E	F	G
Avg. Bin Temp	Jan. Hours	Heating System	Heating Sys.	Part Load factor	Run Time	Energy Input
°F	From Bin Data	Load, Btu/hr	Capacity, Btu/hr	PLF	Hrs	Btu
				Eqn. (17.16)	B*C/(D*E)	F*D/(Sys eff.)
		(can't be <0)		(can't be > 1)	(can't be > col. B)	
62	0	0	98700	0.750	0.00	0.00E+00
57	0	3342	98700	0.758	0.00	0.00E+00
52	0	8952	98700	0.773	0.00	0.00E+00
47	0	14562	98700	0.787	0.00	0.00E+00
42	4	20172	98700	0.801	1.02	1.18E+05
37	25	25782	98700	0.815	8.01	9.30E+05
32	52	31392	98700	0.830	19.94	2.32E+06
27	93	37002	98700	0.844	41.32	4.80E+06
22	99	42612	98700	0.858	49.82	5.78E+06
17	86	48222	98700	0.872	48.18	5.59E+06
12	88	53832	98700	0.886	54.15	6.29E+06
7	81	59442	98700	0.901	54.17	6.29E+06
2	71	65052	98700	0.915	51.16	5.94E+06
-3	57	70662	98700	0.929	43.93	5.10E+06
-8	41	76272	98700	0.943	33.59	3.90E+06
-13	29	81882	98700	0.957	25.13	2.92E+06
-18	11	87492	98700	0.972	10.04	1.17E+06
-23	6	93102	98700	0.986	5.74	6.67E+05
-28	1	98712	98700	1.000	1.00	1.16E+05
					January Energy=	5.19E+07

Problem 17.11 Bin Calculation, Tucson Weather Data, Constant Output Furnace

(UA)effective= Btu/hr	1122	This is the effective (UA)-value for the building
t_i, °F=	68	Indoor temperature
internal gain= Btu/hr	9000	This is the sum of internal gains expected
Sys. Capacity= Btu/hr	98700	This is the constant output of the furnace at steady-state
D_c=	0.25	This is the part load degredation coefficient
Sys. efficiency=	0.85	This is the steady-state furnace efficiency
t_z, °F=	60.0	This is the calcuated Zero-load temp.

A	B	C	D	E	F	G
Avg. Bin Temp °F	Annual Hours From Bin Data	Heating System Load, Btu/hr (can't be <0)	Heating Sys. Capacity, Btu/hr	Part Load factor PLF Eqn. (17.16) (can't be > 1)	Run Time Hrs B*C/(D*E) (can't be > col. B)	Energy Input Btu F*D/(Sys eff.)
62	802	0	98700	0.750	0.00	0.00E+00
57	809	3342	98700	0.758	36.12	4.19E+06
52	780	8952	98700	0.773	91.56	1.06E+07
47	618	14562	98700	0.787	115.87	1.35E+07
42	423	20172	98700	0.801	107.92	1.25E+07
37	215	25782	98700	0.815	68.88	8.00E+06
32	81	31392	98700	0.830	31.06	3.61E+06
27	21	37002	98700	0.844	9.33	1.08E+06
22	4	42612	98700	0.858	2.01	2.34E+05
17	1	48222	98700	0.872	0.56	6.50E+04
					Annual Energy=	5.38E+07

Problem 17.12 Bin Calculation, Denver Weather Data, Constant Output Furnace

(UA)effective= Btu/hr	1739	This is the effective (UA)-value for the building	
t_i, °F=	68	Indoor temperature	
Internal gain= Btu/hr	6500	This is the sum of internal gains expected	
Sys. Capacity= Btu/hr	130000	This is the constant output of the furnace at steady-state	
D_c=	0.25	This is the part load degredation coefficient	
Sys. efficiency=	0.88	This is the steady-state furnace efficiency	
t_z, °F=	64.3	This is the calcuated Zero-load temp.	

A	B	C	D	E	F	G
Avg. Bin Temp °F	Annual Hours From Bin Data	Heating System Load, Btu/hr (can't be <0)	Heating Sys. Capacity, Btu/hr	Part Load factor PLF Eqn. (17.16) (can't be > 1)	Run Time Hrs B*C/(D*E) (can't be > col. B)	Energy Input Btu F*D/(Sys eff.)
62	794	3934	130000	0.758	31.72	4.69E+06
57	776	12629	130000	0.774	97.36	1.44E+07
52	739	21324	130000	0.791	153.25	2.26E+07
47	729	30019	130000	0.808	208.41	3.08E+07
42	752	38714	130000	0.824	271.63	4.01E+07
37	724	47409	130000	0.841	313.89	4.64E+07
32	704	56104	130000	0.858	354.15	5.23E+07
27	555	64799	130000	0.875	316.30	4.67E+07
22	394	73494	130000	0.891	249.90	3.69E+07
17	243	82189	130000	0.908	169.19	2.50E+07
12	137	90884	130000	0.925	103.57	1.53E+07
7	84	99579	130000	0.941	68.34	1.01E+07
2	54	108274	130000	0.958	46.94	6.93E+06
-3	22	116969	130000	0.975	20.30	3.00E+06
-8	13	125664	130000	0.992	12.67	1.87E+06
-13	5	134359	130000	1.000	5.00	7.39E+05
-18	3	143054	130000	1.000	3.00	4.43E+05
-23	1	151749	130000	1.000	1.00	1.48E+05
				Annual Energy=		3.58E+08

Gallons Fuel Oil = (3.58E+08)Btu/140,000 Btu/gal = 2557 gal.

Problem 17.13 Bin Calculation, Denver Weather Data, Air to Air Heat Pump Heating

(UA)effective = Btu/hr	1122	This is the effective (UA)-value for the building
t_i, °F=	68	Indoor temperature
Internal gain= Btu/hr	9000	This is the sum of internal gains expected
Heat Pump Capacity=	$Q_{cond} = 17{,}200 + 385 t_o + 2.54 t_o^2$ Btu/hr	
Dc=	0.25	This is the part load degredation coefficient
Power Requirement=	$W_{comp} = 1.28 + 0.0093 t_o - 0.000085 t_o^2$ kW	
t_z, °F=	60.0	This is the calcuated Zero-load temp.

A	B	C	D	E	F	G	H	I
Avg. Bin Temp °F	Annual Hours From Bin Data	Heating System Load, Btu/hr (can't be <0)	Heat Pump Capacity, Btu/hr	Part Load factor PLF PLF eqn (can't be > 1)	Run Time Hrs B*C/(D*E) (can't be > col. B)	Compressor Power kW	Backup Energy kWh (C-D)/3412 (can't be < 0)	Energy Input kWh (F*G)+H
62	794	0	50834	0.750	0	1.53	0	0
57	776	3342	47397	0.768	71	1.53	0	109
52	739	8952	44088	0.801	187	1.53	0	287
47	729	14562	40906	0.839	309	1.53	0	473
42	752	20172	37851	0.883	454	1.52	0	690
37	724	25782	34922	0.935	572	1.51	0	862
32	704	31392	32121	0.994	692	1.49	0	1031
27	555	37002	29447	1.000	555	1.47	1229	2044
22	394	42612	26899	1.000	394	1.44	1814	2383
17	243	48222	24479	1.000	243	1.41	1691	2034
12	137	53832	22186	1.000	137	1.38	1271	1460
7	84	59442	20019	1.000	84	1.34	971	1083
2	54	65052	17980	1.000	54	1.30	745	815
-3	22	70662	16068	1.000	22	1.25	352	380
-8	13	76272	14283	1.000	13	1.20	236	252
-13	5	81882	12624	1.000	5	1.14	101	107
-18	3	87492	11093	1.000	3	1.09	67	70
-23	1	93102	9689	1.000	1	1.02	24	25
							Annual Energy Input in kWh=	14108

Problem 17.14 Bin Calculation, New York Weather Data, Ground Water Heat Pump Heating

(UA)effective = 1122 Btu/hr — This is the effective (UA)-value for the building

t_i, °F = 68 Indoor temperature

Internal gain = 9000 Btu/hr — This is the sum of internal gains expected

Heat Pump Capacity: $Q_{cond} = 17{,}200 + 385 t_o + 2.54 t_o^2$ Btu/hr

D_c = 0.25 — This is the part load degredation coefficient

Power Requirement: $W_{comp} = 1.28 + 0.0093 t_o - 0.000085 t_o^2$ kW

Calcuated t_z, °F = 60.0 — This is the calcuated Zero-load temp.

A	B	C	D	E	F	G	H	I
Avg. Bin Temp °F	Annual Hours From Bin Data	Heating System Load, Btu/hr	Heat Pump Capacity, Btu/hr	Part Load factor PLF	Run Time Hrs	Compressor Power kW	Backup Energy kWh	Energy Input kWh
		(can't be <0)		PLF eqn (can't be > 1)	B*C/(D*E) (can't be > col. B)		(C-D)*B/3412 (can't be < 0)	(F*G)+H
62	827	0	49444	0.750	0	1.53	0	0
57	780	3342	49444	0.767	69	1.53	0	105
52	751	8952	49444	0.795	171	1.53	0	262
47	771	14562	49444	0.824	276	1.53	0	422
42	829	20172	49444	0.852	397	1.53	0	608
37	842	25782	49444	0.880	499	1.53	0	764
32	697	31392	49444	0.909	487	1.53	0	746
27	453	37002	49444	0.937	362	1.53	0	554
22	279	42612	49444	0.965	249	1.53	0	382
17	169	48222	49444	0.994	166	1.53	0	254
12	83	53832	49444	1.000	83	1.53	107	234
7	35	59442	49444	1.000	35	1.53	103	156
2	8	65052	49444	1.000	8	1.53	37	49
-3	1	70662	49444	1.000	1	1.53	6	8
							Annual Energy Input in kWh=	4544

499

17.14 Cont'd

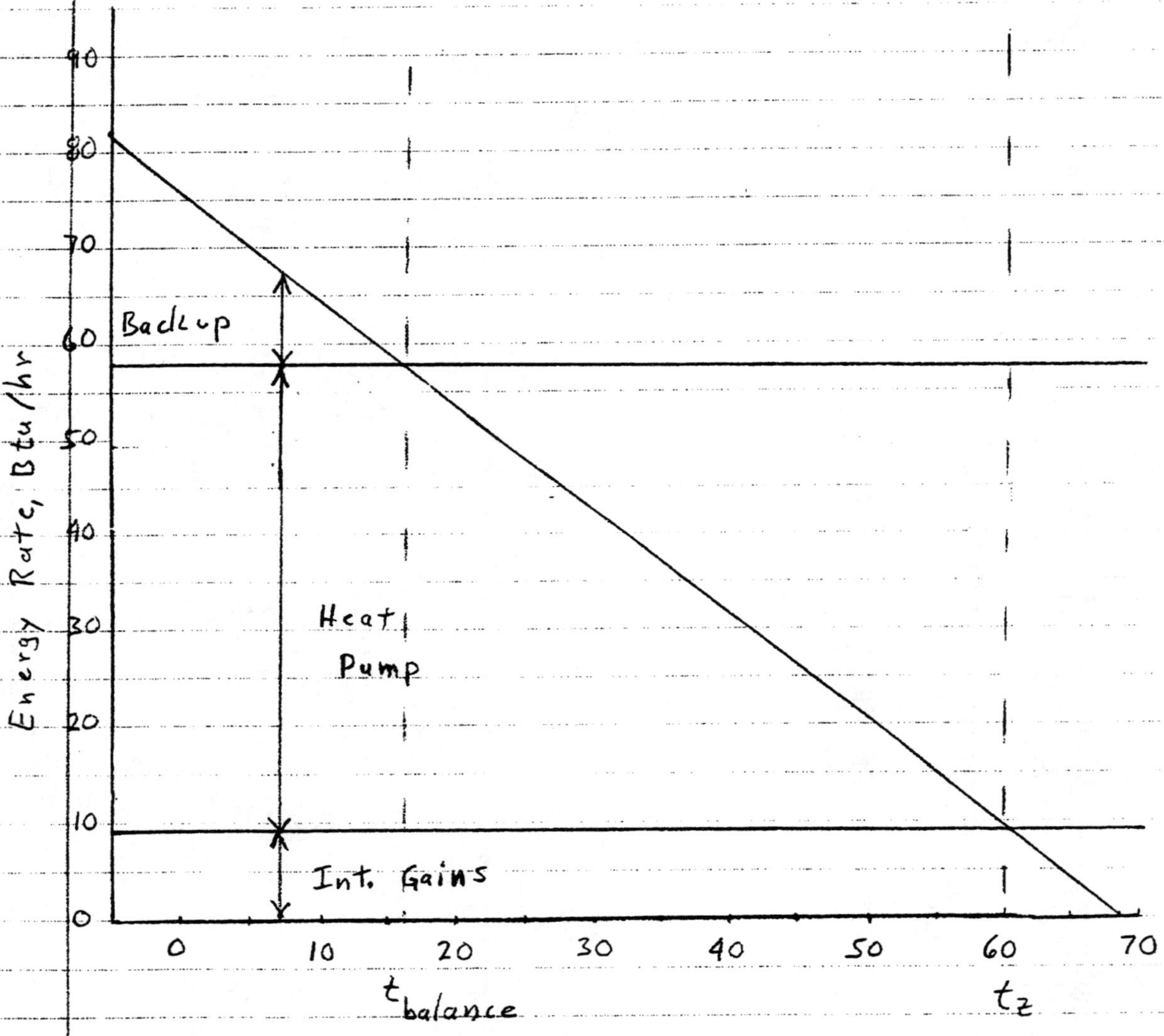
Energy Rate, Btu/hr
90
80
70
60
50
40
30
20
10
0
Backup
Heat Pump
Int. Gains
0
10
20
30
40
50
60
70
$t_{balance}$
t_z

Problem 17.15 Bin Calculation, Denver Weather Data, Air to Air Heat Pump Heating, Occupied

(UA)effective = Btu/hr	3288	This is the effective (UA)-value for the building
ti, °F=	70	Indoor temperature during occupancy
internal gain= Btu/hr	50000	Sum of internal gains during occupancy
Heat Pump Capacity=	51,600+1155to+7.62(to)2	
Dc=	0.25	This is the part load degradation coefficient
Power Requirement=	3.84+0.0279to-0.000255(to)2	
Calcuated tz, °F=	54.8	This is the calcuated Zero-load temp.

A	B	C	D	E	F	G	H	I
Avg. Bin Temp °F	Annual Hours From Table 17.4	Heating System Load, Btu/hr (can't be <0)	Heat Pump Capacity Btu/hr	Part Load factor PLF PLF eqn (can't be > 1)	Occ. Run Time Hrs B*C/(D*E) (can't be > col. B)	Compressor Power kW	Backup Energy kWh (C-D)/B)/3412 (can't be < 0)	Energy Input kWh (F*G)+H
62	273	0	152501	0.750	0	4.59	0	0
57	261	0	142192	0.750	0	4.60	0	0
52	259	9184	132264	0.767	23	4.60	0	108
47	252	25624	122718	0.802	66	4.59	0	301
42	236	42064	113552	0.843	104	4.56	0	473
37	214	58504	104767	0.890	134	4.52	0	608
32	206	74944	96363	0.944	170	4.47	0	759
27	159	91384	88340	1.000	159	4.41	142	843
22	105	107824	80698	1.000	105	4.33	835	1289
17	62	124264	73437	1.000	62	4.24	924	1186
12	35	140704	66557	1.000	35	4.14	761	905
7	23	157144	60058	1.000	23	4.02	654	747
2	14	173584	53940	1.000	14	3.89	491	545
-3	6	190024			0	0	334	334
-8	3	206464			0	0	182	182
-13	1	222904			0	0	65	65
-18	1	239344			0	0	70	70
-23	0	255784			0	0	0	0
						Annual Energy Input in kWh=		8416

Problem 17.15 Bin Calculation, Denver Weather Data, Air to Air Heat Pump Heating, Unoccupied

(UA)effective = Btu/hr	3288 This is the effective (UA)-value for the building
t_i, °F=	65 Indoor temperature during occupancy
Internal gain= Btu/hr	4000 Sum of internal gains during occupancy
Heat Pump Capacity=	$51{,}600 + 1155 t_o + 7.62(t_o)^2$
$D_c=$	0.25 This is the part load degradation coefficient
Power Requirement=	$3.84 + 0.0279 t_o - 0.000255(t_o)^2$
Calcuated t_z, °F=	63.8 This is the calcuated Zero-load temp.

A	B	C	D	E	F	G	H	I
Avg. Bin Temp °F	Annual Unocc Hours From Table 17.4	Heating System Load, Btu/hr (can't be <0)	Heat Pump Capacity Btu/hr	Part Load factor PLF PLF eqn (can't be > 1)	Unocc. Run Time Hrs B*C/(D*E) (can't be > col. B)	Compressor Power kW	Backup Energy kWh (C-D)/B)/3412 (can't be < 0)	Energy Input kWh (F*G)+H
62	522	5864	152501	0.760	26	4.59	0	121
57	515	22304	142192	0.789	102	4.60	0	471
52	480	38744	132264	0.823	171	4.60	0	786
47	477	55184	122718	0.862	249	4.59	0	1141
42	516	71624	113552	0.908	359	4.56	0	1636
37	510	88064	104767	0.960	446	4.52	0	2020
32	498	104504	96363	1.000	498	4.47	1188	3415
27	396	120944	88340	1.000	396	4.41	3784	5529
22	289	137384	80698	1.000	289	4.33	4801	6053
17	181	153824	73437	1.000	181	4.24	4264	5032
12	102	170264	66557	1.000	102	4.14	3100	3522
7	61	186704	60058	1.000	61	4.02	2264	2510
2	40	203144	53940	1.000	40	3.89	1749	1905
-3	16	219584	0		0	0	1030	1030
-8	10	236024	0		0	0	692	692
-13	4	252464	0		0	0	296	296
-18	2	268904	0		0	0	158	158
-23	1	285344	0		0	0	84	84
						Annual Energy Input in kWh=		36399

Problem 17.17a

Problem 17.17 Occupied Hour Distribution								
				Hours				
Mean Temp.	Hour Group % time	I Total	I Occupied	II Total	II Occupied	III Total	III Occupied	Total Hrs Occupied	Total Hrs Unoccupied
		100	9	100	71	100	27		
62		311	28	248	176	286	77	281	564
57		276	25	234	166	263	71	262	511
52		255	23	213	151	241	65	239	470
47		243	22	200	142	222	60	224	441
42		249	22	158	112	201	54	189	419
37		228	21	108	77	135	36	134	337
32		166	15	57	40	80	22	77	226
27		79	7	23	16	32	9	32	102
22		33	3	8	6	10	3	11	40
17		13	1	3	2	7	2	5	18
12		7	1	2	1	0	0	2	7
7		1	0	0	0	0	0	0	1
2		1	0	0	0	0	0	0	1

Table 17.17b

Problem 17.17 Bin Calculation, Atlanta Weather Data, Air to Air Heat Pump Heating, Occupied

(UA)effective = Btu/hr	3288	This is the effective (UA)-value for the building
t_i, °F=	70	Indoor temperature during occupancy
Internal gain= Btu/hr	50000	Sum of internal gains during occupancy
Heat Pump Capacity=	$51,600 + 1155 t_o + 7.62 t_o^2$	
$D_c=$	0.25	This is the part load degredation coefficient
Power Requirement=	$3.84 + 0.0279(t_o) - 0.000255(t_o)^2$	
Calcuated t_z, °F=	54.8	This is the calcuated Zero-load temp.

A	B	C	D	E	F	G	H	I
Avg. Bin Temp °F	Annual Hours From Table 17.17a	Heating System Load, Btu/hr (can't be <0)	Heat Pump Capacity Btu/hr	Part Load factor PLF PLF eqn (can't be > 1)	Occ. Run Time Hrs B*C/(D*E) (can't be > col. B)	Compressor Power kW	Backup Energy kWh (C-D)/B)/3412 (can't be < 0)	Energy Input kWh (F*G)+H
62	281	0	152501	0.750	0	4.59	0	0
57	262	0	142192	0.750	0	4.60	0	0
52	239	9184	132264	0.767	22	4.60	0	100
47	224	25624	122718	0.802	58	4.59	0	268
42	189	42064	113552	0.843	83	4.56	0	379
37	134	58504	104767	0.890	84	4.52	0	380
32	77	74944	96363	0.944	63	4.47	0	284
27	32	91384	88340	1.000	32	4.41	29	170
22	11	107824	80698	1.000	11	4.33	87	135
17	5	124264	73437	1.000	5	4.24	74	96
12	2	140704	66557	1.000	2	4.14	43	52
7	0	157144	60058	1.000	0	4.02	0	0
2	0	173584	53940	1.000	0	3.89	0	0
						Annual Energy Input in kWh=		1862

Table 17.17c

Problem 17.17 Bin Calculation, Atlanta Weather Data, Air to Air Heat Pump Heating, Unoccupied

(UA)effective = Btu/hr	3288	This is the effective (UA)-value for the building
ti, °F=	65	Indoor temperature during occupancy
Internal gain= Btu/hr	4000	Sum of internal gains during occupancy
Heat Pump Capacity=	$51,600 + 1155 t_o + 7.62(t_o)^2$	
Dc=	0.25	This is the part load degradation coefficient
Power Requirement=	$3.84 + 0.0279(t_o) - 0.000255(t_o)^2$	
Calcuated tz, °F=	63.8	This is the calcuated Zero-load temp.

A	B	C	D	E	F	G	H	I
Avg. Bin Temp °F	Annual Unocc Hours From Table 17.17a	Unocc Heating System Load, Btu/hr (can't be <0)	Heat Pump Capacity Btu/hr	Part Load factor PLF PLF eqn (can't be > 1)	Unocc. Run Time Hrs B*C/(D*E) (can't be > col. B)	Compressor Power kW	Backup Energy kWh (C-D)/B)/3412 (can't be < 0)	Energy Input kWh (F*G)+H
62	564	5864	152501	0.760	29	4.59	0	131
57	511	22304	142192	0.789	102	4.60	0	467
52	480	38744	132264	0.823	171	4.60	0	786
47	441	55184	122718	0.862	230	4.59	0	1055
42	419	71624	113552	0.908	291	4.56	0	1328
37	337	88064	104767	0.960	295	4.52	0	1334
32	226	104504	96363	1.000	226	4.47	539	1550
27	102	120944	88340	1.000	102	4.41	975	1424
22	40	137384	80698	1.000	40	4.33	665	838
17	18	153824	73437	1.000	18	4.24	424	500
12	7	170264	66557	1.000	7	4.14	213	242
7	1	186704	60058	1.000	1	4.02	37	41
2	1	203144	53940	1.000	1	3.89	44	48
						Annual Energy Input in kWh=		9745

Problem 17.19 hour distribution

Day	1	2	3	4	5	6	7	8	9	10	11	12	13	14	15	16	17	18	19	20	21	22	23	24
Sunday																								
Monday									A	A	A	A	A	A	A	A	A	B	B	B	B			
Tuesday									A	A	A	A	A	A	A	A	A	B	B	B	B			
Wednesday									A	A	A	A	A	A	A	A	A	B	B	B	B			
Thursday									A	A	A	A	A	A	A	A	A	B	B	B	B			
Friday									A	A	A	A	A	A	A	A	A	B	B	B	B			
Saturday																								
	Hour Group I							Hour Group II									Hour Group III						

Legend:
- Unoccupied
- Occupied A
- Occupied B

Problem 17.19a

Problem 17.19 Occupied Hour Distribution

Mean Temp.	Hour Group % time	I Total 100	I Occ. A 0	I Occ. B 0	II Total 100	II Occ. A 71.43	II Occ. B 12.5	III Total 100	III Occ. A 8.93	III Occ. B 37.5	Total Hrs Occ. A	Total Hrs Occ. B	Total Hrs Unoccupied
62		285	0	0	254	181	32	288	26	108	207	140	480
57		272	0	0	241	172	30	267	24	100	196	130	454
52		262	0	0	237	169	30	252	23	95	192	124	435
47		254	0	0	259	185	32	258	23	97	208	129	434
42		272	0	0	269	192	34	288	26	108	218	142	470
37		298	0	0	249	178	31	295	26	111	204	142	496
32		270	0	0	175	125	22	252	23	95	148	116	433
27		189	0	0	107	76	13	157	14	59	90	72	290
22		125	0	0	65	46	8	89	8	33	54	42	183
17		83	0	0	30	21	4	56	5	21	26	25	118
12		47	0	0	10	7	1	26	2	10	9	11	63
7		23	0	0	4	3	1	8	1	3	4	4	28
2		7	0	0	0	0	0	1	0	0	0	0	8
-3		1	0	0	0	0	0	0	0	0	0	0	1

Problem 17.19 Bin Calculation, New York Weather Data, Constant Output Furnace, Occupancy A

(UA)effective=	8375	Btu/hr	This is the effective (UA)-value for the building
ti, °F=	68		Indoor temperature
internal gain=	120000	Btu/hr	This is the sum of internal gains expected
Sys. Capacity=	700000	Btu/hr	This is the constant output of the furnace at steady-state
Dc=	0.25		This is the part load degredation coefficient
Sys. efficiency=	0.89		This is the steady-state furnace efficiency
Calcuated tz, °F=	53.7		This is the calcuated Zero-load temp.

A	B	C	D	E	F	G
Avg. Bin Temp °F	Annual Hours Occupancy A	Heating System Load, Btu/hr	Heating Sys. Capacity, Btu/hr	Part Load factor PLF	Run Time Hrs	Energy Input Btu
		(can't be <0)		Eqn. (17.16) (can't be > 1)	B*C/(D*E) (can't be > col. B)	F*D/(Sys eff.)
62	207	0	700000	0.750	0.00	0.00E+00
57	196	0	700000	0.750	0.00	0.00E+00
52	192	14000	700000	0.755	5.09	4.00E+06
47	208	55875	700000	0.770	21.56	1.70E+07
42	218	97750	700000	0.785	38.78	3.05E+07
37	204	139625	700000	0.800	50.87	4.00E+07
32	148	181500	700000	0.815	47.10	3.70E+07
27	90	223375	700000	0.830	34.61	2.72E+07
22	54	265250	700000	0.845	24.22	1.91E+07
17	26	307125	700000	0.860	13.27	1.04E+07
12	9	349000	700000	0.875	5.13	4.04E+06
7	4	390875	700000	0.890	2.51	1.97E+06
2	0	432750	700000	0.905	0.00	0.00E+00
-3	0	474625	700000	0.920	0.00	0.00E+00
					Annual Energy=	1.91E+08

Problem 17.19 Bin Calculation, New York Weather Data, Constant Output Furnace, Occupancy B

(UA)effective= Btu/hr	8375	This is the effective (UA)-value for the building
t_i, °F=	68	Indoor temperature
internal gain= Btu/hr	50000	This is the sum of internal gains expected
Sys. Capacity= Btu/hr	700000	This is the constant output of the furnace at steady-state
D_c=	0.25	This is the part load degredation coefficient
Sys. efficiency=	0.89	This is the steady-state furnace efficiency
Calcuated t_z, °F=	62.0	This is the calcuated Zero-load temp.

A	B	C	D	E	F	G
Avg. Bin Temp °F	Annual Hours Occupancy B	Heating System Load, Btu/hr	Heating Sys. Capacity, Btu/hr	Part Load factor PLF	Run Time Hrs	Energy Input Btu
		(can't be <0)		Eqn. (17.16)	B*C/(D*E)	F*D/(Sys eff.)
				(can't be > 1)	(can't be > col. B)	
62	140	250	700000	0.750	0.07	5.24E+04
57	130	42125	700000	0.765	10.23	8.04E+06
52	124	84000	700000	0.780	19.08	1.50E+07
47	129	125875	700000	0.795	29.18	2.30E+07
42	142	167750	700000	0.810	42.02	3.30E+07
37	142	209625	700000	0.825	51.55	4.05E+07
32	116	251500	700000	0.840	49.63	3.90E+07
27	72	293375	700000	0.855	35.30	2.78E+07
22	42	335250	700000	0.870	23.13	1.82E+07
17	25	377125	700000	0.885	15.22	1.20E+07
12	11	419000	700000	0.900	7.32	5.76E+06
7	4	460875	700000	0.915	2.88	2.26E+06
2	0	502750	700000	0.930	0.00	0.00E+00
-3	0	544625	700000	0.945	0.00	0.00E+00
					Annual Energy=	2.25E+08

Problem 17.19 Bin Calculation, New York Weather Data, Constant Output Furnace, Unoccupied

(UA)effective=	8375 Btu/hr	This is the effective (UA)-value for the building
t_i, °F=	68	Indoor temperature
internal gain=	10000 Btu/hr	This is the sum of internal gains expected
Sys. Capacity=	700000 Btu/hr	This is the constant output of the furnace at steady-state
D_c=	0.25	This is the part load degredation coefficient
Sys. efficiency=	0.89	This is the steady-state furnace efficiency
Calcuated t_z, °F=	66.8	This is the calcuated Zero-load temp.

A	B	C	D	E	F	G
Avg. Bin Temp	Annual Hours	Heating System	Heating Sys.	Part Load factor	Run Time	Energy Input
°F	Unoccupied	Load, Btu/hr	Capacity, Btu/hr	PLF	Hrs	Btu
		(can't be <0)		Eqn. (17.16)	B*C/(D*E)	F*D/(Sys eff.)
				(can't be > 1)	(can't be > col. B)	
62	480	40250	700000	0.764	36.11	2.84E+07
57	454	82125	700000	0.779	68.35	5.38E+07
52	435	124000	700000	0.794	97.01	7.63E+07
47	434	165875	700000	0.809	127.09	1.00E+08
42	470	207750	700000	0.824	169.24	1.33E+08
37	496	249625	700000	0.839	210.78	1.66E+08
32	433	291500	700000	0.854	211.11	1.66E+08
27	290	333375	700000	0.869	158.92	1.25E+08
22	183	375250	700000	0.884	110.97	8.73E+07
17	118	417125	700000	0.899	78.22	6.15E+07
12	63	459000	700000	0.914	45.20	3.56E+07
7	28	500875	700000	0.929	21.57	1.70E+07
2	8	542750	700000	0.944	6.57	5.17E+06
-3	1	584625	700000	0.959	0.87	6.85E+05
					Annual Energy=	1.06E+09

18.1 By Eq. 18.1,

$$EDT = (90°F - 70°F) - \frac{0.07°F}{ft/min}(150-30)\,ft/min$$

$$EDT = 11.6°F$$

18.2 By Eq. 18.1,

$$t_x = t_r + EDT + M(V_x - V_r)$$

$$= 25°C + 20°C + \frac{7.0°C}{m/s}(1 - 0.15)\,m/s$$

$$t_x = 51°C$$

18.3

a) For perfect mixing, $\bar{\tau}_p = \tau_n = \dfrac{1}{ACH} = \dfrac{V}{\dot{V}}$

$$\tau_n = \dfrac{1600 \text{ ft}^3}{320 \text{ ft}^3/\text{min}} = 5 \text{ min}$$

b) Concentration is an exponential decay with a time constant of 5 min

$$C_p(t) = C_o e^{-t/\tau_n}$$

$$= 10 \text{ ppm } e^{-t/5\text{min}}$$

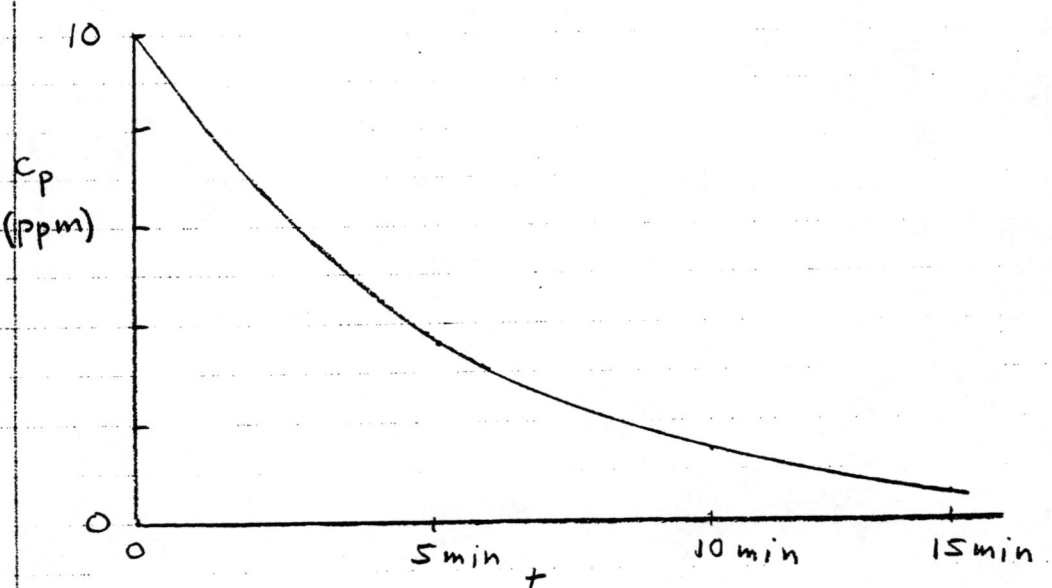

18.4

For ideal piston flow, at the center of the room the concentration retains its initial value (10 ppm) until $t = 5\text{ min}/2 = 2\frac{1}{2}\text{ min}$ when the concentration changes to the inlet value (0 ppm).

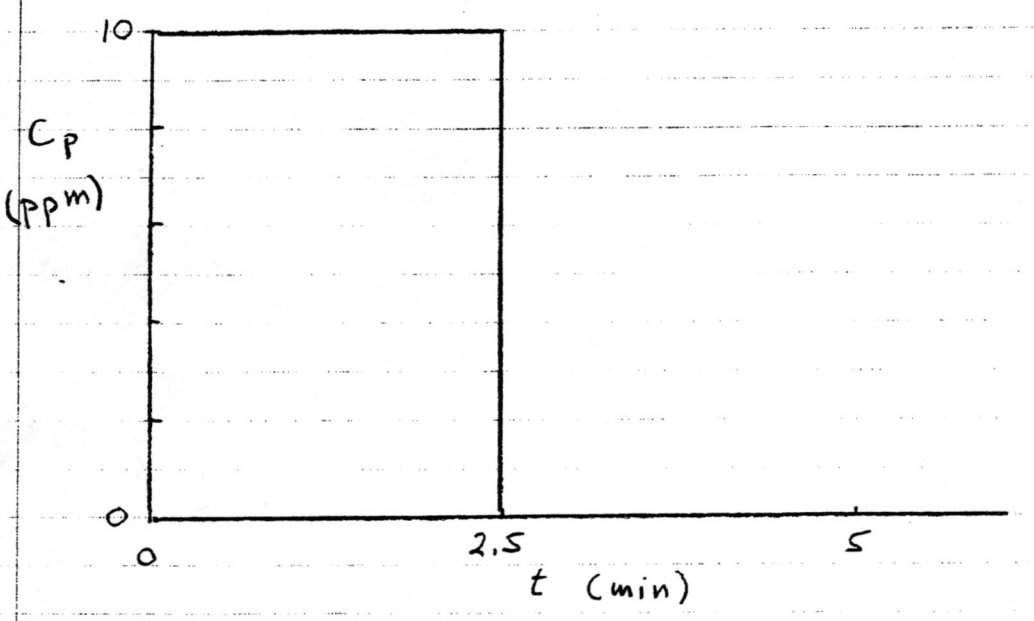

18.5 By Eq. 18.4

$$E_a = \frac{T_n}{2\langle \bar{\tau} \rangle} \times 100$$

for perfect mixing, $T_n = \langle \bar{\tau} \rangle$ so

$$E_a = \frac{100}{2} = 50\%$$

for piston flow, consider a step-down procedure. By Table 18.1

$$\langle \bar{\tau} \rangle = \frac{\dot{v}}{V} \int_0^\infty t \, \frac{C_e(t)}{C(0)} \, dt$$

The concentration at the exhaust will equal $C(0)$ until the time constant is reached, for greater times, the exhaust concentration will be zero. Therefore the integral can be split into two parts:

$$\langle \bar{\tau} \rangle = \frac{\dot{v}}{V} \left[\int_0^{T_n} t \, \frac{C(0)}{C(0)} \, dt + \int_{T_n}^\infty t \, \frac{\cancel{0}}{C(0)} \, dt \right]$$

$$= \frac{\dot{v}}{V} \frac{T_n^2}{2}$$

Substituting into the relation for efficiency,

$$E_a = \frac{T_n}{2 \frac{\dot{v}}{V} \frac{T_n^2}{2}} \times 100 = \frac{V/\dot{v}}{T_n} \times 100 = \frac{V/\dot{v}}{V/\dot{v}} \times 100 = 100\%$$

18.6 By Eqn. 18.5, $Re = \rho L \vec{V}/\mu$

$\vec{V} = 1.016$ m/s, $L = 0.05$ m
By Table A.SSI, $\mu = 1.85 \times 10^{-5}$ kg/m s,
$\rho = 1.180$ kg/m^3

$$Re = \frac{1.180 \text{ kg/m}^3 \cdot 0.05 \text{ m} \cdot 1.016 \text{ m/s}}{1.85 \times 10^{-5} \text{ kg/m s}} = 3240$$

This is turbulent flow as $Re > 2000$

In a full scale room with the same Reynolds number

$$\vec{V}_{room} = \vec{V}_{model} \frac{L_{model}}{L_{room}} = \frac{\vec{V}_{model}}{2}$$

thus $\vec{V}_{room} = \dfrac{1.016 \text{ m/s}}{2} = 0.508$ m/s

18.7

For the model, $ACH = \dot{V}/V$

$$= \frac{1.016 \text{ m/s} \times 0.05 \text{ m} \times W}{(1.95 \text{ m})(1.45 \text{ m}) \times W}\left(\frac{3600 \text{ s}}{\text{hr}}\right) = 64.7 / \text{hr}$$

For full scale room with twice the dimensions and the same jet Reynolds number:

from Problem 18.6, $\vec{V}_{room} = 0.508 \text{ m/s}$

$$ACH_r = \frac{(0.508 \text{ m/s})(2 \times 0.05 \text{ m})(2W)}{(2 \times 1.95 \text{ m})(2 \times 1.45 \text{ m})(2W)}\left(\frac{3600 \text{ s}}{\text{hr}}\right)$$

$$= 16.2 / \text{hr} = ACH_{model} / 4$$

18.8 a) By Eq. 18.7, $Ar = \dfrac{g \beta L \Delta t}{\vec{V}^2}$

For ideal gases, $\beta = 1/T = 1/(27+273.15 \text{ K}) = 3.33 \times 10^{-3}/\text{K}$

$$Ar = \dfrac{(9.8 \text{ m/s}^2)\, 3.33 \times 10^{-3}/\text{K}\, (0.05 \text{ m})\, 27°C}{(1.016 \text{ m/s})^2}$$

$Ar = 0.0427$

b) For the air change rate to be $1/2$ the value in the model, the supply air velocity must be equal to that used in the model (see Prob. 18.7). Thus $\vec{V} = 1.016$ m/s and

$$\Delta t = \dfrac{Ar\, \vec{V}^2}{g \beta L} = \dfrac{(0.0427)(1.016 \text{ m/s})^2}{(9.8 \text{ m/s})\, 3.33 \times 10^{-3}/\text{K}\, (0.1 \text{ m})} = 13.5°C$$

If the walls are at 27°C, the supply air temperature is $27 - 13.5 = 13.5°C$

18.9

By Eq. 18.5, $Re = \rho D \bar{V}/\mu$

Use 15°C to evaluate jet thermophysical properties
By Table A.5SI, assuming standard pressure
$\rho = 1.226$ kg/m³
$\mu = 1.789 \times 10^{-5}$ kg/m·s

$Re = (1.226 \text{ kg/m}^3)(0.4 \text{ m})(0.25 \text{ m/s})/1.789 \times 10^{-5}$ kg/m·s

$Re = 6853$

By Eq. 18.6, $Gr = \rho^2 g \beta D^3 \Delta t / \mu^2$

let $\beta = 1/(15 + 273.15) = 3.47 \times 10^{-3}$ /K

$$Gr = \frac{(1.226 \text{ kg/m}^3)^2 \, 9.8 \text{ m/s}^2 \, (3.47 \times 10^{-3}/K)(0.4 \text{ m})^3 \, 5°C}{(1.789 \times 10^{-5} \text{ kg/m·s})^2}$$

$Gr = 5.11 \times 10^7$

By Eq. 18.7, $Ar = Gr/Re^2 = 5.11 \times 10^7 / (6853)^2 = 1.09$

Neither the jet momentum nor the buoyancy force dominates so the jet direction should be similar to the sketch below.

18.10

$$Re' = \frac{Re\,(0.1\,m/s)}{(0.4\,m/s)} = \frac{6853}{4} = 1713$$

$$Gr' = Gr = 5.11 \times 10^7$$

$$Ar' = Gr'/Re'^2 = 5.11 \times 10^7 / (1713)^2 = 17.4$$

Now the jet buoyancy will dominate the momentum so the resulting jet flow pattern should resemble the following sketch

18.11

~85°F, 1" cylinder, flow at 50 ft/min, 70°F

By Eq. 18.5, $Re = \rho D \bar{V} / \mu$

Use 70°F to evaluate the air thermophysical properties from Table A.5E at standard pressure

$\rho = 0.07493 \text{ lbm/ft}^3$
$\mu = 0.044 \text{ lbm/ft hr}$

$Re = (0.07493 \text{ lbm/ft}^3)(1 \text{ ft})(50 \text{ ft/min})/0.044 \text{ lbm/hr ft} \left(\dfrac{\text{hr}}{60 \text{ min}}\right)$

$Re = 5109$

By Eq. 18.6, $Gr = \rho^2 g \beta D^3 \Delta t / \mu^2$

Let $\beta = 1/(70 + 459.67°R) = 1.888 \times 10^{-3}/R$

$Gr = \dfrac{(0.07493 \text{ lbm/ft}^3)^2 \, 32.2 \text{ ft/sec}^2 (1 \text{ ft})^3 (1.888 \times 10^{-3}/R)\, 15°F}{(0.044 \text{ lbm/ft hr})^2} \left(\dfrac{3600 \text{ sec}}{\text{hr}}\right)^2$

$Gr = 3.43 \times 10^7$

By Eq. 18.7, $Ar = Gr/Re^2 = 3.43 \times 10^7 / (5109)^2$

$Ar = 1.31$

The buoyancy will be slightly more important than the forced flow around this cylinder but the natural and forced convection effects are nearly equal.

18.16

1. Diffuser locations shown on attached figure

2. Circular Ceiling Diffusers selected from Table 18.5 in Text

Compute required flow rates for each room assuming 80% of the load is sensible

$$0.8 \dot{Q}_T = \rho \dot{V} c_p \Delta t$$

std air flow rate

$$\dot{V}_{std} = \frac{0.8 \dot{Q}_T}{\rho_{std} c_p \Delta t} = \frac{0.8 \dot{Q}_T \, BTU/hr}{\frac{0.0765 \, lbm}{ft^3} \cdot \frac{0.24 \, BTU}{lbm \, °F}(75-60)°F} \left(\frac{hr}{60 \min}\right)$$

$$\dot{V}_{std} = \frac{\dot{Q}_T}{20.6} \quad (cfm)$$

recommended $T_{50}/L \approx 0.8$, Table 18.3

Zone I

Room #	Total cfm	No. diffusers	diffuser cfm	L ft	neck size	vel	flow cfm	ΔP_o	NC	Throw
1	485	2	243	7½	8 in	700	243	0.105	26	6
2	485	2	243	7½	8	700	243	0.101	26	6
3	388	2	194	4	8	550	194	0.06	18	4½
4	1456	4	364	5	12	450	364	0.03	<11	5½

Zone II

A	971	4	243	6	8	700	243	0.101	26	6
B	146	1	185	2½	8	400	146	0.033	<15	4
C	1214	3	405	7½	12	525	405	0.048	13	6¼
D	728	2	364	5	12	450	364	0.032	<11	5½
E	728	2	364	4	12	450	364	0.032	<11	5½

Zone III   522

Prob 18.16

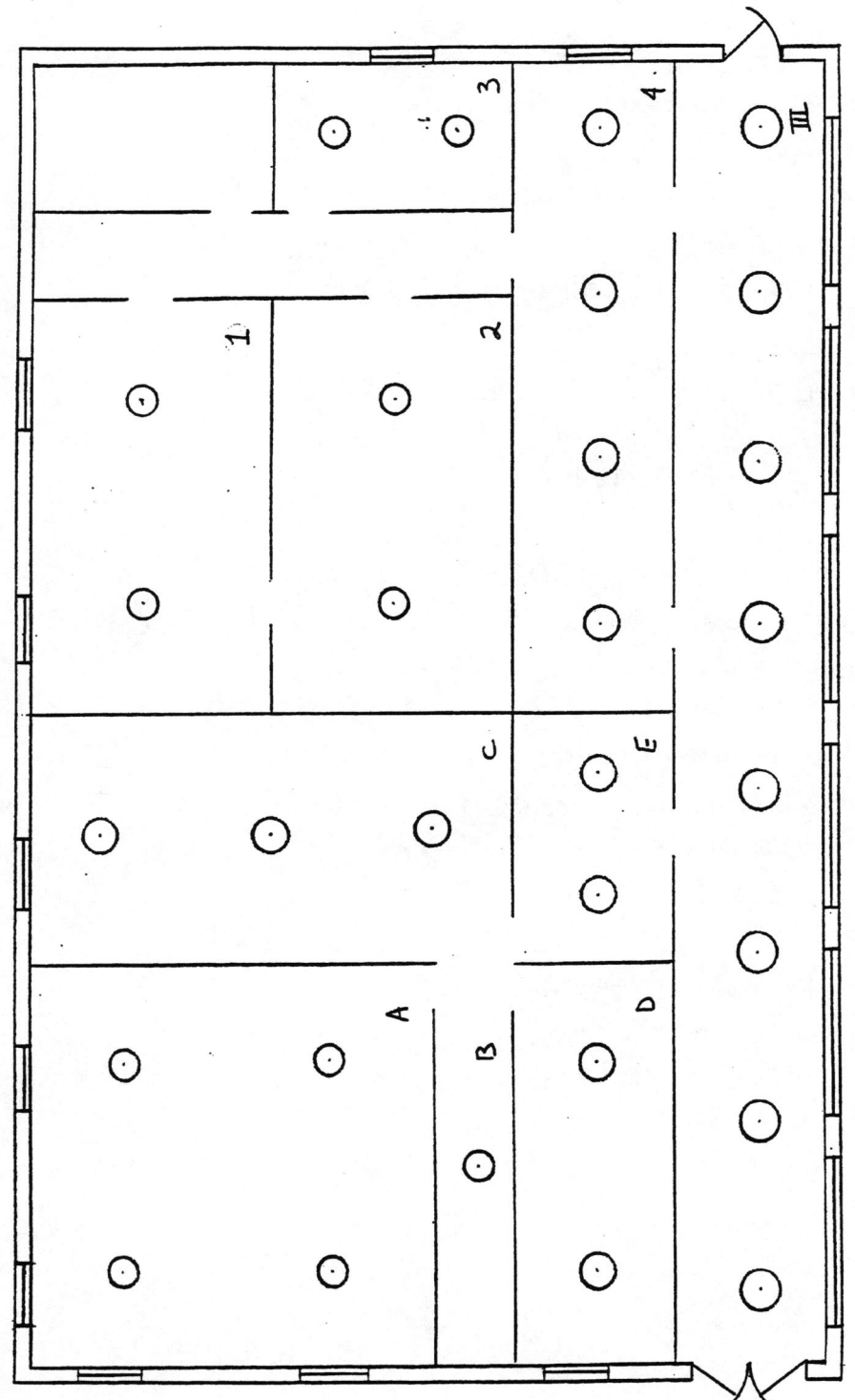

18.17

Procedure follows Example 18.7

a) Fitting ED1-3

$C = 0.22$, $\vec{V} = 4.15$ m/s
$P_v = (4.15/1.29)^2 = 10.35$ Pa
$\Delta P_{0-1} = 0.22 (10.35 \text{ Pa}) = 2.28$ Pa

b) Straight run 1-2

$\Delta P_T/L = 0.9$ Pa/m, $L = 5$ m
$\Delta P_{1-2} = (0.9 \text{ Pa/m}) 5\text{m} = 4.5$ Pa

c) Converging tee, ED5-3

By Table 18.13 with $A_s/A_c = 0.51$, $A_b/A_c = 1.0$
$\dot{V}_s/\dot{V}_c = 200 \text{ L/s} / 800 \text{ L/s} = 0.25$
$C_s = 3.5$ from linear interpolation
In the common duct, $\dot{V} = 1000$ L/s, $D = 35$ cm
By Fig. 18.18b, $\vec{V}_c = 11.5$ m/s, $\Delta P_T/L = 4.0$ Pa/m
$P_v = (11.5/1.29)^2 = 79.5$ Pa
$\Delta P_{2-3} = 3.5 (79.5 \text{ Pa}) = 278$ Pa

d) Straight duct from 3-4

$L = 3$ m
$\Delta P_{3-4} = (4.0 \text{ Pa/m}) 3\text{m} = 12$ Pa

Total Pressure drop from inlet at 0 to 4 is

$2.28 + 4.5 + 278 + 12 = 297$ Pa

18.18

Section 1-2
By Fig. 18.18a, $\vec{V} = 1600$ ft/min, $\Delta P/L = 0.14$ in water/100 ft

$\Delta P_{1-2} = (0.14 \text{ in water}/100 \text{ ft}) \, 150 \text{ ft} = 0.21$ in water

SD5-10 straight
$A_s/A_c = (20/24)^2 = 0.69$, $\dot{V}_s/\dot{V}_c = (3000/5000) = 0.6$

By Table 18.8, $C_s = 0.14$

By Eq. 18.12a, $P_v = (1600/4005)^2 = 0.16$ in water

$\Delta P_{2-3} = 0.14 \, (0.16 \text{ in water}) = 0.02$ in water

Section 3-4
By Fig. 18.18a, $\vec{V} = 1400$ ft/min, $\Delta P/L = 0.14$ in water/100 ft

$\Delta P_{3-4} = (0.14 \text{ in water}/100 \text{ ft}) \, 130 \text{ ft} = 0.18$ in water

SD5-10 branch
$A_b/A_c = (16/24)^2 = 0.44$, $\dot{V}_b/\dot{V}_c = (2000/5000) = 0.4$

By Table 18.8, $C_b = 0.81$

$\Delta P_{2-5} = 0.81 \, (0.16 \text{ in water}) = 0.13$ in water

Section 5-6
By Fig. 18.18a, $\vec{V} = 1420$ ft/min, $\Delta P/L = 0.18$ in water/100 ft

$\Delta P_{5-6} = (0.18 \text{ in water}/100 \text{ ft}) \, 100 \text{ ft} = 0.18$ in water

CD3-10
By Table 18.7, $C_o = 0.07$

$P_v = (1420/4005)^2 = 0.13$ in water

$\Delta P_{6-7} = 0.07 \, (0.13 \text{ in water}) = 0.01$ in water

Section 7-8
$\Delta P_{7-8} = (0.18 \text{ in water}/100 \text{ ft}) \, 130 \text{ ft} = 0.23$ in water

18.18 cont'd

Total Pressure Drop 1-2-3-4

$$\Delta P_{1-4} = 0.21 + 0.14 + 0.18 = 0.53 \text{ in water}$$

$$P_4 = 10 - 0.53 = 9.47 \text{ in water}$$

Total Pressure Drop 1-2-5-6-7-8

$$\Delta P_{1-8} = 0.21 + 0.13 + 0.18 + 0.01 + 0.23 = 0.76 \text{ in water}$$

$$P_8 = 10 - 0.76 = 9.24 \text{ in water}$$

18.19

Elbow w/o turning vanes, CR3-1
 By Table 18.10, $0.13 \leq C_p \leq 1.53$

Elbow with turning vanes, CR3-11
 By Table 18.11, $C_o = 0.15$

Nearly all values of loss coefficient without turning vanes are higher than the value with turning vanes, as much as a factor of ten higher. Turning vanes should be used whenever there is a significant difference between the two coefficients to reduce fan power requirement.

18.20

Pressure drops will be determined for each fitting or straight section separately and then combined to determine the various discharge pressures.

1-2, SR7-13

By Table 18.15, $C_1 = 0.13$

By Fig. 18.19, $D_{e_1} = \dfrac{1.3(30 \times 20)^{5/8}}{(30+20)^{1/4}} = 26.6$ in.

By Fig. 18.18a (6000 cfm, 26.6 in.), $\vec{V} = 1580$ ft/min

$\Delta P/L = 0.115$ in. water / 100 ft

By Eq. 18.12a $P_{v_2} = \left(\dfrac{1580}{4005}\right)^2 = 0.16$ in. water

$\Delta P_{1-2} = 0.13(0.16 \text{ in water}) = 0.02$ in. water

2-3, 10' straight duct

$\Delta P_{2-3} = \left(\dfrac{0.115 \text{ in. water}}{100 \text{ ft}}\right) 10 \text{ ft} = 0.01$ in. water

3-4, SR5-1

By Table 18.12, $\dot{V}_s / \dot{V}_c = 4000/6000 = 0.66$

$A_s/A_c = \dfrac{(20 \times 18)}{(30 \times 20)} = 0.6$, $A_b/A_c = \dfrac{(16 \times 18)}{(30 \times 20)} = 0.48$

$C_s = 0.05$ from linear interpolation

$\Delta P_{3-4} = 0.05(0.16 \text{ in. water}) = 0.01$ in. water

4-5, 30' straight duct

By Fig. 18.19, $D_{e_4} = \dfrac{1.3(20 \times 18)^{5/8}}{(20+18)^{1/4}} = 20.7$ in.

By Fig. 18.18a, (4000 cfm, 20.7 in), $\vec{V} = 1730$ ft/min

$\Delta P/L = 0.19$ in. water / 100 ft

$\Delta P_{4-5} = 30 \text{ ft} \left(\dfrac{0.19 \text{ in water}}{100 \text{ ft}}\right) = 0.06$ in. water

18.20 Cont'd

5-6, SRS-1
By Table 18.12, $\dot{V}_s/\dot{V}_c = 2000/4000 = 0.5$
$A_s/A_c = \frac{(16 \times 18)}{(20 \times 18)} = 0.8$, $A_b/A_c = 0.8$
$C_s = 0.03$ from linear interpolation
By Eq. 18.12a, $P_{v_s} = \left(\frac{1730}{4005}\right)^2 = 0.19$ in. water

$\Delta P_{5-6} = 0.03(0.19 \text{ in. water}) = 0.01$ in. water

6-7, 30 ft straight duct
By Fig. 18.19, $D_{e_6} = \frac{1.3(16 \times 18)^{5/8}}{(16+18)^{1/4}} = 18.5$ in
By Fig. 18.18a (2000 cfm, 18.5 in), $\vec{V} = 1080$ ft/min
$\Delta P/L = 0.085$ in. water / 100 ft
$\Delta P_{6-7} = 30 \text{ ft}(0.085 \text{ in. water}/100 \text{ ft}) = 0.03$ in. water

5-12, SRS-1
By Table 18.12, $\dot{V}_b/\dot{V}_c = 2,000/4000 = 0.5$
$A_b/A_c = A_s/A_c = 0.8$
$C_b = 0.95$ from linear interpolation
$\Delta P_{5-12} = 0.95(0.19 \text{ in. water}) = 0.18$ in. water

12-13, 30 ft straight duct
$\Delta P_{12-13} = \Delta P_{6-7} = 0.03$ in. water

13-14, CR3-10
By Table 18.11, $C_o = 0.12$
By Eq. 18.12a, $P_{v_{13}} = \left(\frac{1080}{4005}\right)^2 = 0.07$ in. water

$\Delta P_{13-14} = 0.12(0.07 \text{ in. water}) = 0.01$ in. water

18.20 Cont'd.

14-15

$$\Delta P_{14-15} = \Delta P_{6-7} = 0.03 \text{ in. water}$$

3-8, SR5-1

By Table 18.12, $\dot{V}_b / \dot{V}_c = 2000/6000 = 0.33$
$A_b/A_c = 0.48$, $A_s/A_c = 0.6$, $C_b = 0.33$
$$\Delta P_{3-8} = 0.33(0.16 \text{ in. water}) = 0.05 \text{ in. water}$$

8-9, 60 ft straight duct
$$\Delta P_{8-9} = 60 \text{ ft}(0.085 \text{ in. water}/100 \text{ ft}) = 0.05 \text{ in. water}$$

9-10, CR3-10
$$\Delta P_{9-10} = \Delta P_{13-14} = 0.01 \text{ in. water}$$

10-11, 60 ft straight duct
$$\Delta P_{10-11} = \Delta P_{8-9} = 0.05 \text{ in. water}$$

Total Pressure Calculations

$$P_7 = P_1 - \Delta P_{1-2} - \Delta P_{2-3} - \Delta P_{3-4} - \Delta P_{4-5} - \Delta P_{5-6} - \Delta P_{6-7}$$

$$= 8 \text{ in. water} - (0.02 + 0.01 + 0.01 + 0.06 + 0.01 + 0.03) = 7.86 \text{ in. water}$$

$$P_{15} = P_1 - \Delta P_{1-2} - \Delta P_{2-3} - \Delta P_{3-4} - \Delta P_{4-5} - \Delta P_{5-12} - \Delta P_{12-13} - \Delta P_{13-14} - \Delta P_{14-15}$$

$$= 8 - (0.02 + 0.01 + 0.01 + 0.06 + 0.18 + 0.03 + 0.01 + 0.03) = 7.65 \text{ in. water}$$

$$P_{11} = P_1 - \Delta P_{1-2} - \Delta P_{2-3} - \Delta P_{3-8} - \Delta P_{8-9} - \Delta P_{9-10} - \Delta P_{10-11}$$

$$= 8 - (0.02 + 0.01 + 0.05 + 0.05 + 0.01 + 0.05) = 7.81 \text{ in. water}$$

18.21

By Eq. 18.17

$$\dot{W}_{sh} = \frac{\dot{m}\, \Delta P_t}{\rho\, \eta_s} = \frac{\dot{V}\, \Delta P_t}{\eta_s}$$

$$= \frac{(5\, m^3/s)\, 50\, Pa}{0.75} = 333\, W$$

18.22

From the fan laws given on p. 625

$\dot{V}_2 / \dot{V}_1 = N_2/N_1 = 1.4$

$P_{t_2}/P_{t_1} = (N_2/N_1)^2 = 1.4^2 = 1.96$

$\dot{W}_2/\dot{W}_1 = (N_2/N_1)^3 = 1.4^3 = 2.74$

18.23

From the fan laws given on p. 625

a) $\dot{V}_2 = \dot{V}_1$ as density does not directly affect the volume flowrate through the fan

b) $\dot{W}_2 / \dot{W}_1 = \rho_2 / \rho_1$

let $\rho = P/RT$

$\rho_2 / \rho_1 = T_1 / T_2$

Assume standard air is dry air at 20°C (p. 204)

$T_1 = 20 + 273.15 = 293.15 \, K$

$T_2 = -40°C = -40 + 273.15 = 233.15 \, K$

$\dot{W}_2 / \dot{W}_1 = \dfrac{293.15 \, K}{233.15 \, K} = 1.26$

a 26% increase in power requirement

18.24

Considering Fig. 18.22 and Fig. 18.23 b, the total pressure at zero flowrate will be approximately proportional to the speed squared, and the maximum flowrate will be approximately proportional to the fan speed.

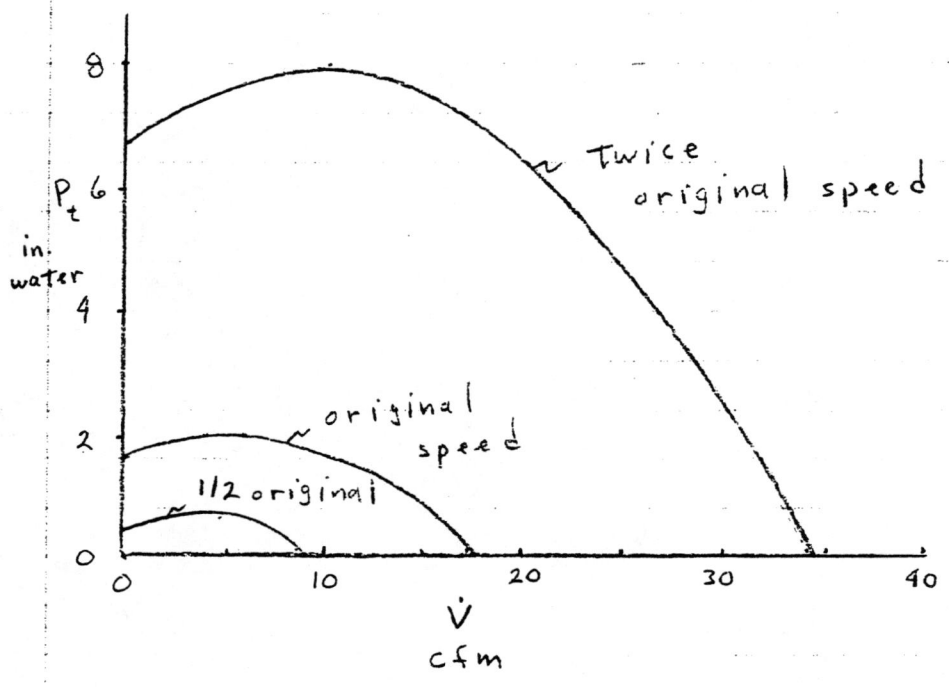

18.25
Duct sizing

By Fig. 18.18b, to the nearest duct diameter shown on the figure:

1-2, 3000 L/s, 630 mm, $\vec{V}$ = 9.6 m/s, $\Delta P/L$ = 1.5 Pa/m
3-6, 1000 L/s, 400 mm, $\vec{V}$ = 8.0 m/s, $\Delta P/L$ = 1.8 Pa/m
7-10, 2000 L/s, 500 mm, $\vec{V}$ = 10.1 m/s, $\Delta P/L$ = 2.1 Pa/m

Pressure Drop Calculations

1-2
$$\Delta P = 20 \text{ m} (1.5 \text{ Pa/m}) = 30 \text{ Pa}$$

2-3
By Table 18.8
$$A_s/A_c = (400/630)^2 = 0.40$$
$$\dot{V}_s/\dot{V}_c = (1000 \text{ L/s})/(3000 \text{ L/s}) = 0.33$$
$$C_s = 0.14, \quad P_v = (9.6/1.29)^2 = 55.4 \text{ Pa}$$
$$\Delta P = 0.14 (55.4 \text{ Pa}) = 7.8 \text{ Pa}$$

3-4 & 5-6
$$\Delta P = (5+30) \text{ m} (1.8 \text{ Pa/m}) = 63 \text{ Pa}$$

4-5
By Table 18.7, $C_o = 0.07$, $P_v = (8.0/1.29)^2 = 38.5 \text{ Pa}$
$$\Delta P = 0.07 (38.5 \text{ Pa}) = 2.7 \text{ Pa}$$

2-7
By Table 18.8
$$A_b/A_c = (500/630)^2 = 0.63$$
$$\dot{V}_b/\dot{V}_c = 2000/3000 = 0.67$$
$$C_b = 0.60$$
$$\Delta P = 0.60 (55.4 \text{ Pa}) = 33.2 \text{ Pa}$$

18.25 Cont'd.

7-8 & 9-10

$\Delta P = (15+25)\text{m} \,(2.1 \text{ Pa/m}) = 84 \text{ Pa}$

8-9

By Table 18.7

$C_o = 0.06, \quad P_v = (10.1/1.29)^2 = 61.3 \text{ Pa}$

$\Delta P = 0.06\,(61.3 \text{ Pa}) = 3.7 \text{ Pa}$

Total Pressure Drop 1-6

$\Delta P_{1-6} = 1-2, 2-3, 3-4, 4-5, 5-6 + \Delta P_{diffuser}$

$= 30 + 7.8 + 63 + 2.7 + 12 = 115.5 \text{ Pa}$

$\Delta P_{1-10} = 1-2, 2-7, 7-8, 8-9, 9-10 + \Delta P_{diffuser}$

$= 30 + 33.2 + 84 + 61.3 + 12 = 220.5 \text{ Pa}$

The Total pressure available at location 1 must be at least 220.5 Pa higher than ambient to overcome the duct losses. The branch from 3 to 6 may require resizing or a balancing damper to balance the system.

18.26

The duct sizes remain the same as in Ex. 18.9, however the pressure drops through the fittings will be determined using the equivalent length approach. Eq. 18.16 is used to convert loss coefficients (C_o) into equivalent lengths (L_{eq}). Values for friction factor (f) are taken from Table 18.17.

0-1, $C_o = 0.44$

$D = 12$ in., $f = 0.019$, $L_{eq\,0-1} = \dfrac{12\,\text{in}\,(0.44)}{0.019\,(12\,\text{in/ft})} = 23$ ft

2-3

$D = 12$ in, $f = 0.019$, $C_o = 0.15$, $L_{eq\,2-3} = \dfrac{12\,(0.15)}{0.019\,(12)} = 7.9$ ft

4-5

$D = 10$ in, $f = 0.022$, $C_o = 0.14$, $L_{eq\,4-5} = \dfrac{10\,(0.14)}{0.022\,(12)} = 5.3$ ft

6-7

$D = 9$ in, $f = 0.022$, $C_o = 0.1$, $L_{eq\,4-5} = \dfrac{9\,(0.1)}{0.022\,(12)} = 3.4$ ft

4-9

$D = 10$ in, $f = 0.022$, $C_o = 1.23$, $L_{eq\,4-9} = \dfrac{10\,(1.23)}{0.022\,(12)} = 47$ ft

2-11

$D = 12$ in, $f = 0.019$, $C_o = 1.11$, $L_{eq\,2-11} = \dfrac{12\,(1.11)}{0.019\,(12)} = 58$ ft

The following table summarizes the pressure drop calculations

18.26 cont'd.

Section	$\Delta P/L$ in.water/100ft	L or Leg ft	ΔP_t in.water
0-1	0.19	23	0.04
1-2	0.19	15	0.03
2-3	0.19	7.9	0.02
3-4	0.3	12	0.04
4-5	0.3	5.3	0.02
5-6	0.22	20	0.04
6-7	0.22	3.4	0.01
7-8	0.22	20	0.04
0-8		107	0.24
4-9	0.3	47	0.14
9-10	0.3	20	0.06
4-10		67	0.20
2-11	0.19	58	0.11
11-12	0.3	20	0.06
2-12		78	0.17

Comparing the pressure drop results with those of Example 18.19 shows that they are not identical but in general agreement.

1/3

18.27

By Fig. 18.18a, determine initial size
Section 1-4, 3000 cfm, D = 17 in, $\Delta P/L$ = 0.28 in. water/100 ft
$$\bar{v} = 1900 \text{ fpm}$$
Sections 9-10 & 11-12, 1000 cfm, D = 11 in, $\Delta P/L$ = 0.3 in. water/100 ft
$$\bar{v} = 1500 \text{ fpm}$$
Section 5-6, 4000 cfm, D = 18 in, $\Delta P/L$ = 0.35 in. water/100 ft
$$\bar{v} = 2200 \text{ fpm}$$
Section 7-8, 5000 cfm, D = 18 in, $\Delta P/L$ = 0.55 in. water/100 ft
$$\bar{v} = 2800 \text{ fpm}$$

Pressure Drop Calculations

1-2 & 3-4 combined
$$\Delta P = (30 + 15) \text{ ft} \left(\frac{0.28 \text{ in. water}}{100 \text{ ft}} \right) = 0.13 \text{ in. water}$$

2-3, CD3-10
By Table 18.7, $C_o = 0.06$
By Eq. 18.12a, $P_v = \left(\frac{1900}{4005} \right)^2 = 0.23$ in. water

$$\Delta P_{2-3} = 0.06 (0.23) = 0.01 \text{ in. water}$$

9-10, 11-12
$$\Delta P = 30 \text{ ft} \left(\frac{0.3 \text{ in. water}}{100 \text{ ft}} \right) = 0.09 \text{ in. water}$$

4-5
By Table 18.13, $\dot{V}_s / \dot{V}_c = 3000/4000 = 0.75$
$A_s/A_c = (17/18)^2 = 0.89$, $A_b/A_c = (11/18)^2 = 0.37$
$C_s = 0.44$
By Eq. 18.12a, $P_{v_c} = \left(\frac{2200}{4005} \right)^2 = 0.30$ in. water

$$\Delta P_{4-5} = 0.44 (0.30 \text{ in. water}) = 0.13 \text{ in. water}$$

18.27 cont'd.

10-5, EDS-3

By Table 18.13, $\dot{V}_b/\dot{V}_c = 1000/4000 = 0.25$
$A_s/A_c = 0.89$, $A_b/A_c = 0.37$, $C_b = -0.17$
$\Delta P_{10-5} = -0.17 (0.30 \text{ in. water}) = -0.05 \text{ in. water}$

5-6

$\Delta P_{5-6} = (15 \text{ ft})\left(\dfrac{0.35 \text{ in. water}}{100 \text{ ft}}\right) = 0.05 \text{ in. water}$

6-7, EDS-3

By Table 18.13, $\dot{V}_s/\dot{V}_c = 4000/5000 = 0.8$
$A_s/A_c = (18/18)^2 = 1$, $A_b/A_c = (11/18)^2 = 0.37$
$C_s = 0.28$
By Eq. 18.12a, $P_v = \left(\dfrac{2800}{4005}\right)^2 = 0.49 \text{ in. water}$
$\Delta P_{6-7} = 0.28 (0.49 \text{ in. water}) = 0.14 \text{ in. water}$

12-7, EDS-3

By Table 18.13, $\dot{V}_b/\dot{V}_c = 1000/5000 = 0.2$
$A_s/A_c = 1$, $A_b/A_c = 0.37$
$C_b = -0.22$
$\Delta P_{12-7} = -0.22 (0.49 \text{ in. water}) = -0.11 \text{ in. water}$

7-8

$\Delta P = 30 \text{ ft} \left(\dfrac{0.55 \text{ in. water}}{100 \text{ ft}}\right) = 0.17 \text{ in. water}$

$\Delta P_{1-8} = \Delta P_{1-2} + \Delta P_{2-3} + \Delta P_{3-4} + \Delta P_{4-5} + \Delta P_{5-6} + \Delta P_{6-7} + \Delta P_{7-8}$

$= 0.13 + 0.01 + 0.13 + 0.05 + 0.14 + 0.17 = 0.63 \text{ in. water}$

18.27

$$\Delta P_{9-8} = \Delta P_{9-10} + \Delta P_{10-5} + \Delta P_{5-6} + \Delta P_{6-7} + \Delta P_{7-8}$$

$$= 0.09 - 0.05 + 0.05 + 0.14 + 0.17 = 0.40 \text{ in. water}$$

$$\Delta P_{11-8} = \Delta P_{11-12} + \Delta P_{12-7} + \Delta P_{7-8} = 0.09 - 0.11 + 0.17 = 0.15 \text{ in. water}$$

The pressure drop from 11-8 is much less than other two branches so a method of flow control is needed to balance the system. One method is to increase the pressure drop between 11-12 by reducing the duct size. This will also affect the fitting loss (gain) between 12-7.

Reduce the diameter between 11-12 from 11 in. to 9 in.
By Fig. 18.18a, $\Delta P/L = 0.83$ in. water /100 ft, $\bar{V} = 2250$ fpm

11-12 (revised)
$$\Delta P = 30 \text{ ft} \left(\frac{0.83 \text{ in. water}}{100 \text{ ft}} \right) = 0.25 \text{ in. water}$$

12-7 (revised)
By Table 18.13, $\dot{V}_b/\dot{V}_c = 0.2$, $A_s/A_c = 1$, $A_b/A_c = \left(\frac{9}{18}\right)^2 = 0.25$
$C_b = 0.21$
$$\Delta P_{12-7} = 0.21 (0.49 \text{ in. water}) = 0.10 \text{ in. water}$$

$$\Delta P_{11-8} = 0.25 + 0.10 + 0.17 = 0.52 \text{ in. water} \quad (\text{revised})$$

The system could be balanced better with balancing dampers or adjustable dampers on the return grilles.

18.28

By Fig. 18.18a, determine equivalent round duct size

Section R1-R4, 4000 cfm, $D_e = 20$ in., $\Delta P/L = 0.22$ in. water/100
$\vec{V} = 1810$ fpm

Section R8-R7, 2000 cfm, $D_e = 14$ in., $\Delta P/L = 0.34$ in. water/100 f
$\vec{V} = 1850$ fpm

Section R5-R6, 6000 cfm, $D_e = 22$ in., $\Delta P/L = 0.30$ in. water/100 f
$\vec{V} = 2300$ fpm

Section 1-2, 6000 cfm, $D_e = 22$ in., $\Delta P/L = 0.30$ in. water/100 f
$\vec{V} = 2300$ fpm

Section 9-10 + 11-14, 2000 cfm, $D_e = 14$ in., $\Delta P/L = 0.34$ in. water/100 ft
$\vec{V} = 1850$ fpm

Section 3-4, 4000 cfm, $D_e = 20$ in., $\Delta P/L = 0.22$ in. water/100 ft
$\vec{V} = 1810$ fpm

Section 5-6, 2000 cfm, $D_e = 14$ in., $\Delta P/L = 0.34$ in. water/100 ft
$\vec{V} = 1850$ cfm

Section 15-16, 1500 cfm, $D_e = 14$ in., $\Delta P/L = 6.2$ in. water/100 ft
$\vec{V} = 1400$ fpm

section 7-8, 500 cfm, $D_e = 8$ in., $\Delta P/L = 0.41$ in. water/100 ft
$\vec{V} = 1410$ fpm

We will size the rectangular duct to the nearest inch with a common height (H) of 12 in. Using the equation for De in Fig. 18.19:

Sections R5-R6 & 1-2, W = 36 in.
Sections R1-R4 & 3-4, W = 30 in.
Sections R8-R7, 5-6, 9-10, 11-14, 15-16, W = 14 in.
Section 7-8, W = 5 in.

Pressure Drop Calculations (Return)
R1-R2 & R3-R4 combined
$$\Delta P = (25+8) \text{ ft} \left(\frac{0.22 \text{ in. water}}{100 \text{ ft}} \right) = 0.07 \text{ in. water}$$

18.28 Cont'd.

R2-R3, CR3-1
By Table 18.10, $H/W = 12/30 = 0.40$
Assume $r/W = 1.0$, $C_p = 0.26$

$$\vec{V} = \frac{4000 \text{ cfm}}{(12 \times 30) \text{ in}^2} \left(\frac{144 \text{ in}^2}{\text{ft}^2}\right) = 1600 \text{ fpm}$$

$$P_{v} = \left(\frac{1600}{4005}\right)^2 = 0.16, \quad \Delta P = 0.26(0.16) = 0.04 \text{ in. water}$$

R7-R8
$$\Delta P = 10 \text{ ft} \left(\frac{0.34 \text{ in. water}}{100 \text{ ft}}\right) = 0.03 \text{ in. water}$$

R4-R-5
By Table 18.14
$A_b/A_c = (12 \times 14)/(12 \times 36) = 0.39$
$A_s/A_c = (12 \times 30)/(12 \times 36) = 0.83$
$\dot{V}_s/\dot{V}_c = 4000/6000 = 0.66$
$C_s = 0.12$

$$\vec{V}_s = \frac{6000 \text{ cfm}}{12 \times 36} \cdot 144 = 2000 \text{ fpm}$$

$$\Delta P = 0.12 \left(\frac{2000}{4005}\right)^2 = 0.03 \text{ in. water}$$

R7-R5
By Table 18.14
$A_b/A_c = 0.39$, $A_s/A_c = 0.83$, $\dot{V}_b/\dot{V}_c = 2000/6000 = 0.33$
$C_b = 0.06$

$$\Delta P = 0.06 \left(\frac{2000}{4005}\right)^2 = 0.02 \text{ in. water}$$

R5-R6
$$\Delta P = 20 \text{ ft} \left(\frac{0.3 \text{ in. water}}{100 \text{ ft}}\right) = 0.06 \text{ in. water}$$

18.28 Cont'd.

Total pressure drops through return

R1-R6

$$\Delta P = \Delta P_{grille} + \Delta P_{R1-R2} + \Delta P_{R2-R3} + \Delta P_{R3-R4} + \Delta P_{R4-R5} + \Delta P_{R5-R6}$$

$$= 0.06 + 0.07 + 0.04 + 0.03 + 0.06 = 0.26 \text{ in. water}$$

R8-R6

$$\Delta P = \Delta P_{grille} + \Delta P_{R8-R7} + \Delta P_{R7-R5} + \Delta P_{R5-R6}$$

$$= 0.06 + 0.03 + 0.02 + 0.06 = 0.17 \text{ in. water}$$

Pressure Drop Calculations (Supply)

0-1

By Table 18.15
$\theta = 20°$, Assume $A_1/A_0 = 2.5$, $C_1 = 0.13$
$V_1 = 2000$ fpm, $P_{v_1} = \left(\frac{2000}{4005}\right)^2 = 0.25$ in. water

$$\Delta P = 0.13 \, (0.25 \text{ in. water}) = 0.03 \text{ in. water}$$

1-2

$$\Delta P = 10 \text{ ft} \left(\frac{0.3 \text{ in. water}}{100 \text{ ft}}\right) = 0.03 \text{ in. water}$$

2-3

By Table 18.12
$A_s/A_c = 30\text{in}/36\text{in} = 0.83$, $A_b/A_c = 14\text{in}/36\text{in} = 0.39$
$\dot{V}_s/\dot{V}_c = 4000 \text{ cfm}/6000 \text{ cfm} = 0.66$
$C_s = -0.04$
$\Delta P = -0.04 \, (0.25 \text{ in. water}) = -0.01 \text{ in. water}$

18.28 cont'd

2-9
By Table 18.12
$A_s/A_c = 0.83$, $A_b/A_c = 0.39$, $\dot{V}_b/\dot{V}_c = 2000/6000 = 0.33$
$C_b = 0.61$
$\Delta P = 0.61 (0.25) = 0.15$ in. water

3-4
$\Delta P = 7 \text{ ft} \left(\dfrac{0.22 \text{ in. water}}{100 \text{ ft}} \right) = 0.02$ in. water

4-5
By Table 18.12
$A_s/A_c = 14\text{in}/30\text{in} = 0.47$, $A_b/A_c = 0.47$,
$\dot{V}_s/\dot{V}_c = 2000\text{cfm}/4000\text{cfm} = 0.5$
$C_s = 0.05$
$\vec{V}_c = \dfrac{4000 \text{cfm}}{(12 \times 30)} \, 144 = 1600 \text{ fpm}$, $P_v = \left(\dfrac{1600}{4005}\right)^2 = 0.16$ in. water
$\Delta P = 0.05 (0.16 \text{ in. water}) = 0.01$ in. water

4-11
By Table 18.12
$A_s/A_c = 0.47$, $A_b/A_c = 0.47$, $\dot{V}_b/\dot{V}_c = 0.5$
$C_b = 0.52$
$\Delta P = 0.52 (0.16 \text{ in. water}) = 0.08$ in. water

5-6
$\Delta P = 7 \text{ ft} \left(\dfrac{0.34 \text{ in. water}}{100 \text{ ft}} \right) = 0.02$ in. water

6-7
By Table 18.12
$A_s/A_c = 5\text{in}/14\text{in} = 0.36$, $A_b/A_c = 14\text{in}/14\text{in} = 1.0$

18.28 cont'd

$\dot{V}_s / \dot{V}_c = 500/2000 = 0.25$, $C_s = 0.4$

$\dot{V}_c = \dfrac{2000 \text{ cfm}}{12 \times 14} \cdot 144 = 1714 \text{ fpm}$, $P_v = \left(\dfrac{1714}{4005}\right)^2 = 0.18 \text{ in. water}$

$\Delta P = 0.4 \, (0.18 \text{ in. water}) = 0.07 \text{ in. water}$

6-15

By Table 18.12

$A_s / A_c = 0.36$, $A_b / A_c = 1.0$, $\dot{V}_b / \dot{V}_c = 1500/2000 = 0.67$

$C_b = 0.36$

$\Delta P = 0.36 \, (0.18 \text{ in. water}) = 0.06 \text{ in. water}$

7-8

$\Delta P = 15 \text{ ft} \left(\dfrac{0.41 \text{ in. water}}{100 \text{ ft}}\right) = 0.06 \text{ in. water}$

9-10

$\Delta P = 25 \text{ ft} \left(\dfrac{0.34 \text{ in. water}}{100 \text{ ft}}\right) = 0.09 \text{ in. water}$

11-12 & 13-14

$\Delta P = (15 \text{ ft} + 20 \text{ ft}) \left(\dfrac{0.34 \text{ in. water}}{100 \text{ ft}}\right) = 0.12 \text{ in. water}$

12-13

By Table 18.10

$H/W = 12 \text{ in}/14 \text{ in} = 0.86$, Assume $r/W = 1.0$

$C_p = 0.22$

$\dot{V} = \dfrac{2000 \text{ cfm}}{12 \times 14} \cdot 144 = 1714 \text{ fpm}$, $P_v = \left(\dfrac{1714}{4005}\right)^2 = 0.18 \text{ in. water}$

$\Delta P = 0.22 \, (0.18 \text{ in. water}) = 0.04 \text{ in. water}$

15-16

$\Delta P = 25 \text{ ft} \, (0.2 \text{ in. water}) = 0.05 \text{ in. water}$

18.28 Cont'd

Total pressure drops through supply

$$\Delta P_{0-10} = 0-1-2-9-10 + \Delta P_{diffuser}$$

$$= 0.03 + 0.03 + 0.15 + 0.09 + 0.04 = 0.34 \text{ in. water}$$

$$\Delta P_{0-14} = 0-1-2-3-4-11-12-13-14 + \Delta P_{diffuser}$$

$$= 0.03 + 0.03 - 0.01 + 0.02 + 0.08 + 0.12 + 0.04 + 0.04 = 0.35 \text{ in. water}$$

$$\Delta P_{0-16} = 0-4-5-6-15-16 + \Delta P_{diffuser}$$

$$= 0.07 + 0.01 + 0.02 + 0.06 + 0.05 + 0.04 = 0.25 \text{ in. water}$$

$$\Delta P_{0-8} = 0-6-7-8$$

$$= 0.10 + 0.07 + 0.06 + 0.04 = 0.27 \text{ in. water}$$

Fan pressure rise must equal the sum of the largest ΔP on the return side and the largest ΔP on the supply side.

$$\Delta P_{fan} \geq 0.26 + 0.35 = 0.61 \text{ in. water}$$

18.29

Supply and return ductwork is sketched on attached floor plan.

1. Flexible duct used to connect diffusers to main duct and branches

2. Ductwork kept fairly simple while trying to achieve equal pressure drop through each diffuser

3. Return is through light fixtures in suspended ceiling.

4. Use space between drop ceiling & roof as return plenum as all three zones are separated.

5. Return ducts are kept as short and simple as possible

6. Assuming separate air handler units for each zone, ducts are shown attached to each AHU in mechanical room.

7. We will use CD3-10 elbows (Table 18.7), and SD5-10 diverging tees (Table 18.8).

8. Duct size reductions occur only near room boundaries to simplify design specifications and ease of computation.

18.29

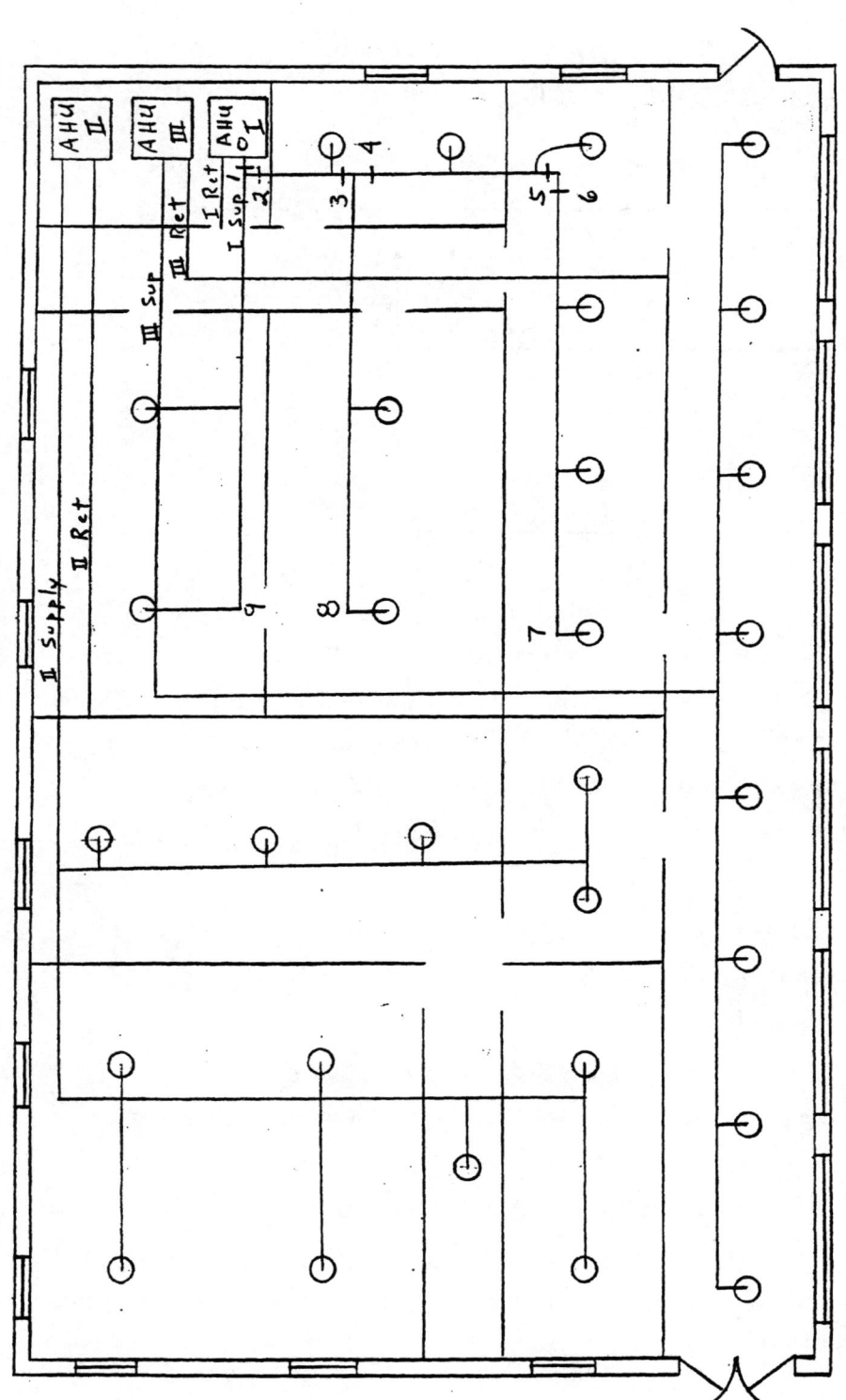

18.30
Zone I

The basic layout is shown below for the supply (not showing flex duct or diffusers)

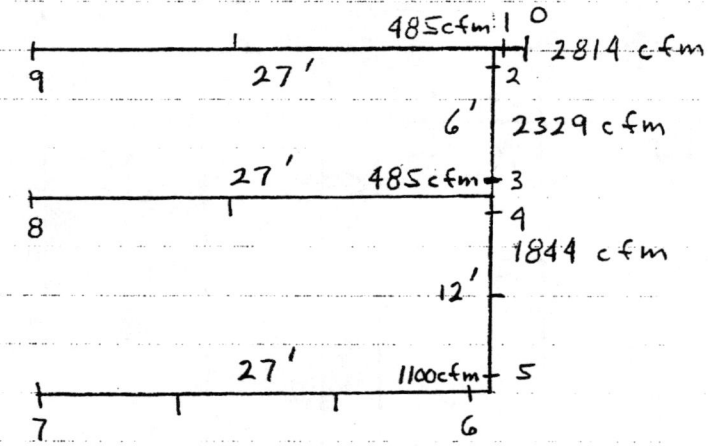

Duct Sizing (see Fig. 18.18a), To nearest 2 in. diameter
0-1 2814 cfm, 18 in., $V = 1600$ fpm, $\Delta P/L = 0.19$ in. water/100 ft
2-3 2329 cfm, 16 in., $V = 1600$ fpm, $\Delta P/L = 0.22$ in. water/100 ft
4-5 1844 cfm, 14 in., $V = 1400$ fpm, $\Delta P/L = 0.26$ in. water/100 ft
6-7 1100 cfm, 12 in., $V = 1400$ fpm, $\Delta P/L = 0.25$ in. water/100 ft

Approximate pressure drop from 0-7 will be computed as this run probably has the largest ΔP. Air flow rate at beginning of each section assumed to flow through entire section for ease of calculation; this provides a conservative estimate for the pressure drop.

We will assume an average value of $\Delta P/L = 0.20$ in. water/100 ft for all straight duct and a velocity of 1500 fpm with $P_v = (1500/4005)^2 = 0.14$ in. water for all fittings.

Assume $C_o = 0.07$ for all elbows (see Table 18.7)

18.30 cont'd

Assume $C_s \cong 0.15$ for all diverging tees

For the branch tee between 1 and 2
$A_b / A_c = (16 \text{ in} / 18 \text{ in})^2 = 0.8$
$\dot{V}_b / \dot{V}_c = 2329 \text{ cfm} / 2814 \text{ cfm} = 0.83$
By Table 18.8, $C_b = 0.61$

Total Pressure Drop 0-7
$\Delta P = 0-1, 2-3, 4-5, 6-7$ (straight)
$= (1 + 6 + 12 + 27) \text{ ft} \, (0.20 \text{ in. water} / 100 \text{ ft}) = 0.09 \text{ in. water}$

Fittings
$\Delta P = 1-2 + 6 \text{ (Tees)} + 1 \text{ (elbow)}$
$= [0.61 + 7(0.15) + (0.07)] (0.14 \text{ in. water}) = 0.24 \text{ in. water}$

Total Supply ΔP, 0-7, $= 0.09 + 0.24 + \Delta P_{diffuser} + \Delta P_{flex}$
$= 0.31 \text{ in. water} + 0.03 + \Delta P_{flex} = 0.34 \text{ in. water} + \Delta P_{flex}$

Return = 18 in (same as section 0-1 in supply)
$L = 4 \text{ ft}$
$\Delta P = $ inlet + straight
assume a conservative square-edged inlet
By Table 18.7, $C_o = 0.5$
$\Delta P = 0.5 (0.14 \text{ in. water}) + 4 \text{ ft} (0.2 \text{ in. water} / 100 \text{ ft}) = 0.08 \text{ in. water}$

Zone I AHU total fan ΔP
$\Delta P_I = 0.08 + 0.34 + \Delta P_{flex}$

$= 0.42 + \Delta P_{flex}$

Use similar procedures for zones II and III

18.31

Section numbers are shown below

```
   a     b       e      f
  □─────────────────────□
         c    d      g    h
```

a-b

$\dot{V} = 40,000$ L/s

By Fig. 18.18b, $d = 1600$ mm, $\vec{V} = 20$ m/s, $\Delta P/L = 1.9$ Pa/m

$P_{v_a} = (20/1.29)^2 = 240$ Pa $= P_{v_b}$

By Table 18.6, $C_o = 0.15$

$\Delta P = 0.15(240 \text{ Pa}) + 35\text{m}(1.9 \text{ Pa/m}) = 102.5$ Pa

b-d

  b-c,

  By Table 18.8

  $\dot{V}_s / \dot{V}_c = 30,000/40,000 = 0.75$

  let $d_c = 1600$ mm, $A_s/A_c = 1.0$, $C_s = 0.15$

  $\Delta P_{b-c} = 0.15(240 \text{ Pa}) = 36$ Pa

  c-d

  $\vec{V} = 14.5$ m/s, $\Delta P/L = 1.0$ Pa/m

  $\Delta P_{c-d} = 40\text{m}(1.0 \text{ Pa/m}) = 40$ Pa

  $P_{v_c} = (14.5/1.29)^2 = 126$ Pa

By Eq. 18.20

$P_{s,b} - P_{s,d} = 36 \text{ Pa} + 40 \text{ Pa} + (126 - 240)\text{Pa} = -38$ Pa

This is reasonably close to zero and is the best solution for the sizes given on Fig. 18.18b.

18.31 Cont'd.

d-f

  d-e

    By Table 18.8

    $\dot{V}_s / \dot{V}_c = 20{,}000 / 30{,}000 = 0.67$

    let $d_e = 1600$ mm, $A_s/A_c = 1.0$, $C_s = 0.15$

    $\Delta P_{d-e} = 0.15(126\ Pa) = 19\ Pa$

  e-f

    $\bar{V} = 10$ m/s, $\Delta P/L = 0.5$ Pa/m

    $\Delta P_{e-f} = 40\text{m}(0.5\ Pa/m) = 20\ Pa$

    $P_{v_f} = (10/1.29)^2 = 60\ Pa$

  By Eq. 18.20

    $P_{s,d} - P_{s,f} = 19\ Pa + 20\ Pa + (60\ Pa - 126\ Pa) = -27\ Pa$

f-h

  f-g

    By Table 18.8

    $\dot{V}_s / \dot{V}_c = 10{,}000 / 20{,}000 = 0.5$

    let $d_g = 1600$ mm, $A_s/A_c = 1.0$, $C_s = 0.19$

    $\Delta P_{f-g} = 0.19(60\ Pa) = 11\ Pa$

  g-h

    $\bar{V} = 5$ m/s, $\Delta P/L = 0.14$ Pa/m

    $\Delta P_{g-h} = 60\text{m}(0.14\ Pa/m) = 8.4\ Pa$

    $P_{v_h} = (5/1.29)^2 = 15\ Pa$

  By Eq. 18.20

    $P_{s,f} - P_{s,h} = 11\ Pa + 8.4\ Pa + (15\ Pa - 60\ Pa) = -26\ Pa$

Duct size should be slightly reduced for best balance.

18.31 cont'd.

Total ΔP between plenum and terminal box at h:
ΔP = a-b, b-c, c-d, d-e, e-f, f-g, g-h

= 102.5 + 36 + 40 + 19 + 20 + 11 + 8.4 = 237 Pa

18.32

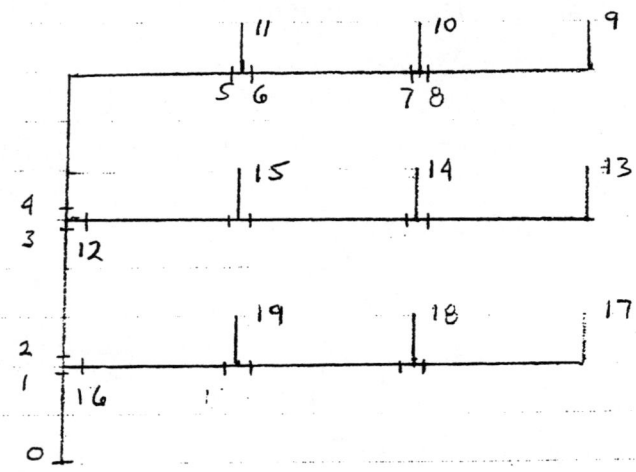

As the branches 4-9, 10 & 11, 12-13, 14, 15 & 16-17, 18, 19 are nearly identical, the longest run from 0-9 will be sized and similar sizes will be used for the other two branches (from 12 & 16)

0-1

45,000 cfm

By Fig. 18.18a, select $d = 45$ in., $\vec{V} = 4000$ fpm,
$\Delta P/L = 0.35$ in. water /100 ft, $P_v = \left(\dfrac{4000}{4005}\right)^2 = 1.0$ in. water

By Table 18.6, $C_o = 0.12$ (inlet), $\Delta P = 0.12$ (1.0 in. water) $= 0.12$ in. water

Straight duct: $\Delta P = 40$ ft $(0.35$ in. water $/100$ ft$) = 0.14$ in. water
$\Delta P = 0.12 + 0.14 = 0.26$ in. water
0-1

1-3

By Fig. 18.18a, $\dot{V} = 30,000$ cfm
let $d_2 = 40$ in., $\vec{V} = 3350$ fpm, $P_v = (3350/4005)^2 = 0.70$ in. water
$\Delta P/L = 0.295$ in. water /100 ft
By Table 18.18a, $\dot{V}_s/\dot{V}_c = 30,000/45,000 = 0.67$
$A_s/A_c = (40/45)^2 = 0.85$, $C_s = 0.13$

18.32 Cont'd.

$\Delta P_{2-3} = 50 \text{ ft} (0.295 \text{ in. water}/100 \text{ ft}) = 0.15 \text{ in. water}$

By Eq. 18.20
$P_{s,1} - P_{s,3} = 0.13 + 0.15 + (0.70 - 1.0) = -0.02 \text{ in. water}$

3-5

$\dot{V} = 15,000 \text{ cfm}$
By Fig. 18.18a, let $d_5 = 32 \text{ in}$, $\bar{V} = 2600 \text{ fpm}$
$P_v = (2600/4005)^2 = 0.42 \text{ in. water}$
$\Delta P/L = 0.245 \text{ in. water}/100 \text{ ft}$
By Table 18.8, $\dot{V}_s/\dot{V}_c = 15,000/30,000 = 0.5$
$A_s/A_c = (32/40)^2 = 0.64$, $C_s = 0.15$
$\Delta P = 0.15 (0.42 \text{ in. water}) = 0.06 \text{ in. water}$

3-4

By Table 18.7 (elbow), $C_o = 0.05$
$\Delta P_{elbow} = 0.05 (0.42 \text{ in. water}) = 0.02 \text{ in. water}$

$\Delta P_{straight} = (50+30) \text{ ft} (0.245 \text{ in. water}/100 \text{ ft}) = 0.20 \text{ in. water}$

By Eq. 18.20 (modified to include elbow)
$P_{s,3} - P_{s,5} = 0.06 + 0.02 + 0.20 + (0.42 - 0.70) = 0.0 \text{ in. wat}$

5-7

$\dot{V} = 10,000 \text{ cfm}$
By Fig. 18.18a, let $d_7 = 28 \text{ in.}$, $\bar{V} = 2300 \text{ fpm}$
$P_v = (2300/4005)^2 = 0.33 \text{ in. water}$
$\Delta P/L = 0.22 \text{ in. water}/100 \text{ ft}$
By Table 18.8, $\dot{V}_s/\dot{V}_c = 10,000/15,000 = 0.67$
$A_s/A_c = (28/32)^2 = 0.77$, $C_s = 0.13$
$\Delta P = 0.13 (0.42 \text{ in. water}) = 0.05 \text{ in. water}$

5-6

$\Delta P_{6-7} = 30 \text{ ft} (0.22 \text{ in. water}/100 \text{ ft}) = 0.07 \text{ in. water}$

18.32 cont'd.

By Eq. 18.20
$$P_{s,5} - P_{s,7} = 0.05 + 0.07 + (0.33 - 0.42) = 0.03 \text{ in. water}$$

7-9
$\dot{V} = 5,000$ cfm
By Fig. 18.18a, let $d = 24$ in., $\vec{V} = 1600$ fpm
$P_v = (1600/4005)^2 = 0.16$ in. water
$\Delta P/L = 0.145$ in. water / 100 ft
By Table 18.8, $\dot{V}_s / \dot{V}_c = 5,000/10,000 = 0.5$
$A_s / A_c = (24/28)^2 = 0.73$, $C_s = 0.16$
$\Delta P = 0.16 (0.33 \text{ in. water}) = 0.05$ in. water

7-8
By Table 18.7 (elbow), $C_o = 0.05$
$\Delta P_{elbow} = 0.05 (0.16 \text{ in. water}) = 0.01$ in. water

$\Delta P_{straight} = (30 + 25) \text{ ft} (0.145 \text{ in. water}/100 \text{ ft}) = 0.08$ in. water

By Eq. 18.20 (modified to include elbow)
$$P_{s,7} - P_{s,9} = 0.05 + 0.01 + 0.08 + (0.16 - 0.33) = -0.03 \text{ in. water}$$

Use 24 in. duct to terminal boxes 10 d 11 also

Total Pressure Drop 0-9
$\Delta P_{0-9} = 0.26 + 0.13 + 0.15 + 0.06 + 0.02 + 0.20 + 0.05 + 0.07 + 0.05 + 0.01 + 0.08$

$\Delta P_{0-9} = 1.08$ in. water

Plenum pressure must be $1.08 + 0.1 = 1.18$ in. water to provide 0.1 in. water at the fartest Terminal box (9).

18.32

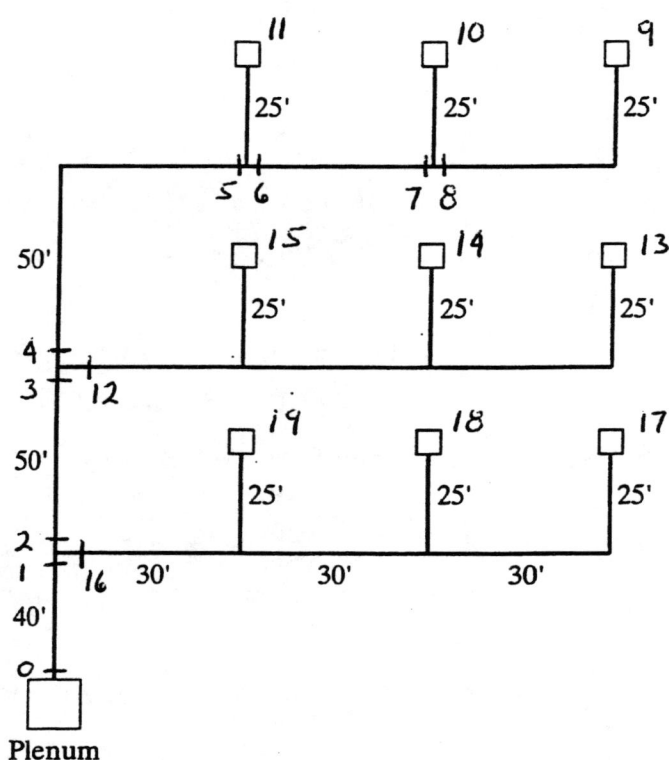

Figure 18.39 Duct-system schematic for Prob. 18.32.

19.1

$$\Delta h_{1-2} = a + b(\dot{V})^2$$

By Eqn. 19.1, $a = (z_2 - z_1)$

$$b = \ell_f / \dot{V}^2$$

By Eqn. 19.2, $b = \dfrac{fL\vec{V}^2}{D\,2g\,\dot{V}^2}$

$$\dot{V}^2 = \left(\dfrac{\pi D^2}{4} \vec{V}\right)^2 = \dfrac{\pi^2 D^4 \vec{V}^2}{16}$$

Thus: $b = \dfrac{fL\vec{V}^2 \cdot 16}{D\,2g\,\pi^2 D^4 \vec{V}^2} = \dfrac{8fL}{g\pi^2 D^5}$

19.2

By Eqns. 19.1 and 19.2

$$\Delta h = \frac{fL\vec{V}^2}{D\,2g} \quad \text{or} \quad \frac{\Delta h}{L} = \frac{f\vec{V}^2}{D\,2g}$$

let $\vec{V}^2 = \left(\frac{\dot{V}}{\pi D^2/4}\right)^2 = \frac{\dot{V}^2\, 16}{\pi^2 D^4}$

Thus $\dfrac{\Delta h}{L} = \dfrac{f}{D\,2g}\left(\dfrac{\dot{V}^2\,16}{\pi^2 D^4}\right) = \dfrac{8f}{g\pi^2 D^5}\dot{V}^2$

on a log-log scale

$$\log\left(\frac{\Delta h}{L}\right) = \log\left(\frac{8f}{g\pi^2 D^5}\right) + 2\log(\dot{V})$$

all head loss curves are parallel

solve for the volumetric flowrate in gpm given $\Delta h/L$,

$$\dot{V} = \frac{(\Delta h/100\text{ft})\, g\, \pi^2 D^5}{8f\quad 100\text{ft}}$$

$$= \left[\frac{(\Delta h/100\text{ft})\, 32.2\text{ft/sec}^2\, \pi^2 D(\text{in}^5)}{8 \times 0.03 \times 100\text{ ft}}\left(\frac{264.2\text{ gal}}{35.32\text{ ft}^3}\right)^2 \left(\frac{3600\text{ sec}^2}{\text{min}^2}\right)\left(\frac{\text{ft}}{12\text{in}}\right)^5\right]$$

$$\dot{V} = \left[(\Delta h/100\text{ft})\, D^5(\text{in})\, 10.72\right]^{1/2} \text{ gpm}$$

let $\Delta h/100\text{ ft} = 0.5$ and 30, $D = 1, 2$ and 4 in

19.2 Cont'd

GPM

$\Delta h / 100\, ft$	D = 1 in	2 in	4 in
0.5	2.32	13.1	74.1
30	17.9	101.4	573.9

Lines of constant velocity

$$\dot{V} = \vec{V} \pi D^2 / 4$$

$$\frac{\Delta h}{L} = \frac{8f}{g\pi^2 D^5}\left(\frac{\vec{V}\pi D^2}{4}\right)^2 = \frac{f\vec{V}^2}{2gD}$$

$$\frac{\Delta h\,(ft)}{100\,ft} = \frac{100\,ft}{2} \frac{0.03\; \vec{V}^2\,(ft/sec)^2}{32.2\,\frac{ft}{sec^2}\;D(in)}\left(\frac{12\,in}{ft}\right) = \frac{0.559\;\vec{V}^2\,(ft/sec)}{D(in)}$$

$$\dot{V} = \frac{\vec{V}(ft/sec)\,\pi\,D^2(in)}{4}\left(\frac{ft^2}{144\,in^2}\right)\left(\frac{264.2\,gal}{35.32\,ft^3}\right)\left(\frac{60\,sec}{min}\right) = 2.45\;\vec{V}(ft/sec)\,D^2$$

let $\vec{V} = 1, 2$ and 4 ft/sec, $D = 1, 2, 4$ in

	D	1 in		2 in		4 in	
$\vec{V}$		$\Delta h/L$	$\dot{V}$	$\Delta h/L$	$\dot{V}$	$\Delta h/L$	$\dot{V}$
1		0.56	2.45	0.28	9.80	0.14	39.2
2		2.24	4.90	1.12	19.60	0.56	78.4
4		8.94	9.80	4.47	39.20	2.23	156.8

561

Problem 19.2

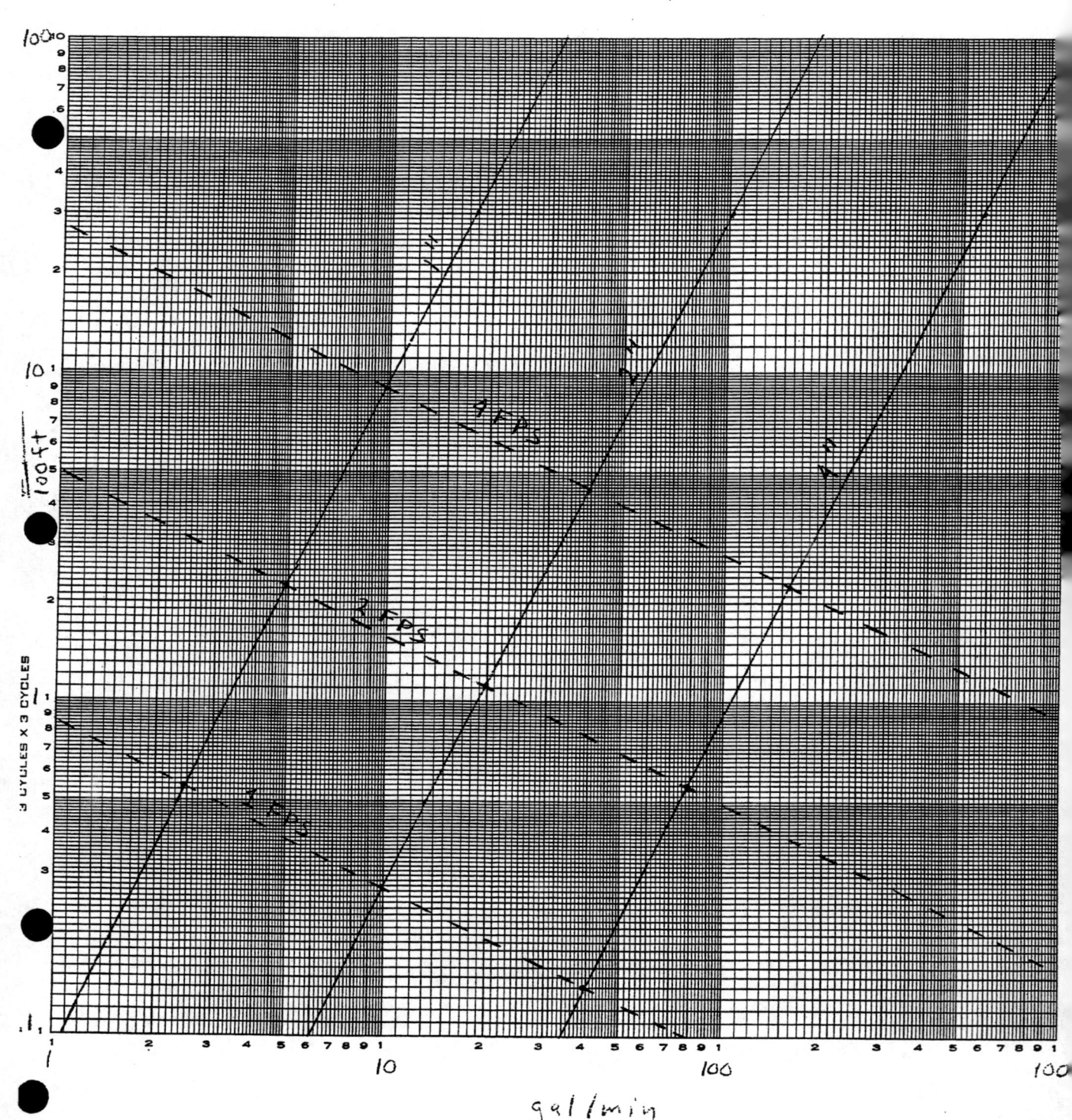

19.3

Compute L_{eq} for the fitting in Ex. 19.2

$L_{eq} = kD/f$, $k = 0.17$

Determine f from straight pipe head loss

By Eq. 19.2 $l_f = f \dfrac{L \vec{V}^2}{D\ 2g}$ or $f = \dfrac{l_f\ D\ 2g}{L\ \vec{V}^2}$

By Fig. 19.2, $l_f/L = 7.1\ \text{ft}/100\text{ft}$ (60 gpm, 2" pipe)

From Ex. 19.2, $\vec{V} = 5.8\ \text{ft/s}$

By Table 19.1, $D = I.D. = 2.067\ \text{in}$

$f = \left(\dfrac{7.1\ \text{ft}}{100\ \text{ft}}\right) \dfrac{2.067\ \text{in}\ \ 2\ (32.2\ \text{ft/sec}^2)}{(5.8\ \text{ft/sec})^2} \left(\dfrac{\text{ft}}{12\ \text{in}}\right) = 0.0234$

$L_{eq} = \dfrac{0.17\ (2.067\ \text{in})}{0.0234} \left(\dfrac{\text{ft}}{12\ \text{in}}\right) = 1.25\ \text{ft}$

19.4

By Eq. 19.2 and 19.4, $L_{eq} = KD/f$

By Table 19.1, for 3 in. schedule 40 pipe, $D = 3.068$ in.
Using screwed pipe fittings
By Table 19.3, $K_s = 0.80$
Using flanged-welded pipe fittings
By Table 19.4, $K_f = 0.34$

Determine the friction factor from Fig. 19.2 and Eq. 19.2 for 3 inch pipe

$$f = \frac{l_f}{L} \frac{D}{(V^2/2g)}$$

$\vec{V} = 10$ ft/s, $l_f/L = 12$ ft water /100 ft

$$f = \left(\frac{12 \text{ ft water}}{100 \text{ ft}}\right) \frac{3 \text{ in.}}{(10 \text{ ft/s})^2} \frac{2(32.2 \text{ ft/sec}^2)}{(12 \text{ in/ft})} = \left(\frac{l_f}{100 \text{ ft}}\right)\left(\frac{16.10}{\vec{V}^2}\right) = 0.019$$

similarly for the other velocity values given on Fig. 19.2

$\vec{V}$	$(l_f/100 \text{ ft})$	f
10	12	0.019
8	8	0.020
6	4.8	0.021
4	2.2	0.022
3	1.3	0.023
2	0.6	0.024

values of f not listed above will be interpolated or extrapolated from these values. The results for L_{eq} are given in the following table

19.4 Cont'd.

For example, $\vec{V} = 10$ ft/s, $f = 0.019$, screwed fitting,

$$L_{eq} = \frac{0.80\,(3.068\,\text{in})}{0.019}\left(\frac{\text{ft}}{12\,\text{in}}\right) = 10.8\,\text{ft}$$

$\vec{V}$	f	L_{eq} (ft) screwed ($k=0.80$)	flanged ($k=0.34$)
10	0.019	10.8	4.6
9	0.019	10.8	4.6
8	0.020	10.2	4.3
7	0.020	10.2	4.3
6	0.021	9.7	4.1
5	0.021	9.7	4.1
4	0.022	9.3	4.0
3	0.023	8.9	3.8
2	0.024	8.5	3.6
1	0.025	8.2	3.5

19.5

Solving for the flow coefficient by combining Eqs. 19.4 and 19.5a

$$C_v = \frac{\dot{V}}{\vec{V}}\sqrt{\frac{2.31 \text{ ft } 2g}{k}}$$

By Fig. 19.2, at $\vec{V} = 10$ ft/sec, $\dot{V} = 230$ gpm (3 in. pipe) for screwed fitting (Table 19.1), $k = 0.8$

$$C_v = \frac{230 \text{ gpm}}{10 \text{ ft/sec}}\sqrt{\frac{2.31 \text{ ft } 2(32.2 \text{ ft/sec}^2)}{0.8}} = 314 \text{ gpm}$$

similar procedure used for other velocities and for flanged fittings ($k = 0.34$). Results are tabulated below.

		C_v (gpm)	
$\vec{V}$ (fps)	$\dot{V}$ (gpm)	screwed ($k=0.8$)	flanged ($k=0.34$)
10	230	314	481
8	185	315	484
6	140	318	488
4	92	314	481
3	70	318	488
2	47	320	492

As seen from the table, the value of the flow coefficient does not depend on velocity and is approximately 315 gpm for the screwed elbow and 485 gpm for the flanged elbow.

19.6

Compute Δh through the branch of a Tee, 1-in schedule 40 steel pipe, 10 gpm, use screwed fitting Table (19.3)

By Eq. 19.4

$$\Delta h_{1-2} = k \left(\frac{\vec{V}^2}{2g} \right)$$

By Table 19.3, $k = 1.8$

By Fig. 19.2, $\vec{V} = 3.7$ ft/sec

$$\Delta h_{1-2} = 1.8 \left(\frac{3.7^2 \, ft^2/sec^2}{(2)\, 32.2 \, ft/sec^2} \right) = 0.40 \text{ ft } H_2O$$

19.7

We will assume a screwed fitting
By Table 19.3, $k = 1.0$ (100 mm, angle valve)
By Fig. 19.3b, $\vec{V} = 2.5$ m/s
By Eq. 19.4

$$\Delta h = k\left(\frac{\vec{V}^2}{2g}\right) = 1.0 \left(\frac{(2.5 \text{ m/s})^2}{2(9.8 \text{ m/s}^2)}\right) = 0.32 \text{ m water}$$

19.8

Combining Eqs. 19.4 and 19.5a

$$C_v = \frac{\dot{V}}{\vec{V}} \sqrt{\frac{2.31 \text{ ft } 2g}{k}} \quad \text{or} \quad k = \frac{2.31 \text{ ft } 2g}{C_v} \left(\frac{\dot{V}}{\vec{V}}\right)^2$$

As an example, consider water flowing through a globe valve screwed into a 1 in. schedule 40 steel pipe with a velocity of 4 ft/sec
By Fig. 19.2a, $\dot{V} = 11$ gpm
By Table 19.3, $k = 9$

$$C_v = \frac{11 \text{ gpm}}{4 \text{ ft/sec}} \sqrt{\frac{2.31 \text{ ft } 2(32.2 \text{ ft/sec}^2)}{9}} = 11.2 \text{ gpm}$$

19.9

From Ex. 19.5, NPSHA = 30.66 ft water

NPSHR = 25 ft

Therefore the pump inlet could be raised

30.66 - 25 = 5.66 ft above its current location

or 2.33 ft above the surface of the pond.

(However this may not be possible as the pump may not start unless a sealed check valve is located at the inlet)

19.10

Rearranging Eq. 19.7 to solve for the velocity:

$$\vec{V} = \sqrt{2g\, NPSHR + 2\left(\frac{P_{sat}(t) - P_{suction}}{\rho}\right)}$$

NPSHR = 15 m

By Table A.1SI, $P_{sat}(30°C) = 0.004246$ MPa
$\rho = 995$ kg/m³

$P_{suction} = P_{atm} = 101.325$ kPa

$$\vec{V} = \sqrt{2\left(\frac{9.8\,m}{sec^2}\right)15\,m + 2\left(\frac{4.246 - 101.325}{995\,kg/m^3}\right)kPa\left(\frac{kg\ 10^3}{m\,sec^2\,kPa}\right)}$$

$\vec{V} = 10.0$ m/sec

19.11

By Fig. 19.6, 6.5" impeller, 100 gpm, 40 ft water head

Power requirement is $1\tfrac{1}{2}$ hp

By Eq. 19.6

$$\eta_{P,t} = \frac{\dot{V}\,\Delta P_t}{\dot{W}} = \frac{\dot{V}\,\rho_w\,g\,\Delta h}{\dot{W}}$$

$$= \frac{100\,\text{gal/min}\,(40\,\text{ft})}{1\tfrac{1}{2}\,\text{hp}}\left(\frac{35.32\,\text{ft}^3}{264.2\,\text{gal}}\right)\left(\frac{62.4\,\text{lb}_m}{\text{ft}^3}\right)\left(\frac{32.2\,\text{ft}}{\text{sec}^2}\right)\left(\frac{1341\,\text{hp}\cdot\text{hr}}{3412\,\text{Btu}}\right)\left(\frac{60\,\text{min}}{\text{hr}}\right)$$

$$\times \left(\frac{\text{hp}\cdot\text{sec}^2}{32.2\,\text{ft}\cdot\text{lb}_m}\right)\left(\frac{\text{Btu}}{778\,\text{ft}\cdot\text{lb}_f}\right)$$

$$\eta_{P,t} = 0.67$$

Figure 19.6 indicates an efficiency of about 68-69% which is in good agreement

19.12

By Fig. 19.6, a pump with a 7" impeller delivering 120 gpm has a head of 44.5 ft, an efficiency of 67% and a power requirement of 2.1 hp.

a) Reduce flowrate to 60 gpm using a valve.
   The pump curve remains the same as shown in Fig. 19.6 for the 7" impeller. At 60 gpm, the pump head is 53 ft water and the power requirement is 1.4 hp.

b) Reduce the flowrate to 60 gpm by reducing the pump speed. By Fig. 19.7, the % head across the pump is 24% at 50% flow, so head = 0.24(44.5 ft water) = 10.7 ft water.
   The power is 12% of the original value or 0.12(2.1 hp) = 0.25 hp.

19.13

2 pumps in parallel, 6.5 in impeller, 100 gpm each

a) By Fig. 19.6, $\eta_{p,t} \approx 69\%$

b) $\dot{W} = 1.5$ hp

c) $\Delta h = 40$ ft water

19.14

One of 2 pumps is turned off. Assume no elevation head

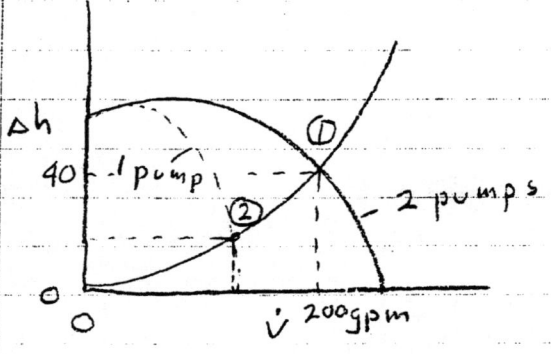

Point ① at 200 gpm, 40 ft

Point ② on single pump curve, same system curve

Need to determine where the system curve intersects the single pump characteristic curve (point 2)

For system curve, $\Delta h_{sys} = C \dot{V}^2$

$40 \text{ ft water} = C \times (200 \text{ gpm})^2$

or $C = \dfrac{40 \text{ ft water}}{4 \times 10^4 \text{ gpm}^2}$

Then $\Delta h_{sys} = \dfrac{40 \text{ ft water}}{4 \times 10^4 \text{ gpm}^2} (\dot{V}^2)$

By Trial: point 2, $\Delta h = 24.3$ ft water, $\dot{V} = 156$ gpm, $\dot{W} = 2$ hp

19.15

The piping system consists of one square-edged inlet, 2-90° elbows and one discharge.

The required water velocity at the discharge:
$$\vec{V}^2 = 2gh = 2(32.2 \text{ ft/sec}^2)(10 \text{ ft}) = 644 \text{ ft}^2/\text{sec}^2$$
or $\vec{V} = 25.4$ ft/sec

By Table 19.2, $d_i = 1.025$ in
$$\dot{V} = \frac{\pi d_i^2}{4} \vec{V} = \frac{\pi}{4}(1.025 \text{ in})^2 \, 25.4 \frac{\text{ft}}{\text{sec}} \left(\frac{\text{ft}^2}{144 \text{ in}^2}\right) = 0.146 \text{ ft}^3/\text{sec}$$

$$= 0.146 \frac{\text{ft}^3}{\text{sec}} \left(\frac{60 \text{ sec}}{\text{min}}\right) \left(\frac{264.2 \text{ gal}}{35.32 \text{ ft}^3}\right) = 65.5 \text{ gpm}$$

Head Requirement (assuming screwed fittings)

Inlet
By Table 19.3, $k = 0.5$
By Eq. 19.4, $\Delta h_{inlet} = 0.5 \left(\frac{(25.4 \text{ ft/sec})^2}{2(32.2 \text{ ft/sec}^2)}\right) = 5.0$ ft water

Elbows
By Table 19.3, $k = 1.5$
$\Delta h_{elbow} = 1.5 (10.0 \text{ ft water}) = 15$ ft water

Discharge, $\Delta h = 10$ ft water

Straight Pipe
By Fig. 19.3a, $\Delta h/L = 200$ ft water/100 ft (extrapolation)
$\Delta h_{straight} = 6 \text{ ft}(200 \text{ ft water}/100 \text{ ft}) = 12$ ft water

Pump head = $\Delta h_{inlet} + 2\Delta h_{elbow} + \Delta h_{discharge} + \Delta h_{straight}$

$= 5.0 + 2(15) + 10 + 12 = 57$ ft water

19.16

Use schedule 40 steel pipe and screwed fittings

a) Size the piping from a to b

for 7 L/s, from Fig. 19.2 (b), the pipe size is larger than 2" (50mm) so the design head loss should be about 0.4 kPa/m or 400 Pa/m. From Fig. 19.2 (b), use 80 mm pipe (3 in) The head loss at 7 L/s is 280 Pa/m

size the remaining piping to each tower

for 3½ L/s from Fig. 19.2 (b), we have a choice between 50 mm and 65 mm pipe, we will choose 65 mm as it provides the lower pressure drop of 230 Pa/m at 3½ L/s

b) Determine the total head loss across the pump

from a to b (piping only)

$$\Delta h_{a-b} = \frac{\Delta h}{L} \times L_{a-b} + \Sigma \Delta h_{fittings}$$

$L_{a-b} = 20 m$

4 90° elbows, by Table 9.3 for screwed fittings
$k = 0.8$
By Fig. 19.2 (b) $\bar{V} = 1.5 m/s$

By Eq. 19.4 $\Delta h_{elbow} = 0.8 \left(\frac{(1.5 m/s)^2}{2 \cdot 9.8 m/s^2} \right) = 0.092 m$

$\Sigma \Delta h = 4 \times \Delta h_{elbow} = 4(0.092 m) = 0.368 m$

19.16 cont'd

$$\Delta h_{a-b} = 280 \frac{Pa}{m} \times 20\,m \left(\frac{4.019 \times 10^{-3}\,in\ water}{1\,Pa}\right)\left(\frac{1\,m}{39.37\,in}\right) + 0.368\,m$$

$$= 0.572\,m + 0.368\,m = 0.940\,m$$

remaining piping to Towers, each side has identical configuration so only one side needs to be analyzed, head loss through spray headers not included

$$\Delta h_{Tower\ pipe} = \Delta h_{straight} + 2(\Delta h)_{branch\ Tee} + \Delta h_{90°\ elbow} + \Delta h_{inlet}$$

$$\Delta h_{straight} = 230 \frac{Pa}{m} \times 10\,m \left(\frac{4.019 \times 10^{-3}}{1}\right)\left(\frac{1}{39.37}\right) = 0.235\,m$$

$$\Delta h_{branch\ Tee} = k\left(\frac{\vec{V}^2}{2g}\right)$$

By Table 19.3, $k = 1.3$
By Fig. 19.2 (b), $\vec{V} = 1.1\,m/s$

$$\Delta h_{branch\ Tee} = 1.3 \left(\frac{(1.1\,m/s)^2}{2 \cdot 9.8\,m/s^2}\right) = 0.080\,m$$

Similarly for the elbow and square edged inlet

$$\Delta h_{elbow} = 0.85(0.062) = 0.052\,m$$

$$\Delta h_{inlet} = 0.5(0.062) = 0.032\,m$$

$$\Delta h_{Tower\ pipe} = 0.235 + 2(0.080) + 0.052 + 0.032 = 0.479\,m$$

19.16 cont'd

Total pump head requirement:

$$\Delta h_{pump} = \Delta h_{a-b} + \Delta h_{condenser} + \Delta h_{Tower\ pipe} + \Delta h_{elevation} + \Delta h_{sprayer}$$

as no information is provided on the spray heads, this will be omitted

inserting values

$$\Delta h_{pump} = 0.940\,m + 3\,m + 0.479\,m + 5\,m = 9.42\,m$$

By Fig. 19.6, we will select a pump with a 6.5" impeller, Δh_{pump} at 7 L/s is about 11.5 m so this should be sufficient when including the spray head pressure loss

$$\dot{W}_{pump} \approx 1.6\ hp, \quad \eta_{P,t} \approx 67\%$$

19.17

We will make an initial assumption that the flow rate remains unchanged.

The head loss from b to a would remain the same, $0.940 + 3 = 3.94$ m water.

The elevation head remains the same, 5 m water.

If all the flow passes through only one tower, a to b, the velocity in the piping doubles so the head increases four times, $0.479 \times 4 = 1.92$ m water.

The total pump head is then
$$\Delta h_{pump} = 3.94 + 5 + 1.92 = 10.9 \text{ m water}$$

However, the increase in system head loss will cause the balance point to shift to a slightly lower total flow rate. This will in turn reduce the system head loss which tends to keep the balance point at a higher flow rate. The net effect in this case, is a flow rate not much different than the original value, perhaps slightly less, such as 6.8 L/s.

19.18

a) Total flowrate from supply to return = 16 gpm
Assume schedule 40 steel pipe with threaded fittings

Split supply & return into 4 sizes:
- I - from supply to 2nd floor, from 7th floor to main return
- II - from 2nd floor to 5th floor on supply and return, 4th to 7th floor on return
- III - from 5th floor to 7th floor on supply and 2nd floor to 4th floor on return
- IV - 1st floor return, 8th floor supply, and all other floors connecting supply and return headers

Sizing - By Fig. 19-2a, approximately 4 ft/sec velocity
- I - 16 gpm, 1 1/4", $\vec{V}$ = 3.5 ft/sec, h_v = 0.19 ft water
- II - 12 gpm, 1", $\vec{V}$ = 4.5 ft/sec, h_v = 0.31 ft water
- III - 6 gpm, 3/4", $\vec{V}$ = 3.75 ft/sec, h_v = 0.22 ft water
- IV - 2 gpm, 1/2", $\vec{V}$ = 2.1 ft/sec, h_v = 0.07 ft water

b) Total head loss should be approximately the same for flow through each heat exchanger. We will take the circuit through the heat exchanger on the second floor.

Supply side
Branch Tee from main supply
By Table 19.3, assuming 4" main supply, k = 1.1
$\Delta h_{b.Tee}$ = 1.1(0.19) = 0.21 ft water

19.18 Cont'd.

Tee-line

By Table 19.3, $k = 0.9$, $\Delta h_{L.Tee} = 0.9(0.19) = 0.17$ ft water

Branch tee to 2nd floor

By Table 19.3, $k = 1.7$, $\Delta h_{b.Tee} = 1.7(0.19) = 0.32$ ft water

Gate Valve (assumed fully open)

By Table 19.3, $k = 0.33$, $\Delta h_{valve} = 0.33(0.07) = 0.02$ ft water

Return side

Branch Tee from 2nd floor to return header

By Table 19.3, $k = 2.4$, $\Delta h_{b.Tee} = 2.4(0.07) = 0.17$ ft water

Tee-line in return

By Table 19.3, $k = 0.9$, $\Delta h_{Tee} = 2(0.9)(0.22) + 3(0.9)(0.31) + 1(0.9)(0.19) = 1.40$ ft water

90° elbows (regular)

By Table 19.3, $k = 1.3$, $\Delta h_{elbows} = 2(1.3)(0.19) = 0.49$ ft water

Branch tee to return main

By Table 19.3, $k = 1.7$, $\Delta h_{Tee} = 1.7(0.19) = 0.32$ ft water

Straight pipe

$1\text{-}1/4''$, $L = 2' + 10' + 10' + 70' + 3' = 95'$

By Fig. 19.2, $\Delta h / L = 4.5$ ft water / 100 ft

$$\Delta h_{1\text{-}1/4''} = 95' \times \left(\frac{4.5 \text{ ft water}}{100 \text{ ft}} \right) = 4.28 \text{ ft water}$$

$1''$, $L = 30'$, $\Delta h / L = 10$ ft water / 100 ft

$$\Delta h_{1''} = 30' \left(\frac{10 \text{ ft water}}{100 \text{ ft}} \right) = 3.0 \text{ ft water}$$

19.18 cont'd.

$3/4"$, $L = 20'$, $\Delta h/L = 10$ ft water / 100 ft

$$\Delta h_{3/4"} = 20' \left(\frac{10 \text{ ft water}}{100 \text{ ft}} \right) = 2.0 \text{ ft water}$$

$1/2"$, $L = 12'$, $\Delta h/L = 5.3$ ft water / 100 ft

$$\Delta h_{1/2"} = 12' \left(\frac{5.3 \text{ ft water}}{100 \text{ ft}} \right) = 0.64 \text{ ft water}$$

Total head loss between main supply & main return

$\Delta h_{Total} = 0.21 + 0.17 + 0.32 + 0.02 + 0.17 + 1.40 + 0.49 + 0.32 + 4.28 + 3.0 + 2.0 + 0.64 + 2.50$ (h.x.)

$\Delta h_{Total} = 15.5$ ft water

19.19

The system should have a head loss of about 15.5 ft water when all valves are open and a flow rate of 16 gpm. The system curve should be parabolic with the following constants

$$\Delta h = 15.5 \text{ ft water} = a\,(16 \text{ gpm})^2$$

or $\quad a = 0.0605 \dfrac{\text{ft water}}{(\text{gpm})^2}$

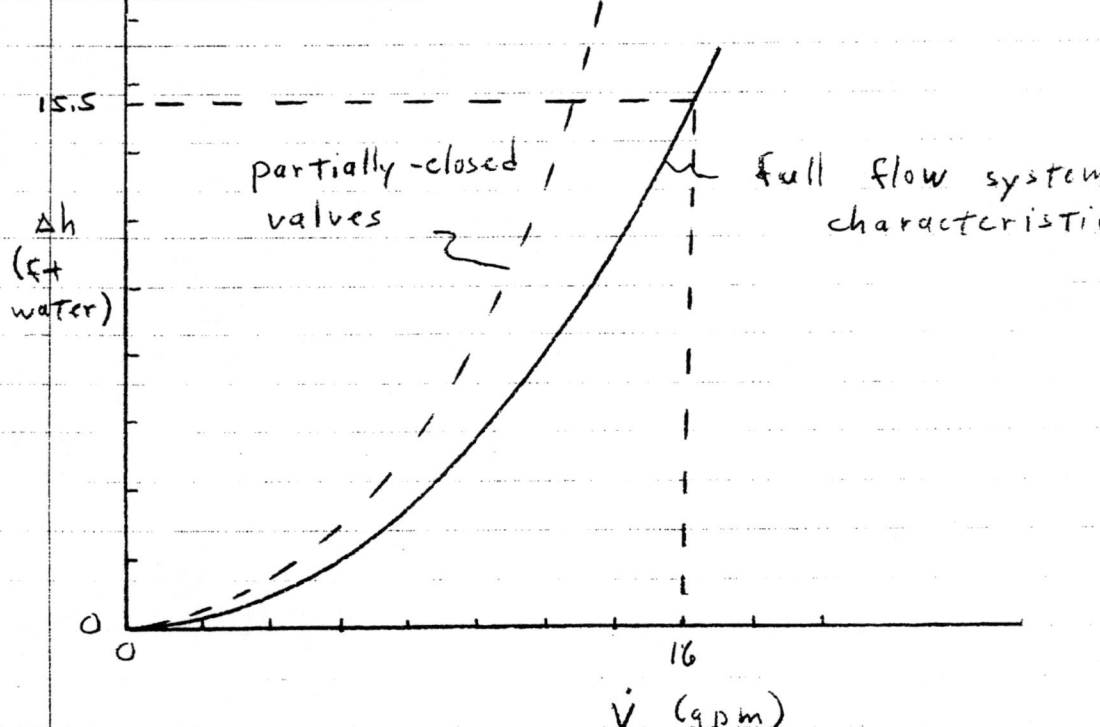

As one or more valves close, the system characteristic will remain parabolic but will shift upward (a becomes larger) as shown by the dashed curve above.

19.20

District heating system providing hot water to 5 bldgs.
(see Fig. 19.13)

19.22

Total head loss through 2-a-3 = 2-b-3 = 10.87 ft water

We need to increase the head loss through branch b by about 3.5 ft water to balance the system.

We can try reducing the pipe size from 2" to 1½"
By Fig. 19.2a, $\Delta h/L = 11$ ft water/100 ft,

$$\Delta h_{pipe} = 4'\left(\frac{11 \text{ ft water}}{100 \text{ ft}}\right) = 0.44 \text{ ft water}$$

The increased velocity will also increase the loss in the gate valves, $\bar{V} \approx 6.2$ ft/sec, and tee branch

$$\Delta h_{valves} = 2(0.19)\left(\frac{6.2^2}{2(32.2)}\right) = 0.23 \text{ ft water}$$

$$\Delta h_{Tee} = 1.6\left(\frac{6.2^2}{2(32.2)}\right) = 0.96 \text{ ft water}$$

$$\Delta h_{2-b-3} = 0.57 + 0.44 + 0.23 + 0.96 + 6.0 = 8.2 \text{ ft water}$$

This is too small, try next smallest pipe, 1¼"
$\Delta h/L = 24$ ft water/100 ft, $\bar{V} = 8.5$ ft/sec

$$\Delta h_{pipe} = 4\left(\frac{24}{100}\right) = 0.96 \text{ ft water}$$

$$\Delta h_{valves} = 2(0.22)\left(\frac{8.5^2}{2(32.2)}\right) = 0.49 \text{ ft water}$$

$$\Delta h_{Tee} = 1.7\left(\frac{8.5^2}{2(32.2)}\right) = 1.91 \text{ ft water}$$

$$\Delta h_{2-b-3} = 0.57 + 0.96 + 0.49 + 1.91 + 6.0 = 9.93 \text{ ft water}$$

This value matches the desired value (10.87 ft water) as closely as possible. Additional balancing should be accomplished using the valves.

19.23

a) At part load, chiller #1 supplies 500 gpm to the load and chiller #2, 100 gpm. Therefore, 400 gpm of 45°F waters flows through the common pipe near chiller #2.

b) The flow into chiller #2 is 400 gpm of 45°F water through the common pipe and 100 gpm of 60°F water from the return. The mixed water temperature entering chiller #2 is

$$t_2 = \frac{400(45°F) + 100(60°F)}{500} = 48°F$$

c) Chiller #2 is at 20% capacity (100 gpm supplied to load vs 500 gpm at full load)

19.24

Assume steel pipe

By Eq. 19.9

$V_w = 1500$ gal

$v_1 = v_f(45°F)$, By Table A.1E, $v_1 = 0.01602$ ft³/lb$_m$

$v_2 = v_f(100°F)$, By Table A.1E, $v_2 = 0.01613$ ft³/lb$_m$

$P_1 = 10$ psig $+ 14.696 = 24.696$ psia, $P_2 = 35 + 14.696 = 49.7$ psia

$\alpha = 6.5 \times 10^{-6}$ in./in. °F

$\Delta t = 100°F - 45°F = 55°F$

$$V_t = \frac{1500 \text{ gal}\left[\left(\frac{0.01613}{0.01602}\right) - 1\right] - 3(6.5 \times 10^{-6})(55)}{1 - (24.696/49.7)}$$

$V_t = 20.5$ gal

19.25
Total System Volume
2½" pipe, $2' + 7' + 5' + 2' + 7' + 7' = 30$ ft
$V = 0.03322$ ft^2 (30 ft) = 1.0 ft^3
2" pipe, $V = 4$ ft (0.0233) = 0.09 ft^3
1½" pipe, $5' + 4' + 5' = 14$ ft
$V = 0.01414$ ft^2 (14 ft) = 0.20 ft^3
Total piping volume = $1.0 + 0.09 + 0.2 = 1.29$ ft$^3 \left(\frac{264.2 \text{ gal}}{53.35 \text{ ft}^3} \right)$
= 6.39 gal

Neglect volume in pump and fittings
Total volume = 6.39 gal + 16 gal = 22.39 gal

By Eq. 19.9
$V_w = 12.4$ gal
$v_1 = v_f$ (40°F), By Table A.1E, $v_1 = 0.01602$ ft^3/lbm
$v_2 = v_f$ (90°F), By Table A.1E, $v_2 = 0.01610$ ft^3/lbm
$P_1 = 15 + 14.696 = 29.696$ psia
$P_2 = 40 + 14.696 = 54.696$ psia
$\alpha = 6.5 \times 10^{-6}$ in./in.·°F
$\Delta t = 90 - 40 = 50$°F

$$V_t = \frac{12.4 \text{ gal} \left[\left(\frac{0.01610}{0.01602} \right) - 1 \right] - 3(6.5 \times 10^{-6}) 50}{1 - (29.696/54.696)}$$

$V_t = 0.13$ gal

To allow for pump and fittings, the volume specified should be at least 0.2 gal.